Sustainable Phosphorus Management

Roland W. Scholz · Amit H. Roy
Fridolin S. Brand · Deborah T. Hellums
Andrea E. Ulrich
Editors

Sustainable Phosphorus Management

A Global Transdisciplinary Roadmap

 Springer

Editors
Roland W. Scholz
Natural and Social Science Interface
ETH Zürich
Zürich
Switzerland

and

Fraunhofer Project Group Materials
　Recycling and Resource
　Strategies IWKS
Alzenau
Germany

Amit H. Roy
Deborah T. Hellums
International Fertilizer
　Development Center
Muscle Shoals, AL
USA

Andrea E. Ulrich
Fridolin S. Brand
Natural and Social Science Interface,
　Institute for Environmental Decisions
ETH Zürich
Zürich
Switzerland

ISBN 978-94-007-7249-6   ISBN 978-94-007-7250-2 (eBook)
DOI 10.1007/978-94-007-7250-2
Springer Dordrecht Heidelberg New York London

Library of Congress Control Number: 2014932665

© Springer Science+Business Media Dordrecht 2014
This work is subject to copyright. All rights are reserved by the Publisher, whether the whole or part of the material is concerned, specifically the rights of translation, reprinting, reuse of illustrations, recitation, broadcasting, reproduction on microfilms or in any other physical way, and transmission or information storage and retrieval, electronic adaptation, computer software, or by similar or dissimilar methodology now known or hereafter developed. Exempted from this legal reservation are brief excerpts in connection with reviews or scholarly analysis or material supplied specifically for the purpose of being entered and executed on a computer system, for exclusive use by the purchaser of the work. Duplication of this publication or parts thereof is permitted only under the provisions of the Copyright Law of the Publisher's location, in its current version, and permission for use must always be obtained from Springer. Permissions for use may be obtained through RightsLink at the Copyright Clearance Center. Violations are liable to prosecution under the respective Copyright Law.
The use of general descriptive names, registered names, trademarks, service marks, etc. in this publication does not imply, even in the absence of a specific statement, that such names are exempt from the relevant protective laws and regulations and therefore free for general use.
While the advice and information in this book are believed to be true and accurate at the date of publication, neither the authors nor the editors nor the publisher can accept any legal responsibility for any errors or omissions that may be made. The publisher makes no warranty, express or implied, with respect to the material contained herein.

Printed on acid-free paper

Springer is part of Springer Science+Business Media (www.springer.com)

Preface

This book is a key product of the first 2 years of the Global TraPs project. The chapters incorporate prevailing views on *critical questions* and *issues* related to current phosphorus management practices. These views have been elaborated among more than 200 key stakeholders of the phosphorus supply–demand chain. For each node, *Exploration, Mining, Processing, Use,* and *Dissipation and Recycling* as well as the cross-cutting issue: *Trade and Finance,* the reader will find *state-of-the-art knowledge, transdisciplinary processes* (i.e., forms of science-practice collaboration) and *topical case studies.* This may help to develop robust orientations on how food security may be achieved and how the current low use efficiency may be increased by improved utilization strategies and the development of new technologies. A closure of the anthropogenic phosphorus cycle may help to avoid eutrophication, hypoxia, and other negative impacts on ecosystems and promotes resources conservation. Finally, the book takes a global perspective on phosphorous and reveals the different use patterns of different types of farmers and countries.

The Global TraPs project and the writing of this book was a consultative process and included participation of representatives from industry and trade, scientists from various disciplines and numerous universities, public agencies and international organizations, Non Governmental Organizations (NGOs), farmers and user associations. This consultative process included four workshops during 2010 and 2012 and the first Global TraPs World-conference in Beijing, China, June 18–20, 2013. We, as the leaders of the project, are very impressed with how well this process worked. The members of the Global TraPs project unselfishly shared their knowledge and time to develop a comprehensive understanding of phosphorus use. Most remarkable was the willingness to listen to each other. Thus, authentic process of mutual learning and knowledge integration took place. We want to thank all authors and reviewers, and the participants of this—presumably first—large-scale transdisciplinary process. This book is an important milestone of this process.

The chapters present a widely shared blueprint of on current phosphorus use and how it may be improved for developing orientations for sustainable phosphorus management. Yet as it is typical for transdisciplinary multistakeholder discourse, the discussion of different chapters revealed the complexity and multilayeredness of the supply–demand chain and identified different and incoherent data, perspectives and valuations that asked for integration. This complexity challenged a re-examinations, re-assessment, and rethinking of key conclusions. A final round of review of all chapters was initiated at the 2013 Beijing conference. For supplementing the current view on sustainable phosphorus management, spotlights were written for explaining key concepts or for introducing nonconventional views. The introductory chapter now includes both a comprehensive and coherent blueprint of an actor- or agent-based phosphorus flows view and outlines the transdisciplinary process, i.e., the specific science-practice collaboration which is needed to foster its sustainable use.

The book thus goes far beyond the mere description of physical phosphorus flows and their impact. As expressed by the subtitle "Global Transdisciplinary Roadmap," the chapters provide a schedule of how critical questions may be answered, in particular by transdisciplinary case studies. The vision is to accomplish mutual learning and consensus building among the key stakeholders of the phosphorus supply–demand chain. This may be valuable not only for the members of the Global TraPs project or those who are interested in sustainable phosphorus management, but also for scientists and key stakeholders who are interested in sustainable resources management.

Amit H. Roy
Roland W. Scholz

Contents

1 **Sustainable Phosphorus Management:
   A Transdisciplinary Challenge** 1
   Roland W. Scholz, Amit H. Roy and Deborah T. Hellums

2 **Exploration: What Reserves and Resources?** 129
   David A. Vaccari, Michael Mew, Roland W. Scholz
   and Friedrich-Wilhelm Wellmer

3 **Mining and Concentration: What Mining to What Costs
   and Benefits?** 153
   Ingrid Watson, Peter van Straaten, Tobias Katz and Louw Botha

4 **Processing: What Improvements for What Products?** 183
   Ludwig Hermann, Willem Schipper, Kees Langeveld
   and Armin Reller

5 **Use: What is Needed to Support Sustainability?** 207
   Robert L. Mikkelsen, Claudia R. Binder, Emmanuel Frossard,
   Fridolin S. Brand, Roland W. Scholz and Ulli Vilsmaier

6 **Dissipation and Recycling: What Losses, What Dissipation
   Impacts, and What Recycling Options?** 247
   Masaru Yarime, Cynthia Carliell-Marquet, Deborah T. Hellums,
   Yuliya Kalmykova, Daniel J. Lang, Quang Bao Le, Dianne Malley,
   Leo S. Morf, Kazuyo Matsubae, Makiko Matsuo, Hisao Ohtake,
   Alan P. Omlin, Sebastian Petzet, Roland W. Scholz,
   Hideaki Shiroyama, Andrea E. Ulrich and Paul Watts

7 **Trade and Finance as Cross-Cutting Issues in the Global
   Phosphate and Fertilizer Market** 275
   Olaf Weber, Jacques Delince, Yayun Duan, Luc Maene,
   Tim McDaniels, Michael Mew, Uwe Schneidewind
   and Gerald Steiner

**Erratum to: Sustainable Phosphorus Management:
A Transdisciplinary Challenge**............................. E1
Roland W. Scholz, Amit H. Roy and Deborah T. Hellums

Figures

Chapter 1

Fig. 1	Phosphate acid was seen as remedies for many diseases (*Source* Hulman & Co)	7
Fig. 2	Phosphorus may have positive and, in certain forms, negative effects.	7
Fig. 3	The Berne Convention of 1906 banned the highly toxic *white phosphorus* from matches.	9
Fig. 4	Shares of phosphate rock (*PR*) use for main purposes according to Prud'homme (2010)	12
Fig. 5	Fertilizer (*left*) and phosphorus dynamics [*right*, based on (FAO 2010b)] show different trends in different countries	13
Fig. 6	Potential trends of phosphorus (*P*) application and trends of phosphorus application and uptake in the croplands of **a** western Europe, **b** Africa and **c** Asia (Sattari et al. 2012), **d** presents the P use efficiency [for 1965–2007 based on historical data, figure **d** provided by (for 1965–2007 based on historical data, figure **d** provided by Sattari 2013)].	15
Fig. 7	The evolution of phosphorus fertilizer use, where mineral phosphorus plays an increasingly important key role (phosphorus data—80 to 85 % for fertilizer use—from USGS, presented as moving mean with a 3-year sliding window; rough estimation of manure data based on literature, see Sect. 4.8.2; population data from (USCB 2009); manure data extrapolated from different data on annual manure production, see text)	16
Fig. 8	Dynamic precision crop rotation asks that at any time t_0 the decision of what amount and type of fertilizer should be added does not only depend on the soil status S_0 and the specific crops C_0 but also depends on the prospective crop grown at that time $C_{n+1}, ..., C_{n+k}$	27
Fig. 9	The coating of the sugar beet seedling is successfully applied (figure taken from Leubner 2013)	28

Fig. 10	Linkage of the Herfindahl-Hirschmann Index (HHI) (figure according to Deutsche-Rohstoffagentur/BGR 2011, present version taken from Scholz and Wellmer 2013, p.15) with the Weighted World Governance Index (WWGI, data from Jasinski 2009, 2010, 2011a, 2012; Kaufmann et al. 2011) for production. "*" Presents the HHI for phosphorus reserves..................................	29
Fig. 11	Price dynamics for rock phosphate, food, commodity index, etc. in US$ per metric ton (graph with the courtesy of Olaf Weber).......................................	34
Fig. 12	Illustration of the "not too little–not too much" principle......	38
Fig. 13	Trends in total fertilizer consumption, distinguishing different world regions (sum of N, P, and K expressed as N, P_2O_5, and K_2O; Sutton et al. 2013; based on (Davidson 2012; Sattari et al. 2012) and trends of world cereal and meat production (*Source* Our Nutrient World, Sutton et al. 2013)....................................	41
Fig. 14	*Box*-and-*whiskers plot* showing median grid-cell P balances (for all grid cells with >5 % cropland area).........	43
Fig. 15	Cereal yields per hectare by region for 2011/2012 (VFRC 2012b, based on FAO data).....................	43
Fig. 16	The five stages of the supply–demand chain...............	47
Fig. 17	Mining, beneficiation, and routes taken as described in 2007 data (Prud'homme 2010)	53
Fig. 18	There are historic trade-offs to the use of dung as fertilizer or fuel (picture taken in Tibet, R. W. Scholz)...............	58
Fig. 19	A flotilla of manure boats on Soochow creek, Shanghai, prepare to transport excreta to cultivated fields (taken from King 1911, p.195)	59
Fig. 20	The process schematic of the anaerobic digestion of biowaste (BTA International 2013)	62
Fig. 21	A blueprint of global phosphorus flows along the steps of the supply–demand chain (non-incorporating virtual flows) referring to the phosphate rock production in 2011 including an equivalent of 25 Mt P year^{-1} according to USGS (2012) (Fig. 21 is electronically accessible by (http://dx.doi.org/10.1007/978-94-007-7250-2_1)	71
Fig. 22	Annual application of mineral fertilizer and manure per hectare (kg P ha year^{-1}) between 1965 and 2007 and phosphorus uptake in cropland and simulation for future phosphorus uptake for the MEA scenarios [Sattari et al. (2012); *AM* Adapting Mosaic, *OS* Order by Strength, *TG* Techno garden, *GO* Global Orchestration, see (MacDonald et al. 2009; Smith et al. 2008)]	73

Fig. 23	Urine separation and separated dry collection and processing are used by the Guatemalan Mopan Mayas in the region of Peten (*Photo* R. W. Scholz)	77
Fig. 24	The dynamics of an ideal transdisciplinary project (Scholz 2011a, p. 375)	86
Fig. 25	Organizational chart of the Global TraPs project (Jan 2013)	87
Fig. 26	Some of the planned and realized case studies (*gray-shaded boxes*) and topics for case study research (*black boxes*) of the Global TraPs project	93
Fig. 27	Development of phosphorus and algal biomass concentration in Lake Constance (There is no clear explanation for the 2011–2012 peak of biomass, changes in the phytoplankton could be observed	97
Fig. 28	Different trends of phosphorus use in different periods (*x*-axis: Mt P, data from USGS Mineral Commodity Summaries; the graph is generated by unweighted moving average statistics to smooth annual fluctuations using a five-year time window)	119
Fig. 29	Disciplines, interdisciplinarity, multi-stakeholder discourses, and transdisciplinarity	120

Chapter 2

Fig. 1	Trends of grades of beneficiated P [in general, which departs P mines after beneficiation; *source* IFDC (2010)]	137
Fig. 2	Per capita global phosphate rock production (PR rock/capita)	145
Fig. 3	Projected PR production based on fixed per capita rate coupled with UN projections of population growth	146
Fig. 4	Scenario for PR production and consumption (Scholz and Wellmer 2013, historical data, and exit strategy after Cordell et al. 2009)	147

Chapter 3

Fig. 1	A schematic of a trailing suction hopper dredge (TSHD) determined to be the optimal method by which the Sandpiper deposit can be developed (Midgley 2012)	156
Fig. 2	The proposed dredging cycle (Midgley 2012)	156
Fig. 3	Global deliveries of phosphate rock (IFA 2012)	158
Fig. 4	Production of phosphate rock in the top four producing countries (IFA 2012)	158
Fig. 5	Phosphate mining and primary processing recovery (Prud'homme 2010)	161
Fig. 6	Phosphate rock concentrate—trend of grade (Prud'homme 2010)	162

Fig. 7　The Total Resource Box illustrates the interrelationship and dynamics of reserves, resources, and geopotential; the area included by the dashed line outside the reserves box marks the resources and geopotential that will be converted to reserves next. 177

Chapter 5

Fig. 1　The primary pathways of P extraction, use, and loss. Opportunities exist at each step for improved P use efficiency 210

Fig. 2　Technical use of P. 240

Chapter 6

Fig. 1　Substance flows of P in the Japanese economy (2005). *Source* (Matsubae et al. 2011) 251

Fig. 2　Virtual P ore requirement of the Japanese economy (2005). *Source* (Matsubae et al. 2011) 252

Fig. 3　Simplified illustration of P balance in Switzerland. *Source* (AWEL 2008) 253

Fig. 4　Locations of hypoxic areas, areas of concern, and areas of recovery. *Source* (World Resources Institute 2012) 255

Fig. 5　Diagram of complete P recovery from sewage sludge and incineration ashes. *Source* (Petzet and Cornel 2010; Petzet and Cornel 2011). 259

Fig. 6　Possibilities of P recycling in agricultural and industrial sectors. *Source* (Ohtake 2010) 267

Chapter 7

Fig. 1　Global fertilizer trade and production (*Source* ICIS) http://www.icis.com/resources/fertilizers/trade-flow-map-2014/. 278

Fig. 2　Phosphate rock prices from January 1960 to April 2012 [*Source* World Bank (www.econ.worldbank.org)] 280

Fig. 3　Fertilizer price index from January 1960 to April 2012 [*Source* World Bank (www.econ.worldbank.org)] 281

Fig. 4　Annualized volatility of real prices of fertilizer from 1960 to 2011 and standardized around 2000 (*Source* Calculation by European Commission/Agrilife Unit; data World Bank) 281

Fig. 5　Prices for commodities used for fertilizer, other commodities, and the food price index 285

Fig. 6　Development of PR, commodity prices, and the stock market. . . 285

Fig. 7　Production costs of phosphate rock in different world regions and price of purchase rock (*Source* FERTECON Research Center). 287

Fig. 8　Factors influencing the price of mineral fertilizers. 290

Tables

Chapter 1
Table 1 Cropland area, phosphorus (*P*) input per ha of cropland, P uptake, and average P consumption in two periods, according to Sattari et al. (2012) 14
Table 2 IFA and IFDC recording of "ore residues" (for a definition of "ore residues" see text) based on the USGS (2012) with linear extrapolation (assuming the same ore grade in the processed phosphorus and the ore residues)...... 50
Table 3 Phosphorus contents in different animal and crop products (Lamprecht et al. 2011, most data refer to Swiss or German reference values) 63
Table 4 Estimates of macronutrient content in crops, crop residues, and inorganic fertilizer in terms of MT year^{-1} (taken from Smil 1999)............................. 64
Table 5 Average N, P, and K content of different types of manure in % of dry matter from different livestock (Schnug et al. 2011) 64
Table 6 Key events and focus-guiding themes of the seven key meetings of Global TraPs[1] to be determined).......... 88
Table 7 Principles of mutual learning sessions.................. 94
Table 8 Regional settings of the two study areas 126

Chapter 4
Table 1 Illustrating type of case study, function/role of transdisciplinary process and potential guiding question 194

Chapter 5
Table 1 Overview of phosphate containing applications (non-exhaustive) *References* Budenheim (2013), Emsley (2000), Phosphate Facts (2013), Prayon; Villalba et al. (2008)............................... 239

Chapter 6
Table 1 P content in manure 258

Chapter 7
Table 1 Regression coefficients and *p*-values from Wald statistic to study association between raw materials and fertilizer prices in two different time periods 287

Appendices

Chapter 1
Appendix: Spotlight 1	Fertilizers Change(d) the World. Amit H. Roy, Deborah T. Hellums, Roland W. Scholz, and Clyde Beaver....................	114
Appendix: Spotlight 2	A Novice's Guide to Transdisciplinarity. Roland W. Scholz and Quang Bao Le	118
Appendix: Spotlight 3	The Yen Chau—Hiep Hoa Case Study: Avoiding P Fertilizer Overuse and Underuse in Vietnamese Smallholder Systems. An Example of How a Transdisciplinary Case Study in the Use Node May be Developed. Quang B. Le and Roland W. Scholz	123

Chapter 3
Appendix: Spotlight 4	Phosphorus Losses in Production Processes Before the "Crude Ore" and "Marketable Production" Entries in Reported Statistics. Roland W. Scholz, Friedrich-Wilhelm Wellmer, and John H. DeYoung, Jr................	174

Chapter 4
Appendix: Spotlight 5	Options in Processing Manure from a Phosphorus Use Perspective. Diane F. Malley......................	204

Chapter 5
Appendix: Spotlight 6	Health Dimensions of Phosphorus. James J. Elser	229
Appendix: Spotlight 7	Phosphorus in the Diet and Human Health. Rainer Schnee, Haley Curtis Stevens, and Marc Vermeulen	232

Appendix: Spotlight 8	Technological Use of Phosphorus: The Non-fertilizer, Non-feed and Non-detergent Domain. Oliver Gantner, Willem Schipper, and Jan J. Weigand	237
Appendix: Spotlight 9	Phosphorus in Organic Agriculture. Bernhard Freyer	243

Chapter 7

Appendix: Spotlight 10	Phosphorus and Food Security from a Greenpeace and Indian Smallholder Farmer View. Olaf Weber, Jacques Delince, Yayun Duan, Luc Maene, Tim McDaniels, Michael Mew, Uwe Schneidewind, and Gerald Steiner.	295

Contributors

Clyde Beaver International Fertilizer Development Center, Muscle Shoals, AL, USA

Claudia Binder Department of Geography, University of Munich, Munich, Germany

Louw Botha Foskor, Phalaborwa, South-Africa

Fridolin S. Brand Institute for Environmental Decisions, ETH Zürich-Natural and Social Science Interface, Zürich, Switzerland

Cynthia Carliell-Marquet School of Civil Engineering, University of Birmingham, Birmingham, UK

Jaques Delince DG Joint Research Centre, European Commission, Brussels, Belgium

John H. DeYoung U.S. Geological Survey, National Minerals Information Center, Reston, VA, USA

Changqun Duan Institute of Environmental Science and Restoration Ecology, Yunnan University, Kunming, People's Republic of China

James J. Elser School of Life Sciences, Arizona State University, Tempe, AZ, USA

Emmanuel Frossard Institute of Agricultural Sciences, Group of Plant Nutrition, ETH Zürich, Zürich, Switzerland

Oliver Gantner Resource Strategy, University of Augsburg, Augsburg, Germany

Debbie T. Hellums International Fertilizer Development Center, Muscle Shoals, AL, USA

Ludwig Hermann Outotec GmbH, Oberursel, Germany

Yuliya Kalmykova Department of Civil and Environmental Engineering, Chalmers University of Technology, Gothenborg, Sweden

Tobias Katz Institute of Mining Engineering I, RWTH Aachen, Aachen, Germany

Daniel J. Lang Chair for Transdisciplinary Sustainability Research, Leuphana University of Lüneburg, Lüneburg, Germany

Kees Langeveld ICL Fertilizers Europe, Amsterdam, The Netherlands

Quang Bao Le Institute for Environmental Decisions, Natural and Social Science Interface, ETH Zürich, 8092 Zürich, Switzerland

Luc Maene International Fertilizer Industry Association, Paris, France

Diane F. Malley Faculty of Agricultural and Food Sciences, Department of Soil Science, PDK Projects, Inc., University of Manitoba, Winnipeg, Canada

Kazuyo Matsubae-Yokoyama Graduate School of Environmental Studies, Tohoku University, Tohoku, Japan

Makiko Matsuo Graduate School of Public Policy, University of Tokyo, Bunkyo, Japan

Tim McDaniels School of Community and Regional Planning, and Institute for Resources, Environment and Sustainability, University of British Columbia, Vancouver, Canada

Michael Mew CRU Fertecon Research Centre Ltd, Twickenham, UK

Robert Mikkelsen International Plant Nutrition Institute, Merced, CA, USA

Leo Morf Office of Waste, Water, Energy and Air, Zürich, Canton, Switzerland

Hisao Ohtake Department of Biotechnology, Graduate School of Engineering, Osaka University, Suita, Japan

Alan P. Omlin Graduate School of Frontier Sciences, Graduate Program in Sustainability Science, University of Tokyo, Tokyo, Japan

Sebastian Petzet Institute IWAR, Technische Universität Darmstadt, Darmstadt, Germany

Armin Reller Resource Strategy, University of Augsburg, Augsburg, Germany

Amit H. Roy International Fertilizer Development Center, Muscle Shoals, AL, USA

Willem Schipper Willem Schipper Consulting, Middelburg, The Netherlands Formerly, Thermphos International, Vlissingen, The Netherlands

Uwe Schneidewind Wuppertal Institute for Climate, Environment and Energy, Wuppertal, Germany

Roland W. Scholz Fraunhofer IWKS, Zürich, Germany; Institute for Environmental Decisions, Natural and Social Science Interface, ETH Zürich, Zürich, Switzerland

Hideaki Shiroyama Graduate School of Public Policy, University of Tokyo, Bunkyo, Japan

Gerald Steiner Weatherhead Center for International Affairs, Harvard University, Cambridgs, MA, USA

Haley Curtis Stevens International Food Additives Council, Atlanta, GA, USA

Christopher Thornton Global Phosphate Forum, Boutgoin Jallieu, France

Reyes Tirado Greenpeace International, University of Exeter, Exeter, Great Britain, UK

Andrea E. Ulrich Institute for Environmental Decisions, ETH Zürich-Natural and Social Science Interface, Zürich, Switzerland

David A. Vaccari Departmemt of Civil, Environmental and Ocean Engineering, Stevens Institute of Technology, Hoboken, NJ, USA

Peter van Straaten Department of Land Resource Science, University of Guelph, Guelph, ON, Canada

Marc Vermeulen European Chemical Industry Council-Phosphoric Acid and Phosphates, Brussels, Belgium

Ulli Vilsmaier Center for Methods, Leuphana University of Lüneburg, Lüneburg, Germany

Krishnan Vijoo All India Kisan Sabha (AIKS: All India Peasants' Union), Patna, India

Ingrid Watson Centre for Sustainability in Mining and Industry, Johannesburg, South Africa

Paul Watts Institute of Arctic Ecophysiology, Churchill Canada and DALUHAY, Mandaluyong City, Phillippines

Olaf Weber Faculty of Environment, Enterprise and Development (SEED), Export Development Canada Chair in Environmental Finance, School of Environment, University of Waterloo, Waterloo, Canada

Jan J. Weigand Department of Chemistry and Food Chemistry, TU Dresden, Dresden, Germany; Institute of Organic and Analytic Chemistry, University of Münster, Münster, Germany

Friedrich-Wilhelm Wellmer Formerly Bundesanstalt für Geowissenschaften und Rohstoffe (BGR) Geozentrum, Stilleweg 2, direct mail Neue Sachlichkeit 32, 30655, Hannover, Germany

Masaru Yarime Graduate School of Public Policy, University of Tokyo, Tokyo, Japan

Nomenclature

mg	Milligramme (10^{-3} g)
kg	Kilogramme (10^3 g)
kt	Thousand tonnes (10^9 g)
Mt	Megatons = Million tonnes (10^{12} g)
Gt	Gigatons = Billion tonnes (10^{15} g)
K	Potassium
N	Nitrogen
P	Phosphorus
PR	Phosphate Rock
DAP	Diammonium Phosphate
NP	Nitro Phosphate
MAP	Monoammonium Phosphate
SSP	Single Superphosphate
TSP	Triple Super Phosphate
MFA	Material Flow Analysis
SFA	Substance Flow Analysis

Basic *conversion factors* among phosphate rock, phosphorus-pentoxide, and phosphorus as used in common annual statistics:

1 t PR contains according to common notation about 300 kg (i.e., 30 %) P_2O_5
1 t P_2O_5 contains a mass of 436 kg (i.e., 43.6 %) P
1 t PR includes 130 kg (i.e., 13 %) P

Basic *data* on phosphorus production and reserved in 2011 (in parenthesis [] in 2012) according to the reports of USGS (2012, [2013[1]]):

---

[1] All references from this page are referring to http://minerals.usgs.gov/minerals/pubs/commodity/phosphate_rock/. Please note that USGS (2012) reports that 191 Mt PR have been produced in 2011. This has been re-adjusted to 198 Mt PR in USGS 2013 report. Both data are used in the different section of this book. Please check the year of reference which is taken for USGS publication.

PR production in 2011: 191 Mt [198 Mt]
P_2O_5 production in 2011: 57.3 Mt [59.4]
P consumption from phosphate rock ore in 2011: 25.0 Mt [2011 = 25.9 Mt]
PR production in 2012[2]: 210 Mt [2011 = 191 Mt]
P reserves amount to 71 Gt P [67 Gt P][3]

The cumulatively world production of P from 1900 till 2012 according to USGS (2013) amounts to 7.6 Gt P.

---

[2] USGS (2013).

[3] The lower number is due to the fact that the reserves for Iraq, (which have been first recorded in 2012) was adjusted to a lower number because they were based on a Russian classification of reserves than the common US classification applied for other countries (Jasinski, personal communication, 5. Feb. 2012).

Chapter 1
Sustainable Phosphorus Management: A Transdisciplinary Challenge

Roland W. Scholz, Amit H. Roy and Deborah T. Hellums

Abstract This chapter begins with a brief review of the history of phosphorus, followed by a description of the role of phosphorus in food security and technology development. It is then followed by discussions on critical issues related to sustainable phosphorus management, such as phosphorus-related pollution, the innovation potential of phosphate fertilizers and fertilizer production, uneven geographical distribution of phosphate resources, transparency of reserves, economic scarcity, and price volatility of phosphate products. In order to identify the deficiencies in the world's phosphorus flows, we utilize the "not too little–not too much" principle (including the Ecological Paracelsus Principle), which is essential to understanding the issues of pollution, supply security, losses, sinks and efficiency of phosphorus use, and the challenges to closing the phosphorus cycle by recycling and other means. When linking the supply–demand (SD) chain view on phosphorus with a Substance or Material Flux Analysis, the key actors in the

---

Electronic supplementary material Supplementary material is available in the online version of this chapter at (http://dx.doi.org/10.1007/978-94-007-7250-2_1) contains supplementary material, which is available to authorized users.

An erratum to this chapter is available at 10.1007/978-94-007-7250-2_8

R. W. Scholz (✉)
Fraunhofer Project Group Materials Recycling and Resource Strategies IWKS,
Brentanostrasse 2, 63755 Alzenau, Germany
e-mail: roland.scholz@isc.fraunhofer.de

ETH Zürich, Natural and Social Science Interface (NSSI), Universitaetsstrasse 22, CHN J74.2, 8092 Zürich, Switzerland

A. H. Roy · D. T. Hellums
International Fertilizer Development Center (IFDC), P.O. Box 2040, Muscle Shoals, AL 35662, USA
e-mail: aroy@ifdc.org

D. T. Hellums
e-mail: dhellums@ifdc.org

global phosphorus cycle become evident. It is apparent that sustainable phosphorus management is a very complex issue that requires a global transdisciplinary process to arrive at a consensus solution. This holds true both from an epistemological (i.e., knowledge) perspective as well as from a sustainable management perspective. To gain a complete picture of the current phosphorus cycle, one requires knowledge from a broad spectrum of sciences, ranging from geology, mining, and chemical engineering; soil and plant sciences; and all facets of agricultural and environmental sciences to economics, policy, and behavioral and decision science. As phosphorus flows are bound to specific historical, sociocultural, and geographical issues as well as financial and political interests, the understanding of the complex contextual constraints requires knowledge of related sciences. The need for transdisciplinary processes is equally evident from a sustainable transitioning perspective. In order to identify options, drivers, and barriers to improving phosphorus flows, one requires processes in; capacity building that may be changed and consensus building on the phosphorus use practices that must be changed and maintained, along with recognition of how changes in phosphorus use in the current market may be framed. The latter is illustrated by means of the Global TraPs (Global Transdisciplinary Processes for Sustainable Phosphorus Management) project, a multi-stakeholder initiative including key stakeholders on both sides of the phosphorus SD chain which includes mutual learning between science and society.

Keywords Sustainable phosphorus management • Supply–demand chain analysis • Food security • Environmental impacts

Contents

1	New Perspective on Phosphorus Management	3
	1.1 What's New About This Book?	3
	1.2 Learning from Phosphorus History: Light, Fertilizers, and a Conflict of Interest	5
2	The Role of Phosphorus in Food Security and Technology Development	9
	2.1 Increasing Demands for Phosphorus in the Future	10
	2.2 Different Phosphorus Demand Trends in Different Parts of the World	12
	2.3 Increasing Efficiency: "Save and Grow" for Food Security?	16
	2.4 Phosphorus and Biofuel	19
	2.5 Virtual Phosphorus Flows	19
	2.6 Phosphorus and Technology Development	20
3	Critical Issues of Phosphorus Management: Phosphorus as a Case of Biogeochemical Cycle Management	20
	3.1 What is Critical?	20
	3.2 Phosphorus as a Pollutant	22
	3.3 Human Health and Phosphorus Food Additives	24
	3.4 Innovation Potential of Fertilizers	25
	3.5 Geographical Distribution of Phosphorus Reserves and Resources	28

3.6	Phosphorus Scarcity: Physical or Economic?	30
3.7	Inefficient Use	33
3.8	Price Volatility	34
4	What is Wrong with the World's Phosphorus Flows?	36
4.1	The Dose Matters: The Ecological Paracelsus Principle	37
4.2	Finiteness: Securing Long-Term Provision of a Public Good	39
4.3	Different Phosphorus Input and Balances in Various Parts of the World	40
4.4	Do We Have Low or High Nutrient Efficiency?	46
4.5	"Losses" and "Sinks" of Phosphorus from a Supply–Demand Chain Perspective	46
4.6	Runoff and Erosion	54
4.7	Changing Lifestyles Increase Anthropogenic Phosphorus Flows	57
4.8	Insufficient Phosphorus Recycling from Manure and Crop Residues	58
4.9	Phosphorus Recycling from Animal Carcass and By-products	64
4.10	Phosphorus Recycling from Sewage	64
4.11	Food Waste	66
4.12	How May We Define Sustainable Phosphorus Use?	66
5	CLoSD Chain Management	68
5.1	The Vision of Closing Anthropogenic Material Flows	68
5.2	A Blueprint of Global Phosphorus Flows	69
5.3	Actor-Based MFA for Changing Flows	80
5.4	Supply–Demand Analysis for Improving Technologies	81
6	The Global TraPs Project: Goals, Methodology, Organization, and Products	83
6.1	What is Transdisciplinarity?	83
6.2	How is the Global TraPs Project Organized?	85
6.3	Knowledge Integration and Mutual Learning as Components of the Global TraPs Project	88
6.4	Mutual Learning Sessions and Dialogue Sessions as Instruments of Transdisciplinary Processes	89
6.5	Transdisciplinary Case Studies	91
7	The Challenge of Increasing Efficiency, Avoiding Environmental Pollution, and Providing Accessibility	92
7.1	Efficiency as an Indicator of Unsustainable Phosphorus Use	94
7.2	Avoiding Pollution from Phosphorus	95
7.3	Securing Access to Phosphorus	97
7.4	A Transdisciplinary Roadmap Toward a Demand-Based Peak Phosphorus	98
References		100

1 New Perspective on Phosphorus Management

1.1 What's New About This Book?

Many books, theses, and papers have been written from various perspectives on phosphorus. The present book targets *global sustainable phosphorus management*. *Transdisciplinarity*, whose core is *the integration of knowledge between practice and science*, is seen as a means by which a sustainable transition toward global efficiency may be achieved. This is a new concept, and the subject of a comprehensive project, the Global TraPs (Global Transdisciplinary Processes for Sustainable Phosphorus management) project, which was officially launched on 6 February 2011.

This book explains what *sustainable phosphorus management* may mean, why we need transdisciplinarity (as defined in the context of the Global TraPs project) and why phosphorus (P) is so distinctive that it may serve as a learning case for any *global biogeochemical cycle management*. In the case of P, sustainable biogeochemical cycle management includes *pollution prevention, resource conservation, technology development,* and *knowledge generation* in a way that future generations may have efficient access to P. Closing the fertilizer loop from the mining of phosphate rock to its use, at least to some extent, is definitely one means. However, this may only be accomplished if we have a clear view of how losses and sinks of phosphorus are related to the actions of the key stakeholders. Clearly, phosphorus atoms do not disappear from earth. We denote those fractions that have been excluded from the value chain by human action (such as phosphorus in mining waste, sewage, and manure) as losses. This is why we take a supply–demand chain perspective. Here, demand is explicitly mentioned because phosphorus is essential, and human life is inextricably linked to the use of considerable amounts of P, particularly for food production. Chapters 2–6 of this book, like the Global TraPs project itself, are structured to follow the stages of the supply–demand chain, which are *Exploration, Mining, Processing, Use* and *Dissipation and Recycling*. A large share of phosphorus flows is tied to economic transaction. Thus, a chapter *on Trade and Finance*, which addresses critical aspects such as the origins of price peaks, is included in this book.

The conceptual vision is elaborated in the *Closed-Loop Supply–Demand* Chain Management of the *anthropogenic* portion of (CloSD Chain Management) phosphorus flows in this chapter. The concept makes reference to ideas in industrial ecology such as "loop closing" (Lifset and Graedel 2002), or "from cradle to cradle" (McDonough et al. 2003, SI 7), stressing the need for recycling. Special emphasis is given to the economic perspective. This is also indicated by including a demand perspective. As Scholz and Wellmer (2013) point out, phosphorus is a demand-driven market rather than a supply-driven market. There is a steady but—as phosphorus is essential and mineral fertilizers a key element of current food supply security, see Spotlight 1—limitedly adaptable demand function and, compared with other minerals and metals, rather abundant resources. Thus, we face a demand-driven market and must understand how the demand side may be affected by technology, population growth, lifestyles, etc.

Though CloSD Chain Management may be considered a necessary condition of sustainable phosphorus management, it is by no means a sufficient one. Sustainability goes beyond the environmental, economic or technological dimensions and includes *social* (Brundtland 1987) and *equity* (Laws et al. 2004) dimensions. The *social dimension* is certainly the most difficult and challenging, but is of major importance. We can easily illustrate this dimension by reviewing the case of sub-Saharan Africa's smallholder farmers. Most of these smallholders live in countries whose soils have the highest need for fertilizers. *Smallholder farms* in this region constitute 80 % of African agrarian land (IFAD 2011), yet they are the most disadvantaged with respect to soil fertility and other factors (i.e., erosion). As a result, the percentage of undernourished within the African rural population is about 16 % (FAO 2010b). In addition to having a large share of highly weathered soils with low nutrient content, some soils in sub-Saharan Africa tend to bind phosphorus (significantly reducing phosphorus available to plants). Gaining access to P, therefore, is fundamental to improving the productivity and livelihoods of smallholder farmers. As documented in 2009, Africa's soils received on the average (including countries with large-scale agroindustrial plantations) only 2.48 kg P ha^{-1} year^{-1}, whereas the soils of Europe and North America, which have higher loads of soil phosphorus from centuries of fertilizer use received 18.7 and 19.12 kg P ha^{-1} year^{-1}, respectively, during the same period (FAO 2012b). Thus, African farmers, and many others who do not practice balanced fertilization, are removing a larger portion of one or more nutrients from the soil (through harvested crops) than is being added (through organic amendments and mineral fertilizers) on an annual basis. Clearly, many smallholder farmers in developing countries are neither able to access manufactured fertilizers, nor do they have access to technologies that promote efficient fertilizer use. In many developing countries, providing access to phosphorus and other nutrients is essential to improving food security.

This book, in support of the Global TraPs project mission, states that we must not only learn from the different stakeholders about their knowledge and cultural backgrounds, but we also must learn from history to better understand the role of phosphorus in biotic and abiotic processes. Ultimately, we must use this knowledge to change current use practices. To that end, we ask the reader to review the brief history of phosphorus that follows.

1.2 Learning from Phosphorus History: Light, Fertilizers, and a Conflict of Interest

In ancient times, the planet Venus was referred to as "phosphorus" by the Greeks (Wisniak 2005). As indicated by its etymological meaning (phosphorus: light bearer; from phos "light" [related to phainein, "to show, to bring to light": see phantasm] + phoros "bearer," (OED 2012)]), the earliest interest in phosphorus was as a lighting element. History indicates that in 1669, the alchemist Henning Brand (c.1630–c.1710) "rediscovered" the element and the procedure to generate phosphorus (Krafft 1969), which entailed boiling silver pieces in urine, drying the silver, mixing it with sand (silica), heating the mixture and collecting the resulting yellowish mass in a condenser. This mass caught fire easily when exposed to air (at ambient temperature). He was believed to have discovered a "black" substance, a "Prima material," or "elemental 'fire,'" i.e., "one of the four Aristotelan 'elements,' earth, water, air, and fire" (Krafft 1969). Giants of the history of science such as Gottfried Wilhelm Leibniz (1646–1716)—who wrote the "Historia inventionis phoshori" (1710)—and Christiaan Huygens (1629–1695) were involved in documenting the procedure, which was run with "a full ton of urine" (Leibniz 1710). Through the work of Lavoisier (1743–1794, Lavoisier 1776), phosphorus became the 13th element in the history of the discovery of elements (Emsley 2000a).

Phosphorus was of *commercial interest* from the very beginning. Alchemists rigorously explored the element, and pharmaceutical companies found uses for it soon after its discovery (Richmond et al. 2003). Eben Norton Horsford (1818–1893), chemist working on phosphorus and a scholar of Justus von Liebig and Professor at Harvard of the Application of Science to the Useful Arts, was the inventor of baking powder. He became cofounder of Rumsford Chemical Works and promoted the selling of phosphate acid for medical purposes (Jackson 1892, see Fig. 1). Large-scale match production began in the early nineteenth century (see Fig. 2) and phosphorus bombs became warfare agents.

After Brand's discovery, "for a century, urine was the only source from which phosphorus was attained" (Färber 1921). But in 1769, Carl Wilhelm Scheele (1742–1786) and Johan Gottlieb Gahn (1745–1818) discovered phosphoric acid in animal bone and many other animal parts (Färber 1921; Petroianu 2010). Théodore de Sassure stated in 1804 that, "we had no means to believe that plants can exist without phosphorus" (Färber 1921, p. 11). These statements were later proven by the emerging experimental "Animal and Vegetable Chemistry" (Dumas and Boussingault 1844), which provided insight into the metabolic nature of plant physiology (Liebig 1840) and set the foundation of *nutrient balance*, which suspected that deficiency of phosphorus was the limiting factor in plant growth (Liebig's Law of the Minimum; see Paris 1992).

But farmers knew about phosphorus long before the modern scientific community. Phosphorus has been used in agriculture, even if unknowingly, since prehistoric times. In fact, archeologists use phosphorus as a tracer element for human settlements (Schlezinger and Howes 2000), and fertilizer use can be traced back to at least the third millennium B.C. (Wilkinson 1982). The Inca civilization used guano as fertilizer (de la Vega 1609/1990). Roman agriculture included manures for crops in the first century (Lelle and Gold 1994), and the use of bones as fertilizer was reported by Walter Blithe (1605–1654, Brand 1937).

Without understanding its scientific properties, farmers utilized the phosphorus and other macronutrients in manure, excrements, and bones as fertilizer. Eventually, scientists learned from these ancient practices, and farmers adapted quickly. As an example, desperate farmers were said to have raided Napoleonic battlefields such as Waterloo (1805) and Austerlitz (1815) to collect human bodies for their phosphorus contents [Hillel, 1991; cited in Foster (1999)]. In his book, *Farmers of forty centuries: organic farming in China, Korea and Japan,* Franklin H. King provides us with another example: "Manure of all kinds, human and animal, is religiously saved and applied to the fields in a manner which secures an efficiency far above our own practices" (King 1911/2004). "This was not done directly, but potential fertilizer such as river mud" was often dried and pulverized before being carried back and used on the fields as makeshift fertilizers (p. 8).

In 1804, Alexander von Humboldt (1769–1859) observed that Peruvian fields were fertilized with guano. He took samples to Europe where chemists noticed high levels of nitrogen (N) and phosphorus (von Pier 2006). In the period between 1857 and 1867, about 50,000 metric tons (mt) of guano were imported annually by Europe (Färber 1921). But technological progress opened other options.

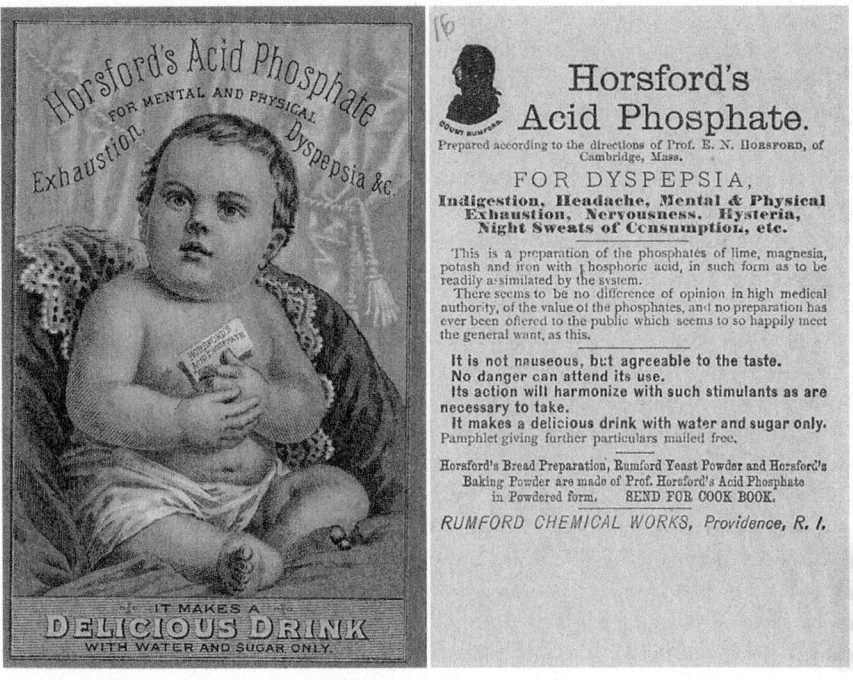

Fig. 1 Phosphate acid was seen as remedies for many diseases (*Source* Hulman & Co)

Fig. 2 Phosphorus may have positive and, in certain forms, negative effects. This picture shows a former employee of the Reliable Match Company of Ashland, Ohio, who died in 1912 due to exposure to *white P*. The disease, which was labeled "phossy jaw," recently reappeared with patients who were treated with *bisphosphonate*, a P-based medicine (Body 2006). Thus, the new term, "bis-phossy jaw," emerged (Hellstein and Marek 2004). Picture taken from the "A last victim," *The Survey* 28, no. 2, 1921 after a reproduction of Moss (1994)

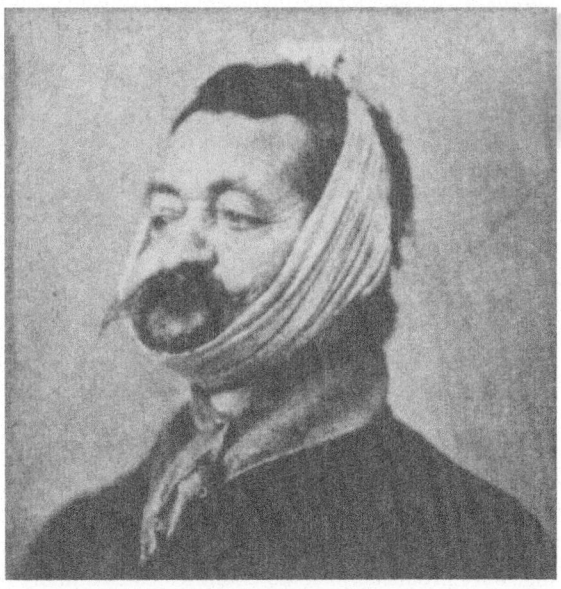

The history of phosphorus fertilizer (organophosphate) invention, and essentially the beginning of the fertilizer industry, was written primarily by three icons of their times, Sir James Murray (1788–1871), Sir John Bennet Lawes (1814–1900), and Baron Justus von Liebig (1803–1873). When referring to experiments in converting "bones to biophosphate of lime as fertilizer, using sulfuric acid," Lawes is noted to have carried out the first field experiments mixing wastes, compost, and manure in 1817. He then experimented with different mixtures of nutrients (Alford and Parkes 1953; Childs 2000). In 1842, trials were carried out to compare the new fertilizer with manure (Childs 2000). Here, superphosphate, a composition of calcium hydrogen phosphate and calcium sulfate was applied. Practically, rock phosphate was treated with sulfuric acid. Lawes bought patents from Murray and Liebig and, in 1843, founded the Rothamsted Experimental Station in an area of the United Kingdom that was extremely deficient in nutrients due to centuries of nutrient-extractive agriculture. In France, Boussignault demonstrated the synergetic effects of P, N, and other minerals. Quickly, the chemical and manure industries developed to fill the growing agricultural need (Daly 1984) to avert famine.

The development of scientific knowledge in the nineteenth century may be considered a history of errors, as even the greatest knowledge of that time was incomplete. This point may be highlighted with the earliest of Liebig's seminal contributions. Initially, Liebig's patented fertilizer proved to be a failure, as it contained no N or potassium (K), and phosphorus was present in an unavailable form. Liebig corrected the latter idea—that soluble forms of phosphorus would be washed away from the soil by rainwater—when he realized that soluble phosphorus was essential for plant growth (Emsley 2000b; Oertli 2008).

Seventy-five years ago, the executive secretary of The US National Fertilizer Association wrote:

> Not so many years ago, the fertilizer industry was largely a waste-products industry. The bone, blood, and tankage of the packaging industry, the fleshings and scraps of the leather industry, the slops of the beet sugar industry and the meal residues of the vegetable oil industry made up the greater part of mixed fertilizer." And he noted that buyers had been "more impressed ... by ... odor than by chemical composition or guarantee of plant food content. (Brand 1937)

The idea to solubilize phosphorus in bones by sulfuric acid (transforming slow release calcium phosphate to superphosphate) was also transferred to phosphate rock. Single superphosphate, triple superphosphate (monocalcium phosphate), and diammonium phosphate became the pillars of the phosphate industry.

But historically, fertilization has been only one of the uses for P. Boyle discovered in 1680 that when sulfur and phosphorus were rubbed together, they caught fire. It took about 140 years until "Lucifers," the original name for contemporary "strike anywhere" matches (Battista 1947), were invented. Excessive exposure to *White* (also called yellow) phosphorus (P_4O_{10})-containing matches that were produced in some countries caused many diseases such a "phossy jaw," a variant of bone cancer. This particular disease was first diagnosed in Vienna

Fig. 3 The Berne Convention of 1906 banned the highly toxic *white phosphorus* from matches. Whereas no biomass may emerge without P, matches without *white phosphorus* and household detergents without phosphorus could be produced to avoid critical collateral impacts (*left* picture taken from Andrews (1910) after a reproduction by Moss (1994), *right* picture after courtesy of the South East Regional Centre for Urban Landcare, Brisbane, Australia)

(Moss 1994), where its P-related etiology was proven (Marx 2008). Young girls who carried matchboxes on their heads became bald (Datta 2005). The import and sale of matches containing white phosphorus were banned by many European countries under the Berne Convention of 1906. "In the United States, nearly all interested parties supported legal abolition, but… no state wanted to be the first to act (for the fear of driving industry from its borders), and the federal government lacked the power to regulate intrastate economic activity …" (Moss 1994). The necessity that global phosphorus management should advocate for international action may be well-learned from this case.

White phosphorus (P_4) is still in use for match production in developing countries and is permitted for contemporary epidemiological studies (González-Andradea et al. 2002). The lethal dose is about 1.0 mg/kg weight in adults (Gossel and Bricker 1994). The critical toxicity of *White* phosphorus can be demonstrated in a new form of "phossy jaw," the "bis-phossy jaw," which is observed in people who are treated with bisphosphonate to combat bone necrosis (about 10 % of the human bone is P, see Fig. 3).

It should be noted that *phosphorus* does not appear in a pure form in nature, but rather, is generally observed in the oxidized form of *phosphate* (PO_4) which becomes *organophosphate* such as DNA if it is bound with organic compounds. This book deals with the chemical element phosphorus, which is denoted as P, though occasionally phosphorus also denotes phosphate in the context of this publication.

2 The Role of Phosphorus in Food Security and Technology Development

"Producing enough food for the world's population in 2050 will be easy." This is the first sentence of a recent Editorial in *Nature* (2010) in a series on world food systems. Given the possible doubling of the food demand in the next 50 years (Tillman et al. 2002), *Nature* undoubtedly makes an extremely optimistic statement. This optimism is linked to what many envision as a second *Green Revolution* and the "sustainable intensification of global agriculture." But under what constraints is this possible? And what role does nutrient management play in general, and phosphorus management in particular, in this vision? The optimistic view has been criticized as far too simplified, or even naïve, due to a singular focus on technology that does not take into account the social dimension, resource dynamics or environmental issues. The authors of this chapter do not believe that the challenge of feeding the world in 2050 will be easy, but it will be possible if a proper system view is taken, which allows us to better understand the demands for phosphorus posed by human systems.

2.1 Increasing Demands for Phosphorus in the Future

Phosphorus is an *essential* element for any living organism, as it cannot be substituted by another element. DNA, the basic building block of life itself, consists of carbon, hydrogen, oxygen, nitrogen, and phosphorus. Phosphorus is also the key component of the "workhorse" molecule, adenosine triphosphate (ATP), which provides the energy to keep cells alive and active. But phosphorus plays many other roles in the body as well; phosphorus is a component of the lipids that make up cell membranes. Phosphorus deficiency is often the *limiting factor* of plant growth (de Vries 1998). Thus, phosphorus is a *critical* element in food security: a shortage of phosphorus in any agrosystem results in low agricultural productivity that, in many cases, may cause undernourishment and, in extreme cases, famine (Ragnarsdottir et al. 2011; Sanchez and Swaminathan 2005). The impetus for the *Green Revolution* that began in the 1940s was a world *population increase* that required an exponential increase in world food supplies. This new age of agriculture relied heavily on mineral fertilizers and other agrotechnological innovations such as new higher-yielding seeds (including the breeding of modern varieties; see Evenson and Gollin 2003), expansion of irrigation systems (with higher groundwater depletion and energy costs), pesticide and herbicide development and application, more efficient agricultural machinery, intensification of crop and grazing land areas, and better means of education.

Today, humankind faces a new challenge as it makes its way into the twenty-first century. The United Nations (2009) projected the population could increase by more than 30 % to 9.2 billion people, by 2050. This estimate reflects more than a tripling of the population in 90 years as there were 3 billion people in 1960. According to latest projections, total population in 2050 may be 9.6 billion (UN 2013). If the population growth trend over the last five decades continues, there may be more than 1 billion undernourished people, considering that 1 billion have been attained in 2009 (FAO 2009). In percentage terms, the number of people starving has declined in the last centuries. A firm projection is difficult as we are witnessing opposing trends, of reduction (e.g., in China or Vietnam) and increase (e.g., India and Pakistan) in the number of undernourished people (Cuesta 2013; Fan et al. 2013). During the next decades, there will be additional demands for phosphorus due to an increased demand for meat and other dietary changes (particularly as emerging nations become more developed). As phosphorus is a cornerstone of food production, world phosphorus management may be challenged in the coming decades.

There were 25 megatons (Mt) phosphorus (corresponding to 191 Mt of phosphate rock [PR]) produced and recorded in the Mineral Commodity Summary of USGS in 2011 worldwide (Jasinski 2012). The importance of the availability of chemical fertilizer for today's world food system may be taken from the following data: in the year 2000, there were 14.2 Mt of phosphorus fertilizer used (given a total production of 18.1 Mt P [calculated from 139 Mt total production of phosphate rock concentrate, according to Jasinski, 2001]) compared with 9.6 Mt of manure produced for crop production. Thus, about 13 kg P ha^{-1} year^{-1} was used on farmland including pastureland (MacDonald et al. 2011a).

On average, each person consumes the equivalent to about 31 kg of phosphate rock per year (Scholz and Wellmer 2013). We may take from Fig. 7 that the average PR demand for a world citizen decreased slightly after the decline of the Soviet Union around 1990 but is currently forecasted to increase. Sustainable phosphorus management seeks to more efficiently use phosphorus and to reduce the relative (kg PR consumed per person annually by means of increasing efficiency) and the absolute consumption (i.e., decreasing the Mts of mineral phosphorus which are inserted into the ecosystems, see Sects. 3.2 and 4.1) per year. Fertilizer is the main segment of phosphorus use (see Fig. 4). But between 10 and 15 % are used for other purposes. In 2011, from a total production of 25 Mt P, the category of food and dark soft drinks accounted for 2 % (0.50 Mt P), while animal feed additives amounted to 7 % (1.74 Mt P). In addition, there is sodium triphosphate (STPP). Most STPP is used for detergents and a broad set of cleaning products. Prud'homme (2010) estimates that phosphorus in STPP amounts to 8 % [4 Mt P_2O_5 year^{-1} out of 49.5 Mt P_2O_5 year^{-1} which would correspond to 1.7 Mt P year^{-1} for STPP out of 21.6 Mt P year^{-1} used as reference in Prud'homme (2010), see Fig. 4]. These 8 % include a wider range of industrial uses of STPP. An estimate of Shinh (2012) of STPP is much smaller estimating about 0.93 Mt P year^{-1} for STPP in 2011. Shinh states that the use of STPP for detergents which historically made up a major share reduced by half between 2007 where he reports a production of 1.23 Mt P year^{-1} (see Sect. 5.2.8) and in 2011. The EU has recently adopted the detergent regulations that call for a ban of phosphorus in laundry detergents as of June 2013 and automatic dishwater detergents beginning in January 2017 (EU 2012b).

Finally, phosphorus is used in other industries such as in lighting or electronics. While the amount of industrial use of phosphorus is limited, it is very beneficial for many technical processes (see Spotlight 8, Gantner et al. 2013). However, in principle, phosphorus could be substituted with other minerals in this instance and in other industrial applications.

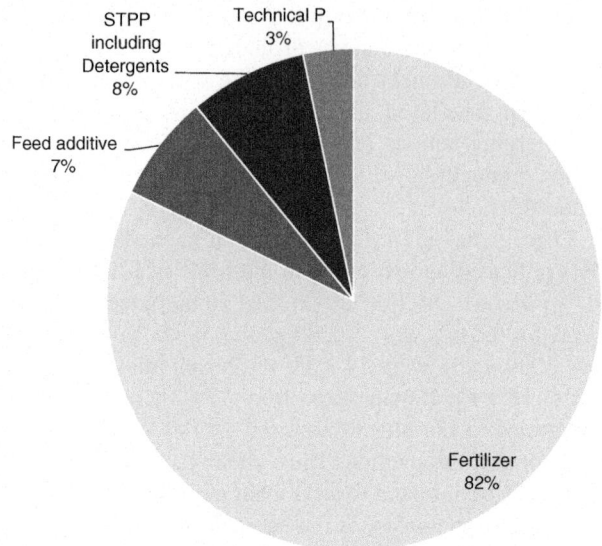

Fig. 4 Shares of phosphate rock (*PR*) use for main purposes according to Prud'homme (2010)

There is an increasing pressure on P availability under *biofuel* demands. Predictions suggest that in 2020, individuals in the developed world will use, on average, 150 kg of maize per person, per year (Rosegrant et al. 2008; FAO 2011). This number may easily demonstrate the trade-off between food and fuel (Jansa et al. 2010). Even using conservative estimates (e.g., assumptions about weight of wet and dry corn), about a quarter of the recommended daily calories (2,000–2,500 calories) of edible vegetable is used for biofuel (Kelly 2012). There are many critical arguments against biofuel, both with respect to bioethanol and biodiesel, because of trade-offs with food (and potential price increase), supplementary land use, etc. (Alexandratos and Bruinsma 2012). From a phosphorus resources management perspective, we have to acknowledge that most of the phosphorus of the plants used for bioethanol and biodiesel production remains in the biofuel co-products (e.g., oilcake and microalgae slurry) may be fed to livestock (Zhang and Caupert 2012) or even processed as organic fertilizer.

2.2 Different Phosphorus Demand Trends in Different Parts of the World

An important lesson to be learned is that we are facing completely different histories, constraints (with respect to soil, crops, etc.), and prospects with respect to phosphorus demand in different regions and countries of the world. A first impression of the different trends may be gained from Fig. 5. We must acknowledge that there are different trends in agricultural phosphorus use per

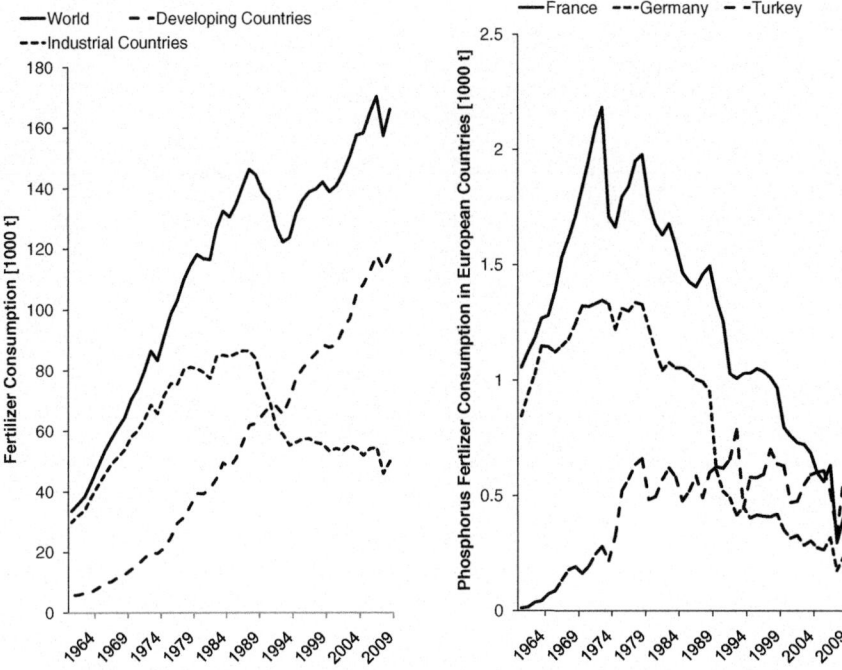

Fig. 5 Fertilizer (*left*) and phosphorus dynamics [*right*, based on (FAO 2010b)] show different trends in different countries

hectare (ha) in many parts of the world. Industrial nations have experienced decreasing demands since the 1980s. This is primarily due to an accumulation of residual phosphorus, which was not taken up by annual crops, but rather, is bound to soil particles and is available for subsequent crops. The decrease is also reflective of the increasing use of livestock manures from concentrated feeding operations. The balanced fertilizer application practice, which has been adopted by farmers in most developed nations, has been one factor which contributed to a moderate reduction in the use of agricultural land. In contrast, many developing countries are increasing demand (Fig. 5 left) for phosphorus though some countries—such as China—are now promoting efficiency and thus may flatten or decrease future demand projections.

An important study (MacDonald et al. 2011a) calculated that 10 % of croplands, mainly in South America (especially Argentina and Paraguay), northern United States, and eastern Europe, had deficits of phosphorus (-3 to -39 kg P/ha) in the year 2000; whereas another 10 % (East Asia cropland, large areas of western and southern Europe, the coastal United States, and southern Brazil) had large surpluses (13–840 kg P/ha). Large surpluses of phosphorus occurred in less than 2 % of cropland in Africa, existing particularly in North Africa or on large plantations that export their crops. MacDonald et al. elaborate that there is a large potential for more efficient use, which may increase world food production (see

Table 1 Cropland area, phosphorus (P) input per ha of cropland, P uptake, and average P consumption in two periods, according to Sattari et al. (2012)

Year(s) region	Cropland (10^6 ha)		P input by fertilizer and manure (kg ha^{-1} year^{-1})		P uptake rate (kg ha^{-1} year^{-1})		Average P input (Mt)	
	1965	2007	1965	2007	1965	2007	1965–2007	2008–2050
World	1390	1520	7.6	16.6	3.2	7.6	18.6	29.1
Western Europe	107	94	23.8	17.2	4.9	9.9	2.6	1.2
Eastern Europe	231	199	6.1	4.7	2.6	3.9	2.1	1.0
North America	230	225	8.7	11.4	3.9	8.8	2.5	3.3
Latin America	112	170	4.4	20.8	3.1	8.9	1.6	3.9
Asia	446	541	6.4	27.3	3.5	10.0	7.9	15.5
Africa	173	247	1.9	4.1	1.8	3.1	0.8	3.8
Oceania	41	46	14.8	16.0	1.1	2.5	5.7	7.8

Calculations are based on the mean area of 1965 and 2007, based on a linear extrapolation of area increase/decrease as between 1965 and 2007

Spotlight 1). We will take a closer look at this issue further on, but we can surmise from Fig. 5 and Table 1 that there are varying trends. Overall, we see an increase in nutrient and phosphorus production (see Chap. 5, Fig. 3). We take a more differentiated view of these trends in Sect. 4.3.

In order to understand fertilizer use trends and the pressure for recycling, we must also acknowledge the dramatic change in cropland availability per person. From 1960 to today, there has been a population increase in about 230 % (from 3 to 7 billion). Almost proportionally, there were 217 % more ha of unutilized cropland available in 1960 than today (Pimentel et al. 2010, all data in those paragraphs are taken from this source). Currently, the available cropland per capita differs dramatically across the globe. China has only 0.08 ha of available land per capita. This is reflected in the consumption rates. Whereas US citizens consume 1,481 kg year^{-1} of agricultural products per capita, the Chinese consume only 785 kg year^{-1}. As the United States still has 0.5 ha per capita of available cropland, it may provide large amounts of grain to China and other countries with fewer arable land resources. It is clear that the pressure to overuse fertilizers—and the subsequent pollution and land degradation—is more pronounced in China, Vietnam, and other highly populated, resource-poor countries. More detail on this issue is provided in the Use Node, Chap. 5.

A recent study by Sattari et al. (2012) provides further insight to the different continental trends of phosphorus use. This study utilized data on phosphorus use by different continents from 1965 to 2007 and provided simulations for another 43-year window from 2008 to 2050. This model is based on the dynamic input–output of phosphorus by plants based on a "labile pool" and a "stable pool" (see also Dumas et al. 2011), including fertilizer and manure as input and runoff, erosion and uptake by plants as output. We should also note that Sattari et al. are using as specific definition for use efficiency that is also applied in this chapter. The ratio of nutrient output (in grain) to nutrient input is defined as use efficiency (see Fig. 6d).

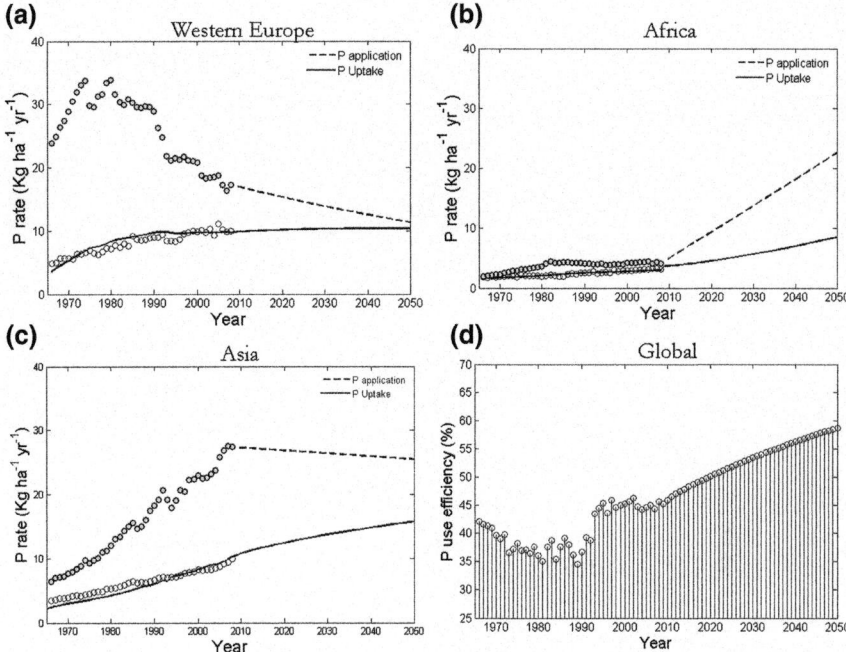

Fig. 6 Potential trends of phosphorus (*P*) application and trends of phosphorus application and uptake in the croplands of **a** western Europe, **b** Africa and **c** Asia (Sattari et al. 2012), **d** presents the P use efficiency [for 1965–2007 based on historical data, figure **d** provided by (for 1965–2007 based on historical data, figure **d** provided by Sattari 2013)]

One message of this chapter is that the phosphorus demand will not decrease, but rather increase in many parts of the world. This assertion is in line with estimations by Dutch researchers (Bouwman et al. 2009, 2012) who provide estimates of the future use of global phosphorus between 26 and 31 Mt P annually in 2050 including mineral fertilizer and manure. Though this model may be considered to contain early, rough estimates, it nevertheless suggests that African countries may quintuple their use of the input "from 4 kg ha^{-1} year^{-1} in 2007 to 23 kg ha^{-1} year^{-1} in 2050" (Sattari et al. 2012). This increase is due to the necessity to build up phosphorus stocks in agricultural soils to increase and sustain productivity in some countries, but in others is related more too balanced fertilization. Figure 6 shows quite well three dynamics of phosphorus use. The western Europe simulations propose a sort of equilibrium of 10 kg ha^{-1} year^{-1}. The Africa graph presents the backlog demand before a level of 10 kg ha^{-1} year^{-1} is attained. And the Asian graph may indicate that overloading of phosphorus in the soils may be close to an end, and that there will be a decreasing demand per ha in the future.

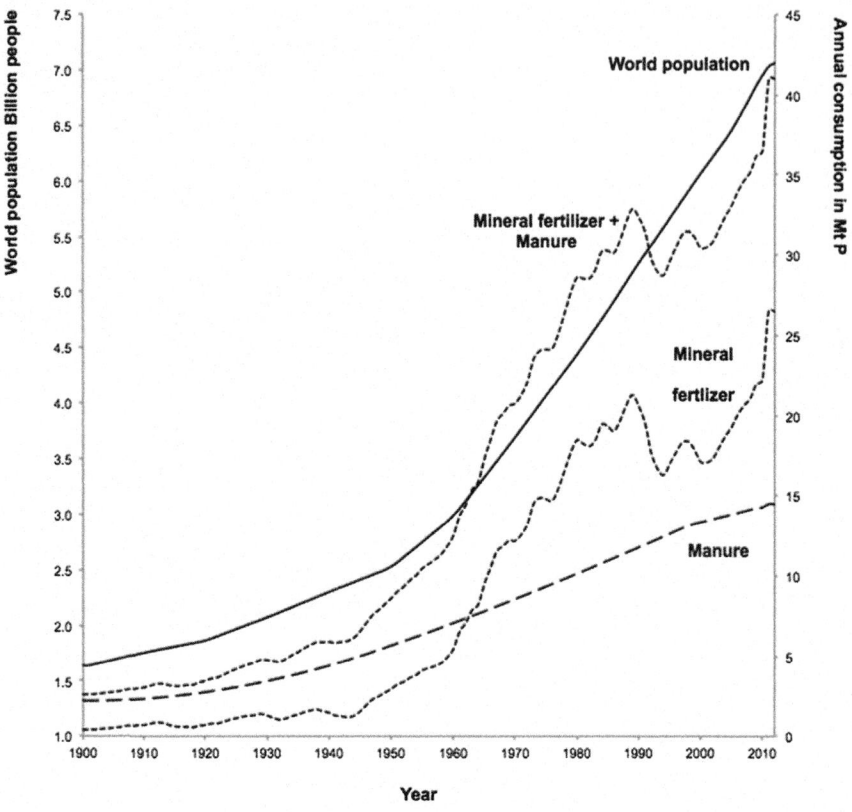

Fig. 7 The evolution of phosphorus fertilizer use, where mineral phosphorus plays an increasingly important key role (phosphorus data—80 to 85 % for fertilizer use—from USGS, presented as moving mean with a 3-year sliding window; rough estimation of manure data based on literature, see Sect. 4.8.2; population data from (USCB 2009); manure data extrapolated from different data on annual manure production, see text)

2.3 Increasing Efficiency: "Save and Grow" for Food Security?

There is no doubt that food security to match the anticipated world population growth is unthinkable without mineral fertilizer. Figure 7 presents the trends in the use of different types of fertilizer and the world population. If we assume that the world population will increase to 9.2 billion people by 2050, we may face different options or scenarios. One is "business as usual," which means that we produce more phosphorus in a "linear fashion," with the same amount of phosphorus required per capita. However, a closer analysis of historical per capita consumption reveals a strong increase leading up to the 1970s followed by a reduction in demand. This demand was due to the increasing efficiency of agriculture in the developed world and other factors (the decline of the demand) at the end. The temporary decline of consumption between 1990 and 2003 due to the decline of the Soviet Union will be dealt with in other sections (see e.g., Fig. 26). The idea of increasing efficiency has been adopted by FAO (2011) in "a policymaker's guide" which—almost exclusively—focuses on agrotechnology. Let us review the potential and limits of such an approach.

Efficiency is an input–output relationship. Phosphate fertilizer use efficiency in the domain of cropland agriculture is simply defined by how much phosphorus of the phosphate fertilizer added to cropland is actually taken up by plants. Here, in the first instance, the uptake of plants is of interest. If we denote the uptake of phosphorus as u, denote the uptake by plants which received phosphorus fertilizer as u_P, denote plants that did not receive phosphate fertilizer as u_o and the input of phosphorus simply as P, we may calculate the *uptake efficiency*: $\text{eff}_u = (u_P - u_0)/P$. There are other forms of efficiency such as "agronomic efficiency," which refers to the yield, where the increase in yield is by phosphorus input, i.e., $\text{eff}_y = (y_P - y_0)/P$ is considered, whereas y_P is the yield with phosphorus, and y_0 is the yield without phosphorus. Sutton et al. (2012) distinguish between nutrient efficiency for food crops, feed crops, and animal uptake, followed by nutrient use efficiency (NUE) of the *food supply* (with follow-up recycling food efficiencies of NUE in sewage and manure recycling). Clearly, as demonstrated in field experiments, the method of measuring the phosphorus use efficiency is in the study of its interaction with the other mineral nutrients (primary and secondary), the availability of hydrogen (H), oxygen (O), carbon (C), and many other factors such as soil texture (see Chap. 7). However, in practice, this has not been properly acknowledged.

Here, we focus different viewpoints on the questions, "how substantial is the NUE (measured by different parameters) for the case of phosphorus? And, how big are the losses?" Answering these questions on a global level is difficult because the regional differences in soils, climates, plants, agricultural technologies, etc. must be considered and integrated. Also, one must distinguish between real losses (e.g., by erosion and losses to the sea) and by temporary "virtual" losses when phosphorus binds with soil particles but becomes plant-available over extended periods of time. Just the difference between estimated soil erosion of 10 t ha^{-1} year^{-1} on cropland in the United States compared with China at 40 t ha^{-1} year^{-1} reveals potentially vast differences. Recent estimates of soil erosion on the African continent indicate an increase by a factor of 30 in the last three decades (Pimentel et al. 2010).

However, we may roughly identify an *optimist* and a *pessimist camp* (Pimentel et al. 2010). The optimists argue that the chemical phosphorus fertilizer in the soil is not lost. Syers et al. (2008)—based on long-term trials, mostly under temperate conditions and with relatively low erosion losses—provide the following pronounced statement:

> The main conclusion of this report is that the efficiency of fertilizer P use is often high (up to 90 percent) when evaluated over an adequate timescale using the balance method. (Syers et al. 2008)

Others, such as Cordell et al. (2011, whose work has been based on a literature review instead of experimental work), suggest an estimate of 8 Mt P associated *erosion losses* from agricultural soil and pastures out of their estimate of 14 Mt P of mineral fertilizer (plus a further 3 Mt P of losses from crop uptake). Even higher estimates are provided by Liu et al., based on the phosphorus balance of topsoil ("plow layer") contents.

> This gives world phosphorus losses at 19.3 MMT P/yr [MMT means that Mt in the nomenclature of this book] from cropland and at 17.2 MMT P/yr from pastures, respectively, ... (Liu et al. 2008)

Most presumably, the reality is somewhere between the two purported limits. We must also acknowledge that "natural P," i.e., phosphorus in the soil deposited through weathering processes, is lost from terrestrial systems, not only "mineral and organic fertilizer" phosphorus. Otherwise, there would be no (extensive) life in the sea. A critical point for understanding the difference between the two statements is the time range. Whereas one position takes a static view with a one-year window, the other takes a dynamic life cycle view. From the latter perspective, phosphorus stored in soil is not lost (following some dynamics that may be described by differential equations), as plants may gain access to most of the phosphorus in the soil (Dumas et al. 2011; Sattari et al. 2012). Clearly, it is important to get a proper picture here and distinguish between *temporary losses* (which may be retrieved over a couple of decades and centuries) and *real* losses (which may become accessible for terrestrial plants after millions of years).

> A new global effort is needed to reduce nutrient losses and improve overall nutrient use efficiency in all sectors, simultaneously providing the foundation for a Greener Economy to produce more food and energy while reducing environmental pollution. (Sutton et al. 2012)

We agree that efficiency is an important means in transitioning to sustainable phosphorus flows. But we should acknowledge that, logically, efficiency is neither a necessary nor a sufficient reason for sustainability. Former agrarian societies, such as those based on "slash and burn" agricultural extensification, may have shown low NUE (Escueta and Tapay 2010; Kauffman et al. 1995), but may have lived in a steady and productive societal state for a long period of time. Thus, efficiency is not a prerequisite of sustainability. To the contrary, we might imagine a world population that has managed a very high NUE in its agrosystem, but has become vulnerable due to its disregard for other aspects, such as population growth, biodiversity, or climate change, which has rendered its agrosystem unsustainable. Nevertheless, given the current situation of population growth, dietary change, scarcity of arable land, etc. NUE is an important means to avoid further land degradation, avoid the negative environmental impacts of fertilizer (mineral and organic), and foster food security. Given our current knowledge of the demand and the environmental impacts of the supply of phosphorus and other inorganic fertilizers, increasing efficiency is an absolute requirement.

2.4 Phosphorus and Biofuel

Liquid biofuel for transportation has seen a significant rise in many countries over the last decade. Bioethanol comprises 80 % of liquid biofuels, and almost 90 % is produced in Brazil and the United States, which are also two leaders in world food production (FAO 2008). For fear of running out of fossil oil, biofuel has been regarded as an acceptable alternative:

> Biofuels may contribute to the crucial goals of enhancing energy security, energy diversification, and energy access; improving health from reduced air pollution; and boosting employment and economic growth for rural communities. (UN 2006, p. 29)

In 2008, the world's arable land used for liquid biofuel production was approximately 1 %; that number is expected to increase to 3.8 % by 2030. However, this would only lead the "global share of biofuels in transport demand to increase to 10 %" (FAO 2008, p. 21). Though it does not fundamentally act as a substitution for oil in transportation, biofuel does affect the demand on P, in particular, in corn-based bioethanol production, which in 2010 amounted to 118 Mt (see also Sect. 5.2.9). Scholz (2011b) identifies and discusses the trade-offs of large-scale bioethanol production, and concludes that bioethanol may not be considered a renewable energy because of the additional use of non-renewable rock phosphate fertilizer and additional agricultural land extension. According to the International Fertilizer Industry Association (IFA):

> High crude oil prices provide strong incentives for biofuel production. This also pulls up agricultural commodity prices, which stimulates intensification and higher fertilizer applications. (Heffer and Prud'homme 2011)

It should be noted that the trade-off between competing uses of phosphorus and biofuel in various manners is a salient historical conflict. Cow dung may be used as fertilizer or fuel, and grass may be used as biofuel to feed plow- and carthorses or livestock animals.

2.5 Virtual Phosphorus Flows

By referring to virtual or unintended flows of phosphorus, we denote that these flows are included in the production of goods and commodities in which phosphorus is embodied, but phosphorus is not targeted or officially recognized. Research on material flows in Japan (Matsubae-Yokoyama et al. 2009; Matsubae et al. 2011) revealed that the quantity of phosphorus in iron- and steel-making slag is of the same percentage compared with the phosphorus contained in phosphate rock.

> Importantly, the results show that our society requires twice as much phosphorus ore as the domestic demand for fertilizer production. The phosphates in "eaten" agricultural products were only 12 % of virtual phosphorus ore requirement. (Matsubae et al. 2011)

Clearly, "our society" in the last quote refers to "Japanese society," which, among others, has a strong car production-related heavy industry. But there are other nations with a large share of industrial metals and other processes which include "non-accounted" phosphorus flows. Thus, "virtual" industrial use is an important part of the anthropogenic phosphorus cycle. In principle, virtual flows represent a sort of secondary feedback loop (Scholz 2011a). As we engage in the production of steel or another commodity, we severely affect the phosphorus cycle. What this means, what impact these phosphorus flows have on the ecosystem or whether the phosphorus may be used as nutrient at some time is an open question. We may also reflect about virtual flows of phosphorus by trade of commodities and their potential multiple impacts.

2.6 Phosphorus and Technology Development

A small share of mined phosphorus (about 3 %) is used for technical non-food purposes (see Spotlight 8, Gantner et al. 2013). *White* phosphorus (P_4) is the most common intermediate for a wide range of P4-containing products. In addition to the use of phosphorus in lighting, there is a wide scope of technological applications, ranging from superconductivity and energy storage (lithium-ion batteries) to warfare implements such as bombs and nerve gas (see Spotlight 8, Gantner et al. 2013), which are based on different forms, or allotropes, of phosphorus such as red or black phosphorus for batteries (Park and Sohn 2007).

3 Critical Issues of Phosphorus Management: Phosphorus as a Case of Biogeochemical Cycle Management

Sections 3 and 4 address the critical aspects of phosphorus flows, including impacts such as pollution, health-related food security issues, scarcity, innovation demands for phosphorus fertilizer, geographical distribution, and price volatility. Section 4 takes a substance flow view, focusing on sinks and losses, for preparing CloSD Chain management. As we noted in Sect. 2.3, phosphorus atoms are not lost from the earth. But they may be removed from the human value chain *temporarily* or *forever*, e.g., through erosion. We elaborate that there are different types of *losses* and *mobilizations* of phosphorus by human activities that may be considered critical.

3.1 What is Critical?

Technological development has changed human life and increased human wealth and health. Most notably, technology has increased human longevity and, therefore, is a direct contributor to population growth. We have determined from Fig. 7

that the tremendous increase in the use of phosphorus, such as the option of using this element in other fields, is highly related to technology development. Humans have become masters of "digesting at the periodic table" (Johnson et al. 2007), and as such, the global biogeochemical cycle of an increasing number of elements is dominated by anthropogenic activities.

The criticality of minerals has been defined by a criticality matrix with dimensions of importance and availability (National Research Council 2008). Graedel et al. (2012) even suggest a methodology of assessing metal criticality when focusing on *environmental implications, supply risk,* and *vulnerability to supply restriction* (Erdmann and Graedel 2011). These dimensions are assessed differently depending on the scale, i.e., whether a company-, state-, national-, or global-level view is taken.

In this section, *criticality* of a system is linked to a state of a human system that may become subject to undesired, drastic alteration, or negative change dynamics. This may be related to environmental, social, or economic issues both from a cause and from an impact side. We consider the use or management of phosphorus *critical* if a human system is exposed to threats and the use or deficiency of phosphorus will cause adverse or unwanted impacts which endanger the vitality of a system.

Given the current discussion both in risk research and sustainability science, this brings us straight to the concepts of *vulnerability* and *resilience* (Adger 2006; Aven 2011; Scholz et al. 2012; Holling 1973). From a sustainable phosphorus management perspective, it is not only of interest how *sensitive* a human system is that is *exposed* to certain threats but rather how *fast* or with *what efforts* a system may cope or adapt if a negative event (*threat*) has factually occurred. Threats with respect to phosphorus management may be low agricultural yields due to nutrient deficiency or eutrophication of aquatic systems. Traditionally, *exposure* and *sensitivity* are the core concepts which define *risk*, at least from a human and environmental health perspective (Paustenbach 2002). The inclusion of the adaptive capacity (after being exposed to a known threat) transfers *risk* to *vulnerability* assessment (Metzger et al. 2008; Scholz et al. 2012). A challenge may be to assess for a specific human or environmental system how resilient it is with respect to deficiency or abundance of phosphorus.

Vulnerability may be considered as *specified resilience*. A system is denoted as resilient with respect to (a known) risk, if it has the ability to recover to a (acceptable) vital level in a tolerable time. The vulnerability assessment may open new research perspectives for research. We want to note that this type of research will be shaped by quantitative and qualitative analysis. This holds particularly true if the *general resilience* of a system is targeted. General resilience may be described as the ability to cope with the (still) unknown (threats). The challenge is to establish a kind of (environmental) system limit management capability, e.g., to provide access to needed phosphorus inputs and to avoid harmful doses.

Criticality in the following refers to these ideas. Criticality and vulnerability have been widely synonymic used in adaptation to climate change (Bohle 2001). Another interpretation of criticality goes back to network analysis. We are talking about floats or buffers in resource constraint networks and identify critical paths which may harm the system's performance (Bowers 1995). The idea of indispensability was used for defining criticality of metals in the supply chain of technical systems (Reller et al. 2013). This idea, which looks at resources filters and barrier of supply chains (Krohns et al. 2011; Reller 2011), may become of interest also for phosphorus. In the latter notion, criticality may differ from vulnerability as it is more focusing on sensitivity than vulnerability (Scholz 2011a).

This section looks at critical trends in essential domains such as food security or ecosystem health. We also discuss technology lock-in and market imponderability, both of which are barriers to advancement and which may promote vulnerability. In general, this section is intended to properly identify *critical aspects* that should be addressed in sustainable phosphorus management.

3.2 Phosphorus as a Pollutant

Phosphorus is a highly reactive element and may thus function as a "secondary pollutant." Sedimentary phosphate rock may include heavy metals, toxic elements, and other precious elements. This may induce long-term critical contamination of soils (Nicholson et al. 1994). With respect to cadmium (Cd) in phosphate fertilizer, we find contradictory risk assessments. In Europe, strict regulations on the cadmium concentration in fertilizers are under discussion. Concentrations between 20–60 mg Cd/kg P_2O_5 have been discussed as thresholds (Chemicals Unit of DG Enterprise 2004). Fertilizer concentrations above this domain are expected to result in critical long-term soil accumulation (Nziguheba and Smolders 2008). These conclusions were drawn from a precautionary perspective. This is justified by the short-term irreversibility of heavy metal soil contamination, which cannot be reduced in a short term if big areas show a critical cadmium concentration. There is a wide range of Cd concentration in fertilizers, yet most of the sedimentary phosphate rock shows concentration below 100 mg/kg (Roberts in print). During the processing of phosphate rock with sulfuric acid, e.g., for fertilizer production, much cadmium is deposited in gypsum. A comprehensive human risk assessment is difficult to construct and is not yet sufficiently developed. There is a wide range of Cd concentration in fertilizers depending on its origin of phosphate rock concentrates (Roberts in print). Thus a differentiated view on the phosphate rock and the heavy metal concentrations (i.e., the purity of fertilizer) may become a subject which asks for a comprehensive assessment.

In general, an overabundance of phosphorus in the aquatic environment from all sources will cause eutrophication, algal blooms with "dead zones" (i.e., hypoxic zones) and fish die-off in lakes, rivers, and oceans. Similar pollution of water bodies also affects drinking water quality. Here, for instance, the UK Water Supply Regulation (MacDonald et al. 2011a) defines a maximum value of 2,200 mg P/l. This value, however, must be seen relative to estimations of no-effect levels of human uptake. Here, longer-term studies (6 weeks) showed that dosages up to

3,000 mg/day did not elicit adverse effects (EFSA Panel on Dietic Products Nutrition and Allergies 2005). This reference value is currently under re-evaluation.

We argue that water systems are the most sensitive environmental compartment. Diaz and Rosenberg (2008) identified 400 systems throughout the world with hypoxic zones. When assessing the critical load of phosphorus, we should distinguish between the geogene and the anthropogene. Phosphorus in its natural ecological state is released by weathering and transported by runoff and rivers to the sea, where it is a cornerstone of marine life. On the other hand, there are freshwater systems, such as alpine lakes, which have very low natural phosphorus content; these systems are highly sensitive to additional phosphorus input. Against this background, it is difficult to define standards for aquatic systems with respect to phosphorus loads. According to Dutch environmental thresholds, a concentration of phosphorus below 0.1 mg P l^{-1} is considered critical to stave off eutrophication and protect ecosystem health (van der Molen et al. 2012), but this value depends on hydrological and other factors.

Eutrophication, due to perpetual algal blooms, became a problem for the Great Lakes of North America in the 1950s. The blooming algae prevented the sunlight from reaching the lakes' deeper domains. As the algae died, the microbes responsible for the breakdown of decay on the lake bottoms began using the oxygen dissolved in the lakes' water. The lakes became green and malodorous, and an alarming fish die-off occured. At the height of the issue, research was engaged to discover the genesis and impacts of this phenomenon.

Synthetic phosphate products, and in particular, sodium tripolyphosphate (STPP), had been added to detergents on a large commercial scale since 1948. From the 1940s to the 1970s, raw wastewater effluent increased from 3 to 11 mg l^{-1} (Litke 1999). STPP has the property to bind magnesium and calcium ions, and thus the ability to increase the effectiveness of detergents. Environmental chemists provided clear evidence that phosphorus from all sources was a major source of eutrophication (Stumm and Stumm-Zollinger 1972). And detergents contributed significantly. In 1983 and in the United States alone, 2 Mt phosphorus was annually used for detergents.

The scientific community made the first real effort to understand the eutrophication process and problem (Knud-Hansen 1994). Regression models were applied to the contamination of the lakes (Vollenweider 1970). After 1977, with the introduction of the US Environmental Protection Agency's (EPA's) "Detergent Phosphate Ban," several US States and European countries banned phosphorus-containing detergents. Following the ban, the industry voluntarily offered laundry detergents free of phosphate. However, phosphate-containing dishwashing powders remained, as no economic substitute could provide the same performance and satisfy consumer demands. The effect of STPP in detergents is due to its ability to bind minerals and metals and does thus "enable cleaning components of the detergent to act" (Global Phosphate Forum 2012).

From their very inception, the role of detergents in eutrophication has been disputed. For one camp of detractors, detergents seemed "the devil in disguise, and the less we had to do with it the better" (Emsley 2000b, p. 270). Others provided loose calculations such as: "if detergents make 30 % of the total phosphorus of domestic wastewater and the wastewater discharge represents only 25 % of the phosphorus load to the lake, a ban would provide a reduction in only 7.5 %" (Lee and Jones 1986). If we were to introduce (full) phosphorus precipitation or biological extraction in the sewage plants, phosphorus would be an ideal additive to detergent, in particular as zeolite. Zeolites, the primary substitute for STPP, perform better, based on a cradle to grave-based life cycle analysis (LCA) for sites with very advanced wastewater treatment plants (such as those in Scandinavian countries). The LCA recorded roughly the same environmental performance in the UK, which features relatively simple sewage cleaning systems (Köhler 2006). Compared with human excreta, the contribution of phosphorus-based detergents was, and continue to be (in those places where STPP-based products are still used), relatively low. Thus, whether a phosphorus ban on detergents is effective seems to be site-specific. For developing countries, in which only a minor part of the domestic population is connected to sophisticated wastewater treatment plants, a release of detergent phosphorus seems to be instantly feasible and operationally eco-efficient, as it may provide some environmental effect, but with low costs (Scholz and Wiek 2005).

A lesson learned from history is that water pollution is a complex issue that depends on a variety of historical and situational factors. The role of phosphate is certainly important and must be viewed in relation to other eco-toxicants that eliminate natural zooplankton that in turn consume the algae.

3.3 Human Health and Phosphorus Food Additives

Phosphorus is present in almost any food, though a few products, such as industrially processed sugar, do not contain phosphorus. As outlined in Spotlight 6 (Elser 2013), both a deficiency and excess of phosphate in food may cause health issues. In that light, phosphorus-based food additives deserve a special review. Disodium phosphate (DSP) and STPP are used to improve the texture of ham and other meats, and phosphate is used to avoid weight losses in humans. DSP and sodium aluminum phosphate are used in producing processed cheese and to process milk. Trisodium phosphate "is considered too risky to be used in general cleaning products, but it is still used to remove food-poisoning germs from raw chicken" (Emsley 2000b, p. 264). Bacteria such as salmonella on surface fat of the carcass are washed away with a trisodium phosphate solution. This procedure was patented and received approval from the US Food and Drug Administration (FDA) in 1992 (Emsley 2000b).

Regarding phosphate additives in general, and STPP in particular, the critical question is how sensitive the human health system may be. The chemist Emsley provides a relaxed view:

Even when the food additive is STPP, which is the commonly used dishwater and laundry detergent component, it is deemed safe because the body is plentifully equipped with phosphatase enzymes that can break down this complex phosphate into simple phosphate. (2000b)

According to our current knowledge, the use of phosphate suggests that the human metabolic systems are fairly well equipped and robust with respect to the processing of phosphate and phosphate-based acids, at least for most forms in medium doses. In what domains there are limits, and where and whether we may identify more "concealed effects," is yet to be determined. Such has been the case with phthalates, whose (estrogen) health impacts have been identified, but whose real effect is still elusive (Halden 2010)—even after more than two decades of intense research.

The daily uptake of phosphorus between countries widely differs and depends on the diet. European countries show mean dietary intakes of 1,000–1,500 mg P/day (EFSA Panel on Dietic Products Nutrition and Allergies 2005). The recommended intake for adults between 19 and 65 years is 700 mg P/day, the highest recommended value is for adolescents between 10 and 19 years and is 1,250 mg P/day (DACH 2008). Given normal health conditions, there is no evidence for negative health impacts by normal Western diet (Schnee et al. 2013).

3.4 Innovation Potential of Fertilizers

We describe the early history of chemical phosphorus fertilizers in Sect. 1.2. The big gray box of Fig. 21 presents the main wet processes for producing fertilizer, e.g., single superphosphate (SSP), nitrophosphate (NP), monoammonium phosphate (MAP), diammonium phosphate (DAP), triple super phosphate (TSP), the common NPK fertilizers, and others. These wet processes are acid based and are linked to high amounts of waste, energy use, and water consumption. In its basic sense, the wet process is more than 100 years old (see Chap. 6, Herrmann et al. 2013).

Perhaps, because phosphorus fertilizer may be considered a low-cost commodity, technological processing has not yet attained high efficiency with respect to waste and to purity. Zhang et al. (2008) consider the amount of gypsum and other waste such as phosphate slag as one key environmental indicator for improving phosphorus fertilizers. There are about 5–7 mt of phosphogypsum produced for each mt of sulfuric acid-based wet-process phosphate fertilizer via the sulfuric acid route. "There are currently about 1 billion tons of phosphogypsum stacked in 25 stacks in Florida (22 are in central Florida) and about 30 million new tons are generated each year." Phosphogypsum includes high percentages of calcium and sulfur, e.g., 26 % CA and 20 % S in the case of Richard Bay, South Africa, production line. We should note that there are alterative routes of phosphorus fertilizer production via the nitrophosphate route (SSP, which was once the most commonly used fertilizer) that do not generate any gypsum (IPNI 2013).

The amelioration and valorization of gypsum for use in the construction of buildings and roads and for agricultural purposes is another challenge that requires innovation (Hilton and Dawson 2012). However, the market is limited as there is an abundance of flue gas desulfurization gypsum as by-products from coal power plants (FIPR 2013). Another interesting aspect is that some phosphate rocks contain relatively high amounts of uranium and heavy metals (Schnug et al. 1996). Here, the long-term accumulation in the soil and potential toxicity of the food chain is one consideration, but recovery/recycling of these metals presents an ancillary opportunity. This toxicity concern is primarily linked to cadmium, as has been mentioned in Sect. 3.2. Under what constraints what cadmium concentrations in fertilizers may cause adverse health effects is assessed with deterministic risk assessment methods (Woltering 2004). There is no evidence that cadmium in fertilizers may cause health risk in the near future.

Most phosphorus fertilizer is from phosphate rock of sedimentary origin. "The average cadmium content in European fertilizers is 138 mg/kg phosphorus" (Finnish Environment Institute 2000, p. ii). Given the data about cadmium in rock phosphate from different mines in the worked used for fertilizers reported by Roberts (in print), these concentrations seem to be surprisingly high (Chemicals Unit of DG Enterprise 2004). A large share of the phosphorus fertilizers used in Finland, for instance, are from the igneous form of phosphate rock and has a cadmium concentration from 1 to 5 mg/kg P (Finnish Environment Institute 2000).

Given that phosphatic uranium could cover the current rate of uranium consumption for some centuries (Schnug et al. 1996; Hilton and Dawson 2012), the option of mining and meaningfully using this uranium and other heavy metals is a key challenge in sustainable resource management. The above aspects refer to the first stages of the phosphorus supply chain. From a chemical engineering aspect, we may clearly identify the potential for the extraction of heavy metals and uranium by efficient means. Whether this may become economically feasible depends on the accessibility from conventional uranium deposits. In this scenario, however, a sophisticated assessment of the long-term application of fertilizer related to different crops is missing.

If we look at the aspect of use, we can see three or four further opportunities for the innovation of fertilizer. The *first* is the improvement of the "compounds" contained in organic and inorganic fertilizer. Compounds, on one hand, are heavy metals, radionuclides, and pathogens associated with manures and sewage-based products. Here, the economic extraction of heavy metals or radionuclides from phosphate rock with high concentration may be considered an innovation. On the other hand (VFRC 2012), there is an emerging deficiency of micronutrients such as zinc (Zn), manganese (Mn), iron (Fe), sulfur (S), and even boron (B), a situation that may require much more sophisticated fertilizers in the future.

Second, the challenge will not only be the site-specific optimization of the currently available phosphorus. Rather, we see that the relationship among soil–plant/species–fertilizer should be viewed from a prospective, *dynamic perspective* (Zhang et al. 2002, see Fig. 8). At the least, large-scale farming should anticipate

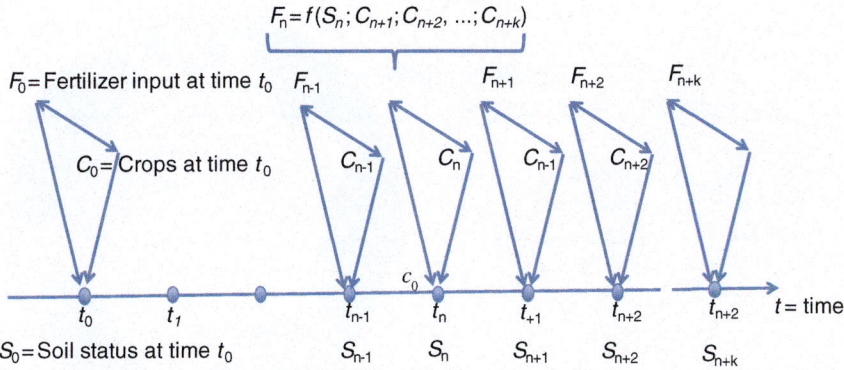

Fig. 8 Dynamic precision crop rotation asks that at any time t_0 the decision of what amount and type of fertilizer should be added does not only depend on the soil status S_0 and the specific crops C_0 but also depends on the prospective crop grown at that time $C_{n+1}, ..., C_{n+k}$

prospective cropping and available phosphorus, and thus, apply prospective balancing. But this problem may require a biotech solution. When water-soluble phosphate fertilizers are added to the soil, a significant portion not taken up by the plant is converted to less plant-available forms that may or may not become available again in the near term. These less soluble forms result from the pH-dependent reactions of phosphate with iron (Fe^{3+}), aluminum (Al^{3+}), calcium (Ca^{2+}) and magnesium (Mg^{2+}) ions that are present in the soil. Some plants, however, possess biological properties (including enzymes) that allow them to utilize phosphorus from these less soluble forms. This capability opens up other opportunities for technological innovations to either produce new fertilizers that are responsive to the soil and plant enzymes or to transfer this capability to other plant species.

Third, we may consider *seed coating and pelleting* as another form of innovation. This idea originated in the 1940s and is common in sugar beet (Vogelsang 1950) and cotton, but may be applicable to other crops (Fig. 9). The advantage is that the fertilizer is optimally placed in the rhizosphere. The pellets may be processed to include fungicides, pesticides, or micronutrients, if necessary. Naturally, we must carefully reflect under what constraints of farming such technologies make sense. Clearly, the coating in itself is not sufficient to improve plant growth. But here, the coating is a concept that has the opportunity to improve efficiency. With respect to nitrogen, the coating may also have positive effects from a use-efficiency perspective. We should mention that in general, the coating of the seed usually only contributes a minor, but important start-up fertilization of the root zone.

To ensure that one has a complete picture of P, one must recognize that though many phosphorus fertilizers are water soluble, phosphorus in the soil is virtually immobile. The leaching of phosphorus to groundwater is not a common occurrence and thus not a major issue (Hesketh and Brookes 2000). Losses are primarily related to erosion of soil particulate matter containing P.

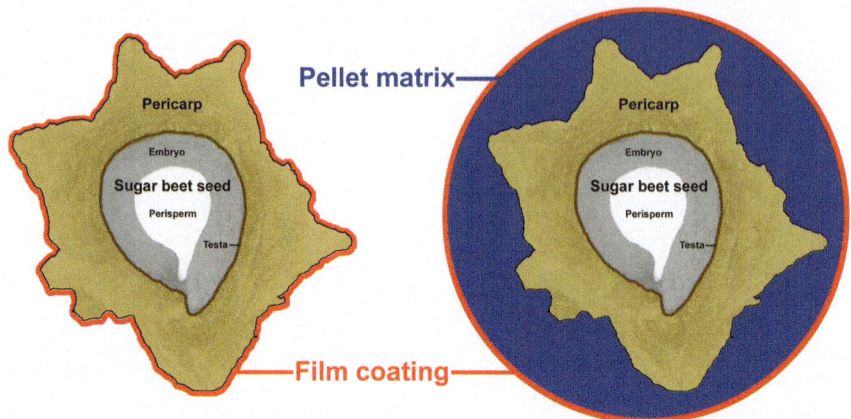

Fig. 9 The coating of the sugar beet seedling is successfully applied (figure taken from Leubner 2013)

Fourth, the amelioration of organic fertilizers from different types of manure, from compost or peat to sewage may require low and high technologies. Here, not only soil-biological knowledge and plant-biological knowledge are required, but also the understanding of comprehensive transdisciplinary processes, including the collaboration of key stakeholders from a given region to assess whether the technology is "sociotechnologically robust." In particular, this may hold true for "smart" manure management.

3.5 Geographical Distribution of Phosphorus Reserves and Resources

The availability and accessibility of phosphorus supply may be endangered by geopolitical risks. A common method of assessing the risks that a country may not be able to deliver (phosphorus) minerals is through the Worldwide Governance Index (WWGI) (Kaufmann et al. 2009). In WWGI calculations, political stability is a critical aspect. A second dimension that is usually considered is the global *supply concentration*, which is measured by the Herfindahl–Hirschmann Index (HHI) (Scholz and Wellmer 2013; Graedel et al. 2012). The first important question when applying the WWGI and HHI is whether the reserves, the beneficiation sites, and the fertilizer production sites are integrated. Figure 10 presents the two aforementioned indices for metals and minerals. As a rule of thumb, values outside of the medium domain are considered risky.

We may take from Fig. 10 that *phosphate production* is in the non-critical domain (within the medium). The *reserves* show a critical tendency in the WWGI

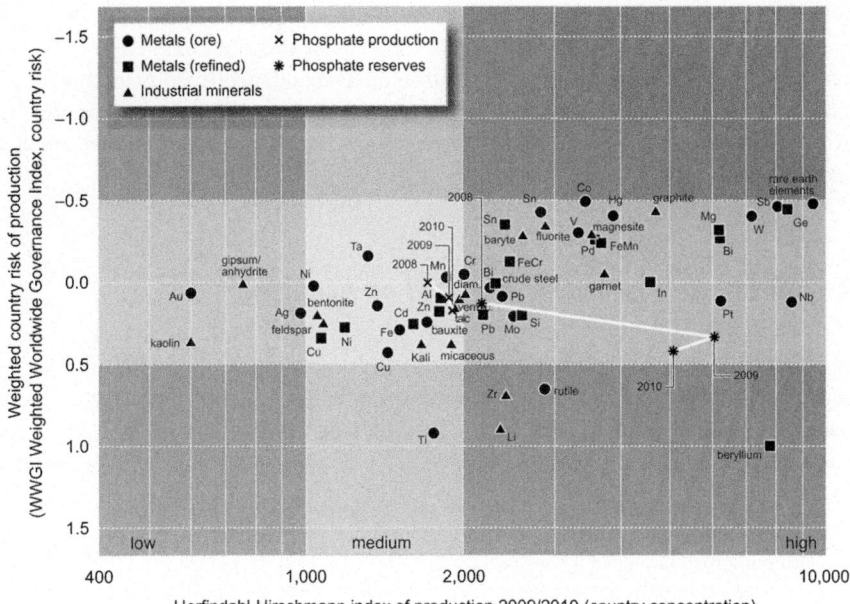

Fig. 10 Linkage of the Herfindahl-Hirschmann Index (HHI) (figure according to Deutsche-Rohstoffagentur/BGR 2011, present version taken from Scholz and Wellmer 2013, p. 15) with the Weighted World Governance Index (WWGI, data from Jasinski 2009, 2010, 2011a, 2012; Kaufmann et al. 2011) for production. "*" Presents the HHI for phosphorus reserves

due to increasing political instabilities in many *countries* in North Africa and the Near East, which began in late 2010. The strong increase in the reserve-based concentration index is due to the increase in the Moroccan data from 5,700 Mt of PR in 2008 to 50,000 Mt in 2011.

We should also note that the essential element potash shows a high concentration with respect to reserves. Here, two countries, Canada and Russia, have a combined 81 % of the documented potash reserves, where Morocco and China represent 80 % of phosphorus reserves (USGS 2012).

Does the HHI index properly represent the *supply* concentration in the case of phosphorus? Does the graphical increase in the HHI with respect to phosphate *reserves* really mean that the world is facing increased supply vulnerability? As we will show, this is not necessarily the case. Let us look at the data. Before 2009, Morocco had only 33 % of the 16 Gt of PR world resources (USGS 2010). In 2011, it had 70 %, or 50 Gt of the 71 Gt of PR world resources. The HHI suggests "increasing geographical concentration," and an increasing dependency on Morocco. Upon further inspection, one may argue that the dependence on Morocco has decreased, because the non-Morocco reserves increased from 11 Gt of PR before 2009 to 21 Gt of PR after 2011. If we roughly calculate with a (current) annual demand of about 0.2 Gt of PR annually, one may argue that the static lifetime without Morocco has increased from 50 to 100 years (which is roughly the 2010 static lifetime of titanium). We should take from this that the HHI index of reserves does not represent the demand-related abundance/scarcity of the commodity and, despite the large reserves in Morocco, there are sufficient options to avoid a dependence on one major provider—even if recycling options are not taken into account.

3.6 Phosphorus Scarcity: Physical or Economic?

Historically, concerns have emerged repeatedly regarding the potential scarcity of phosphorus. US President Franklin D. Roosevelt addressed fundamental aspects of sustainable phosphorus management in his Message to Congress on Phosphates for Soil Fertility, May 20, 1938[1]:

> The necessity for wider use of phosphates and the conservation of our supplies of phosphates for future generations is, therefore, a matter of great public concern. We cannot place our agriculture upon a permanent basis unless we give it heed.
>
> I cannot overemphasize the importance of phosphorus, not only to agriculture and soil conservation, but also to the physical health and economic security of the people of the Nation. Many of our soil types are deficient in phosphorus, thus causing low yields and poor quality of crops and pastures ...
>
> Recent estimates indicate that the removal of phosphorus from the soils of the United States by harvested crops, grazing, erosion, and leaching, greatly exceeds the addition of phosphorus to the soil through the means of fertilizers, animal manures and bedding, rainfall, irrigation and seeds ...
>
> It appears that even with a complete control of erosion, which obviously is impossible, a high level of productivity will not be maintained unless phosphorus is returned to the soil at a greater rate than is being done at present ...
>
> Therefore, the question of continuous and adequate supplies of phosphate rock directly concerns the national welfare ...
>
> It is, therefore, high time for the Nation to adopt a national policy for the production and conservation of phosphates for the benefit of this and coming generations. (Roosevelt 1938)

We can see from this passage how the basic principles of phosphorus management had already been keenly identified by the 1930s. The prevalence of concern over phosphorus can be illustrated by Aldous Huxley, who used the absence of phosphorus recycling as an example of technological entrapments in his very serious 1928 novel, *Point Counter Point*:

> "With your intensive agriculture," he went on, "you're simply draining the soil of phosphorus. More than half of one percent a year. Going clean out of circulation. And then the way you throw away hundreds of thousands of tons of phosphorus pentoxide in your sewage! Pouring it into the sea. And you call that progress. Your modern sewage systems ..." (quoted according to Ulrich 2011)

---

[1] We wish to thank Andrea E. Ulrich for hints in the historic sourcing.

The physical scarcity, the finiteness of rock phosphate reserves, and the dissipation of phosphorus have been recently brought to the fore again by Cordell et al. (2009):

> However, modern agriculture is dependent on phosphorus derived from phosphate rock, which is a non-renewable resource and current global reserves may be depleted in 50–100 years. While phosphorus demand is projected to increase, the expected global peak in phosphorus production is predicted to occur around 2030. (p. 292)

Though this statement has been frequently cited, this alarmist statement does not adequately describe the scarcity issue. As Scholz and Wellmer (2013) conclude, the Cordell et al. analysis is (a) applying an inappropriate mathematical model; (b) ignoring basic geological data; and (c) ignoring the dynamics of resources and reserves which are given in a demand-driven market. In the case of phosphorus, scarcity is an *economic* issue. *Ceteris paribus*, there are enough resources that may become reserves within feasible production cost ranges for phosphate rock for at least some centuries before a purported "peak" in resources occurs. Extraordinarily, high phosphorus prices, production peaks, or the volatility of prices (see Chap. 9, Figs. 3 and 5) are not due to physical scarcity, but rather, due to other reasons such as bubbles in the financial markets, imbalances of supply and demand, unfruitful prospecting efforts in the mining industry, geopolitical effects and many other factors (see Chaps. 2 and 7).

The mathematical model applied by Cordell et al. (2009) is the Hubbert curve fitting (Hubbert 1956; Brandt 2010). This model assumes that for a finite deposit, production may be described as a bell-shaped curve (logistic curve, Gaussian curve, etc.). The curve has been very successfully applied for predicting the "Peak Oil" in US production, though the actual curve more closely resembles a lognormal curve than a normal curve. We acknowledge that the Hubbert curve may show high validity dynamics for US oil production (if we exclude unconventional forms of oil production such as oil shale production). But, as the paper with the indicative title "Peak Nothing" (Rustad 2012) indicates, for most minerals, and even for oil, we are facing different dynamics, including multiple peaks, plateaus, and other issues:

> Although many resources have exhibited logistic behavior in the past, many now show exponential or superexponential growth. (Rustad 2012)

As Scholz and Wellmer (2013) elaborate, Hubbert curve modeling may be applied for limited resources with a supply market structure. Thus, in cases such as nineteenth century guano production or current US oil production, you may assume that (annual) production increases due to increasing exploration, expertise, technology, equipment, etc. until a certain peak is attained. Then, in a second phase, the production declines as the mined resource becomes more difficult to access due to location, decreasing ore grade and other issues. Supply-driven markets show no saturation and—in principle—take everything that is available. This is definitely not the case for phosphorus. If we refer to the recorded reserves by the US Geological Survey (USGS) of 71 Gt P and the annual demand of

200 Mt P year^{-1} (=0.2 Gt P), then look at the data for fertilizer demand, it is easy to see that the market may be saturated for a good length of time.

Cordell et al. (2009) simply take USGS data from 2009, which was 16 Gt P, add the former cumulative production to this value (thus figuring what Hubbert modelers call the Ultimate Recoverable Resource [URR]), and fit a Gaussian curve to past annual production data applying least square technique. Based on this formula, "Peak 2030" and "depleted in 50–100 years" are inferred. However, the documented reserves increased; in 2012, they were 71 Gt P. Cordell et al. (2011, April 4) then revised their 2009 calculations using (USGS 2010) data of 60 Gt P, which resulted in a peak around 2070. It should be noted that usual Hubbert curve modeling does not refer to an estimated URR. Some authors (Déry and Anderson 2007; Ward 2008) provide applications without referring to the URR, which Cordell et al. simply assess by adding the formerly mined (cumulative) phosphate rock with the USGS data.

We also know that some reserves of the USGS survey are underestimated, and large reserves from other countries such as Saudi Arabia, USA, Peru, Kazakhstan (Evans 2012) or the largest reserves in Europe, i.e., Estonia, have not been comprehensively assessed to date though resources of more than 250 Mt P have been already identified in surface rock just for this country of the Baltic Basin (Äikäs 1989).

An estimate without a URR is provided in a paper by Dery and Anderson (2007). This application suggests a URR of 8–9 Gt P, of which, about 6.3 Gt have already been mined. Similar extraordinary underestimation is also reported in a paper by Ward (2008). Here, the application of the Hubbert provided an estimate of 11 % of the mined or known reserves. It should be noted that these papers are mostly published in unreviewed journals.

But the physical scarcity judgments also ignore geological data. There is geological evidence that many current PR resources may become reserves. If one considers the history of exploration of the Western Phosphate Field (WPF) in the United States (Jasinski et al. 2004; Moyle and Piper 2004), or other related literature (such as Volume 8, *Handbook of Exploration and Environmental Geochemistry* (Hein 2004), one can assume that as the price of PR increases, a large portion of these occurrences (along with others) currently classified as resources may become reserves. Since 1904, 70 mines have operated in the WPF; 49 conducted underground mining and stopped when cheaper surface mines began operation or increased production. This illustrates the dynamic boundary between reserves and resources (Jasinski et al. 2004). For example, a price increase by a factor of 2 to 4, accompanied by an improvement in technologies, may make phosphate rock from a discontinued mine more economically viable for a very long time. This point in particular holds true, as phosphorus is a low-cost commodity. Currently, each person consumes about 31 kg of PR per year, with a price of about US $6 per year (Scholz and Wellmer 2013). Given this price level, there is a great deal of flexibility if we simply look at "average global" consumption. However, we should acknowledge that some farmers, and in particular smallholder farmers, are highly sensitive to fertilizer prices. Nevertheless, we may conclude that scarcity with respect to phosphorus, for the foreseeable future, will be an economic issue, though—from a very long time perspective—physical scarcity of mineable ores may emerge.

3.7 Inefficient Use

Efficiency has become a key term in resource management. Simply stated, "efficiency" targets an increase in the input–output relationship. One critical point is that efficiency relates to specific systems (e.g., an individual, company, society, or a world scale) and to a certain time frame. What may be efficient in an accounting cycle may be inefficient from a company life cycle or a societal or ecosystem perspective. How "inefficient use" is defined must be specified case by case. From the chosen system theoretical view on sustainability, "an ongoing inquiry on system limit management in the frame of intra- and intergenerational justice" (Laws et al. 2004), it is clear that a global long-term perspective should include the economic and social perspectives. However, acknowledging the multi-level nature of human-environment systems (Scholz 2011a), it is clear that efficiency must be viewed from many perspectives, from the consumer via smallholder and industrial farmers to high-tech companies, and from mining companies to geological services, to mention a few. All of them are challenged in the systems scale, but single-scale efficiency must be reflected in a contextualized larger scale. In some cases, this may require the framing of phosphorus management, e.g., by national or international rules or (environmental) laws.

We reveal in Sect. 2.3 that efficiency is neither a necessary nor sufficient condition of sustainable transitioning. However, given the challenges of twenty-first century human development, it is clear that sustainable phosphorus management on a global scale is not meaningful without efficiency along all steps of the supply chain, i.e., exploring, mining, processing, use and dissipation and recycling. Exploration should be efficient in the sense that we must reflect on how much manpower and money we have to invest in the assessment of reserves, given current knowledge and future demands. It should be acknowledged that efficiency is not absolute. Rather, it is relative to technologies and context, and also to the options of improving efficiency in other domains. Here, we speak about cross-sectional efficiency, or operational (eco)efficiency (Scholz and Wiek 2005). In practice, this means that the improvement of efficiency at one stage of the phosphorus supply–demand chain (how much improvement does one get for how much investment) must be considered relative to other stages and other elements, for instance, nitrogen.

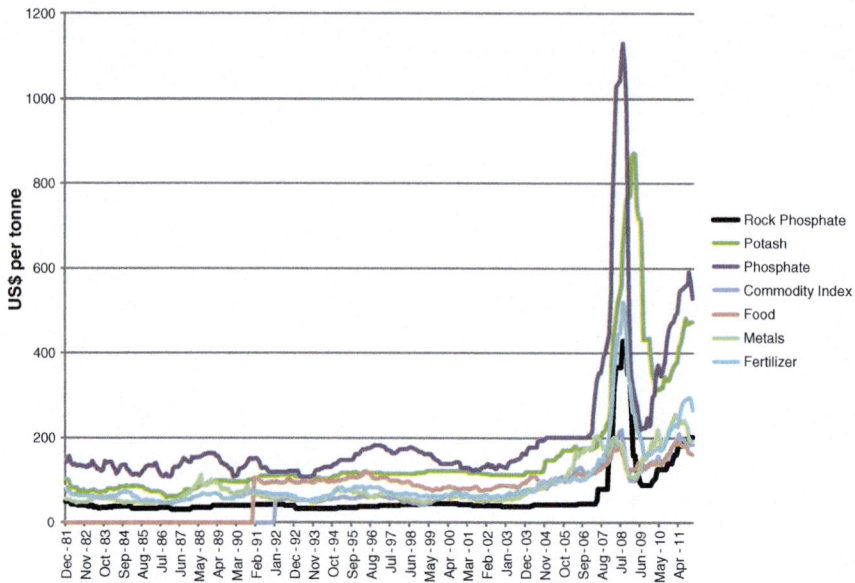

Fig. 11 Price dynamics for rock phosphate, food, commodity index, etc. in US$ per metric ton (graph with the courtesy of Olaf Weber)

3.8 Price Volatility

Price volatility of fertilizer has remained high since 2007/2008. This has strongly affected farmer access to fertilizers, particularly the smallholders in less-developed countries. This is one of the main conclusions in the analysis of Weber et al. (2013), featured in Chap. 7 of this book. In the 2007 peak, prices were closely linked to those of other commodity markets (see Chap. 7, Fig. 5). It is clear that *energy prices* affect phosphorus fertilizer production, in particular those related to nitrogen such as DAP and MAP, and transportation. The level of energy intensity can be traced back to different crops (Mitchell 2008) and has lingering effects on transportation. Figure 11 presents grain, nitrogen, and PR prices over the last five decades. One may observe that there is a step-like increase after the 1973 oil crises, with fairly constant price-level increases over the following three decades. However, around 2002, there is a sharper increase, culminating in a historic peak in 2008. We should acknowledge, however, that econometric analysis reveals that fertilizers experienced much higher (standardized) annualized volatility in 1974 than in 2008 (see Weber et al. 2013). It is important to note that both volatile periods have been linked to world food crises.

In an analysis of the 2008 price peak, specific effects are identified. US fertilizer production declined 42 % between 2000 and 2008 and production capacity could not be quickly adapted. In 2008, dramatic price increases were due to supply–demand imbalances, weather-related crop failures, high energy prices, biofuel mandates, and numerous trade barriers. For example, China, which exported about 3 Mt of urea in 2011 (YARA 2013), imposed an export tariff of 185 % (CNCIC 2008) which contributed to low exports. As a result, world demand could not be met by available supply, affecting global fertilizer price stability. The 2007/2008 food crises may be seen as converging events resulting from this trend.

Given that the demand for food raises crop and livestock prices is a common market effect, the purchase of fertilizer became more attractive, thus increasing fertilizer prices as a result of higher demand.

Often, the argument is made that commodity speculation led by the global financial sector affected the sharp rise in food prices.

> The dramatic rise and fall of world food prices in 2007–2008 was largely a result of speculative activity in global commodity markets, enabled by financial deregulation measures in the United States and elsewhere. (Ghosh 2010)

There is some intuitive evidence to this theory. Speculation in physical markets may be defined as financial activities that are not related to "fundamental" production and commercial activities. Following the collapses of the technology "bubble" and the US housing market, financial investors expectedly shifted to far less risky investments, including commodities. And, as market speculators tend to follow the most active, high-volume trading, it seems plausible that speculation had a hand in the 2007–2008 price spikes. However, this theory cannot be proven by econometric analysis for all commodities (Wright 2011).

Taking an alternative view, we note that according to standard economic theory (Ghosh 2010), speculation—in unbiased markets—can stabilize market prices. This may seem counterintuitive, but it does meet economic theory:

> ... the existence of a derivative market increases in the long term the level of inventory. Indeed, suppose a commercial hedger (for its physical stock) needs a counterpart to hedge it, and there is no hedger accepting the risk (counter position) business... In the opposite, a liquid derivative market will ensure the hedger to find a counterpart, which can be a commercial trade or a speculator... A liquid derivative market leads to larger stock, which in turn lowers the volatility, as it can act as a buffer to mitigate supply (or demand) shocks. Speculators help to increase the liquidity of the market. (Ott 2012)

Thus, there may be both positive and negative effects that financial agents pose in the management of fertilizer price risks. The assertion that speculation had an affect on the 2007/08 commodity price peak could be proven (Sanders and Irwin 2010). Kenkel (2012) points out that there are two main strategies to becoming less exposed to price peaks: *diversification* (e.g., by not being solely dependent on chemical fertilizers), and *hedging* (e.g., buying fertilizer earlier and allowing another party to take the risks inherent in future price volatility), referred to loosely as "insurance." Kenkel estimates that the price volatility of fertilizers represents two-thirds of the price volatility of crop-related commodities.

4 What is Wrong with the World's Phosphorus Flows?

This section is devoted to the review of global phosphorus flows, with a focus on identifying the characteristics that may be viewed as *critical from a* general *sustainability perspective*. We subscribe to a general definition of sustainability, which refers to *sustainable development* as an *ongoing inquiry on system limit management in the frame of intra- and intergenerational justice* (Laws et al. 2004). The phrase "ongoing inquiry" reflects that we do not know exactly what is and is not sustainable, and that we must continue to explore the question. *In order to avoid the negative aspects*—in the context of this book—we must be critical in identifying ways of utilizing phosphorus that may induce unacceptable risk or vulnerability with respect to famines, environmental degradation, economic crises, etc.

One of the other aspects is "intra- and intergenerational justice." This aspect of phosphorus management refers to the "needs component" of the Brundtland (1987) definition. In the European discussion on sustainability, this has been called "sustainability as a regulating idea" (Minsch et al. 1998). We are aware that this includes a normative component. But goals such as "providing access to phosphorus to the poor" or "human beings far into the future should have access to phosphorus" ask for a reference point, which is provided by the idea of intragenerational and intergenerational justice, which are included in the presented definition of sustainability. We should note that this section also lays the foundation for the use and a critical view on the Global Material Flow Analysis, which is presented in the following chapter.

Contrary to other essential elements such as carbon, hydrogen, oxygen, or nitrogen, phosphorus does not freely circulate between the earth's spheres. Phosphorus may not be fixed to the atmosphere, as only a marginal amount of phosphorus (about 0.3 Mt P is in the air at any given time) goes into the atmosphere, mainly as dust and sea spray. Given the scale of human development, there is no noteworthy natural recycling to match human need, as the formation of sedimentary rock phosphate and its uplift to accessible mines took tens or even hundreds of millions of years. About 13 Mt of P $year^{-1}$ are released from rocks through natural weathering each year with an additional human-induced input of up to 5 Mt P $year^{-1}$ (Carpenter and Bennett 2011) to the soil system. Similar amounts pass into rivers, lakes, and seas and are deposited as sediments (Emsley 2000b). Ruttenberg (2003) provides an estimate of 20 Mt $year^{-1}$ of phosphorus as a flux from rocks and sediments to the soil by weathering and erosion, causing soil accumulations. However, there is also the opposite flow by lithification and "deep burial," with a "fuzzy" estimate of 9.3–19.5 Mt $year^{-1}$, indicating that the Emsley estimate may be a reasonable one (for more details, see Sect. 4.6).

Through the natural cycle, phosphorus moves from the land into rivers and the seas. In addition to natural erosion, phosphorus is formed through the decomposition of dead organic material or by the formation of insoluble calcium phosphate that sinks to the seabed as sediment. Fish-eating birds provide some minor recycling from the sea back to the land. Their droppings have built guano deposits on some islands and coasts that represent about 1 % of world reserves.

However, one should acknowledge that, as Smil (2000) stated, there is a "paucity," or lack, of knowledge, as many estimates refer to amounts first published in the 1970s. The inconsistency of some estimates can be illustrated by the estimate of erosion and runoff, considered to be 18–22 Mt P year^{-1} for particulate and 2–3 Mt P of dissolved phosphorus (Smil 2000). This estimate does not align with the Emsley estimate. The Liu et al. estimate (presented in Sect. 2.3) of 36 Mt P year^{-1} from pastures and cropland is even more divergent and shows how critically the existing data should reviewed.

In the following, based on the uncertainty of the data, there will be variances between years, measuring techniques, modeling assumptions, extrapolations, etc. This inconsistency will be a "steady companion." Here we may expect, in many fields, a typical *factor 2 uncertainty* (Fresenius et al. 1995) of quantitative estimations of concentrations in systems which are well known in the specific, but whose extrapolation is linked to uncertainty. For instance, the uncertainties of the estimates of the amount of phosphorus mined each year are certainly below factor 2. Naturally, some data are much more precise, such as data on fertilizer production. And for some data, we may have multiple datasets or sophisticated statistical methods of assessing uncertainty, which may result in much higher accuracy and robustness. But given multiple environmental and economic variability, we must be aware that regional and global data are subject to multiple transformations.

4.1 The Dose Matters: The Ecological Paracelsus Principle

It is difficult to precisely define what is wrong with the phosphorus flows, or to "put it in a nutshell." However, there is no doubt that—independent of the scale of system—we may find a "not too much" principle, expressed by the medieval alchemist Paracelsus (1493–1541) in the following terms: "All things are poison, and nothing is without poison; only the dose permits something not to be poisonous." Paracelsus, who lived during the Renaissance, was a deterministic thinker, rather than a probabilistic one. Thus, he supposed that there is a "threshold" for each living system, beyond which, the exposure to any (chemical) element is negative or toxic. In simple terms, this means that "the dose matters." From an ecological/human health perspective, this means that we may identify for living systems concentrations that are too high for positive performance. This concept aligns with the balanced nutrient supply that any living system should have. Thus, we assume—simply and deterministically—that there is a level of uptake or access $\bar{P}_{crit}$, beyond which the phosphorus use is considered critical, or negative. In addition, we can construct an ecological Paracelsus principle that

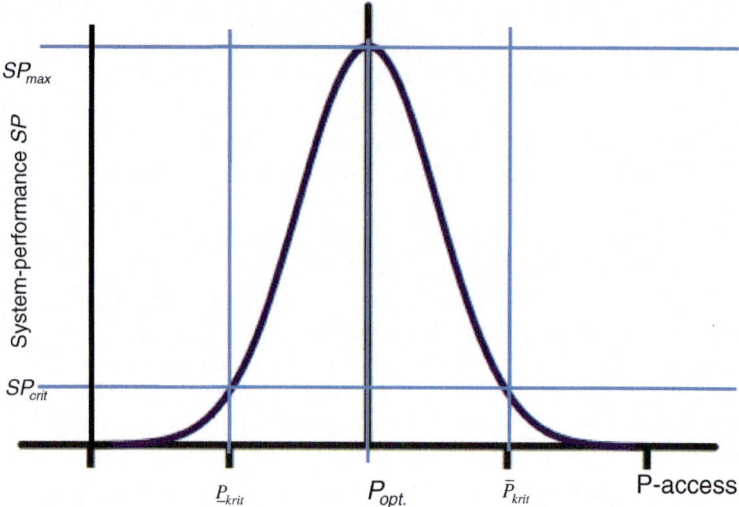

Fig. 12 Illustration of the "not too little–not too much" principle. Phosphorus (P) access (x-axis) with lower ($\underline{P}_{krit}$) and upper ($\bar{P}_{krit}$) critical "thresholds," optimal phosphorus use P_{opt} and y-axis with critical (SP_{crit}) and maximal (SP_{max}) system performance level

defines that which makes the use of phosphorus sustainable. Here, "critical" may refer to an ecological or an economic dimension, or both. Likewise, we have critical intake or access $\underline{P}_{crit}$, under which, the performance may be considered economically or environmentally critical. And there may be an application P_{opt} that is considered optimal. Figure 12 presents the case that we may measure the performance of the system with a one-dimensional parameter SP.

We may speak of an ecological Paracelsus principle, which defines what represents the sustainable use of phosphorus. The uncertainties of the estimates of the amount of phosphorus mined each year are certainly below a factor of 2.

We should acknowledge that from an evolutionary perspective, the tripling of the flows of an essential element such as phosphorus in less than 100 years is considered to be a dramatic and rapid change to the ecosystem grid. Diaz and Rosenberg (2008) conclude that industrialized primary production and the increase in fertilization after the 1940s, and in particular after the 1960s, led to widespread hypoxia. More than 400 "hypoxic zones" with more than 245,000 km^2 have emerged. However, outside of increased nutrient overflow, hypoxia can be caused by a number of other factors, including water stratification due to saline or temperature gradients that prevent mixing of oxygen-rich surface waters with oxygen-poor waters at greater depths. The weight of evidence indicates that in pre-industrialized times, coastal and offshore ecosystems seldom became hypoxic, except in the occasional natural upwelling of cold, nutrient-rich ocean water (Diaz and Rosenberg 2008).

In addition to sea and freshwater systems, an overabundance of phosphorus may also be seen as a threat to terrestrial systems. If we look at biodiversity, we come to more critical judgments on the extraordinary increase in phosphorus flows. Plant ecologists state that, "the use of P fertilizers is unsustainable and may cause pollution" (Hammond et al. 2004). One critical issue is that phosphorus is seen as a cause of the loss of biodiversity, as high phosphorus loads "favor a few species that would competitively displace many other species from a region" (Tilman and Lehman 2001). Or to express this in other terms, "enhanced phosphorus is more likely to be the cause of species loss than nitrogen enrichment" (Venterink 2011). But further evidence is needed to determine under what constraints this may occur (Molina et al. 2009).

Given our current knowledge, the groundwater concentration of phosphorus is not a critical issue if we exclude highly sandy soils. This is obviously due to the very high absorption capacity of soil, which functions as an effective buffer (Smil 2000).

4.2 Finiteness: Securing Long-Term Provision of a Public Good

Though phosphorus is ubiquitous, high-grade igneous and sedimentary PR seams (or layers) are limited. Nevertheless, we may define resource availability as "a sustainable phosphorus cycle if—in the long run—the economically mineable (primary and secondary) reserves of phosphorus increase higher than the losses to sinks (i.e., dissipation), which are not economically mineable" (Scholz and Wellmer 2013).

The dissipative nature of phosphorus is a long-term threat for the sufficient supply of the mineral. In the natural biogeochemical cycle, most of the weathered phosphorus ends its migration into the sea, with only minor amounts returning to the land as guano. The phosphorus reserves that are currently mined cannot infinitely provide high-grade ore phosphate rock.

Though we currently operate with a static lifetime of about 400 years, and we have identified reserves that may be mined at feasible costs over a few thousand years, it seems conceivable that there will be a time for humans in which the mining of phosphorus in conventional form might not be possible. One may argue that one could extract phosphorus from sea water, as the ocean shows a large reservoir of 93,000 Mt P (Scholz and Wellmer 2013). A rough estimate (based on an average concentration of phosphorus in seawater) shows the extraction of phosphorus from seawater not to be a realistic option. The 191 Mt of PR that are recorded by the USGS (2012) per year correspond roughly to 25 Mt P year^{-1}. Considering that the volume of the world's oceans is about 1.4×10^{21} l and recognizing that the concentration of phosphorus in seawater varies strongly (depending on oxygen and carbon content), a rough estimation of the amount of seawater required to extract that amount of phosphorus is 4.2×10^5 km^3 (4.2×10^{17} l). This represents about 0.03 % of the ocean's volume.

From an environmental systems perspective, we are facing different timescales within various human systems. With respect to phosphorus, *individuals* are tasked with trying to ensure consumption of nutritious foods with the appropriate dietary amount of phosphorus on a daily basis. This situation is different for farmers, fertilizer producers and other service providers who may maintain stocks, forcing them to consider longer time frames of several months to a year. If one considers the phosphate rock mining companies, the perspective is much longer, from a 50- to 100-year time frame, similar to the planning horizon of coal mines, hydropower plants or urban sewage or water utilities. An additional consideration is that a large share of current world food production is dependent on chemical or mineral fertilizers (see Roy et al. 2013, Spotlight 1), and that—depending on the soil type (Dumas et al. 2011)—there will be a large reduction in the yield without mineral fertilizer use. Naturally, one may think about organic farming or small-scale farming where one seeks out local produce, but here (as previously seen in the small-scale renewable energy sector), the global community is ultimately dependent on large-scale operations with longer planning horizons. In the case of phosphorus, change on a large scale, e.g., for the recycling of phosphorus from sewage and manure, not only requires a long-term planning horizon but technological and economically robust innovations.

4.3 Different Phosphorus Input and Balances in Various Parts of the World

From a global and regional viewpoint, phosphorus use has a unique history and will have an equally unique future. Amounts of indigenous soil phosphorus, patterns of consumption, levels of crop production and the resulting nutrient balances vary greatly by regions and countries of the world. In developing the regional trends presented below, data were drawn from different statistical sources; in cases of non-availability of complete phosphorus data, occasionally nitrogen/phosphorus/potassium (NPK) data are reported to build a more comprehensive picture.

Figure 13 presents global and regional fertilizer consumption. We see how sharply eastern European (East Europe) fertilizer consumption declined after the collapse of the Soviet Union, and how sharply Asian consumption has increased since the 1970s.

If we look at phosphorus fertilizer consumption, Table 1 provides insight into the regional differences. With near certainty, world demand will be considerably higher than today when we consider the sum of manure and mineral fertilizers. One reason is the continued increase in meat production/consumption per capita. Meat production is remarkably inefficient on both energy and nutrient scales.

However, future trends in phosphorus application differ in different regions of the world. Sattari et al. (2012) predict an increase in total phosphorus consumption by a factor of 2 in Latin America and Asia by 2050, and a quintupling of the

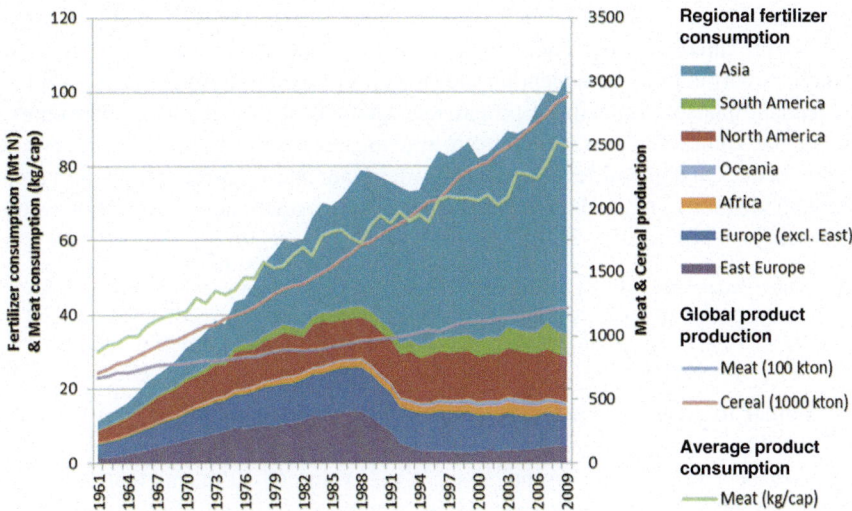

Fig. 13 Trends in total fertilizer consumption, distinguishing different world regions (sum of N, P, and K expressed as N, P_2O_5, and K_2O; Sutton et al. 2013; based on (Davidson 2012; Sattari et al. 2012) and trends of world cereal and meat production (*Source* Our Nutrient World, Sutton et al. 2013)

demand in Africa compared with 1965 consumption rates within a past and future 42-year window. Conversely, the consumption in Europe is expected to decrease by a factor of 2, whereas North America and Oceania are projected to modestly increase consumption.

Table 1 also informs on how the estimation of the average *phosphorus nutrient use efficiency* (phosphorus NUE) changed across the last four decades. Here, we learn that the annual "phosphorus plant uptake ratio" to "phosphorus soil input" increased in Western Europe; this means that the "one-year P nutrient efficiency accounting" dramatically increased. In 1965, the crops in Western Europe extracted 4.9 kg P ha^{-1} year^{-1} given an input of 23.8 kg ha^{-1} year^{-1}. In 2007, the extraction was 9.9 kg P ha^{-1} year^{-1} given an input of 17.2 kg P ha^{-1} year^{-1}. Thus, the *nutrient efficiency for phosphorus* in Western Europe increased from *phosphorus NUE* = 21 % to *phosphorus NUE* = 58 %, although the fertilizer input increased from 1.9 to 4.1 kg P ha^{-1} year^{-1}. Annual fertilizer P input in Western Europe decreased over the time period by 6.6 kg P/ha year^{-1}. The phosphorus use efficiency in Africa was *phosphorus NUE* = 95 % in 1967, which means that the balance was slightly negative. In 2007, Africa experienced a *phosphorus NUE* = 76 %, which indicates that P efficiency decreased in Africa.

MacDonald et al. (2011a) provide a profound analysis of phosphorus imbalances, referring to data for the year 2000. The authors state that cereal crops accounted for more than half of phosphorus removal from soils. Further, in 29 % of global cropland areas, there were phosphorus deficits; in 71 % of the areas, there were phosphorus surpluses. The large surpluses, with a mean value of

26 kg ha^{-1} year^{-1}, are found in East Asia, areas of Western and Eastern Europe, and North America, whereas the most widespread deficits may be found in South America, particularly in Argentina and Paraguay. An important issue for global sustainable phosphorus will be proper management of all streams of phosphorus flows: the natural streams, those from mineral fertilizer, and those from sewage and manure. On a global level, manure is important and has been included in most simulation studies. According to MacDonald et al., manure application shows high variability:

> Manure P alone, exclusively of P fertilizer, exceeded crop use only in 11 % of croplands globally, particularly in areas with high livestock densities but relatively limited cropland areas ... or in regions with relatively low P fertilizer applications and low P surpluses (e. g., across central Africa). (MacDonald et al. 2011a)

In order to understand the differential use of phosphorus fertilizer, we present Fig. 14. The y-axis is a double logarithmic scale of phosphorus NUE expressed in kg P per hectare per year (kg ha^{-1} year^{-1}). A value of 10 means that there is a surplus of 10 kg P per hectare each year; a value of −10 indicates a loss at the same rate. The data are attained by a large-scale simulation (MacDonald et al. 2011a; Potter et al. 2010) for fertilized areas with certain exclusions (i.e., excluding non-cultivated grassland). The uncertainty appears here, as in other data, due to the definition of fertilized cropland for spatial units of 50 × 50 km (which underlies the simulation) and due to the varying quality of national statistical data. Further, the calculations do not include natural system-based losses or gains of phosphorus through runoff or flooding. Thus, the robustness of these simulations is not definite.

We see that—Europe notwithstanding—there are positive phosphorus balances, indicating that (depending on assumptions of phosphorus loss in crop residues (MacDonald et al. 2011b), we may have an annual surplus of about 12 Mt P on a global scale. What happens with this surplus will be discussed hereafter. Note that many data of the simulation refer to the year 2000 and, thus, may not completely coincide with Table 1 or Fig. 14 data.

Often, to illustrate the intensity of mineral fertilizer use, we present the mean annual input of mineral NPK fertilizer in different countries for the years 2007–2009 on arable land in the 2007–2009 window according to the World Bank (The World Bank 2012) records, which are denoted as NPK fertilizer input. All unreferenced data in Sect. 4.3 refer to this source. To complement the efficiency discussion by absolute values, Fig. 15 presents the cereal yield per hectare for the different regions.

Asia, China, and India: We observe in Fig. 14 that Asia occupies the upper two quartiles (the upper 50 % of the Asian grid areas), showing a high P balance surplus. This indicates the extraordinarily high phosphorus consumption in parts of Asia, and in urban horticulture in particular. China holds about 20 % of the world population, yet since consumes since 2006 about 40 % of world nitrogen fertilizers (Gong et al. 2011). And the numbers for phosphorus fertilizers should be similar. On the other hand, we see from the outliers (the lower 5 % of the box-and-

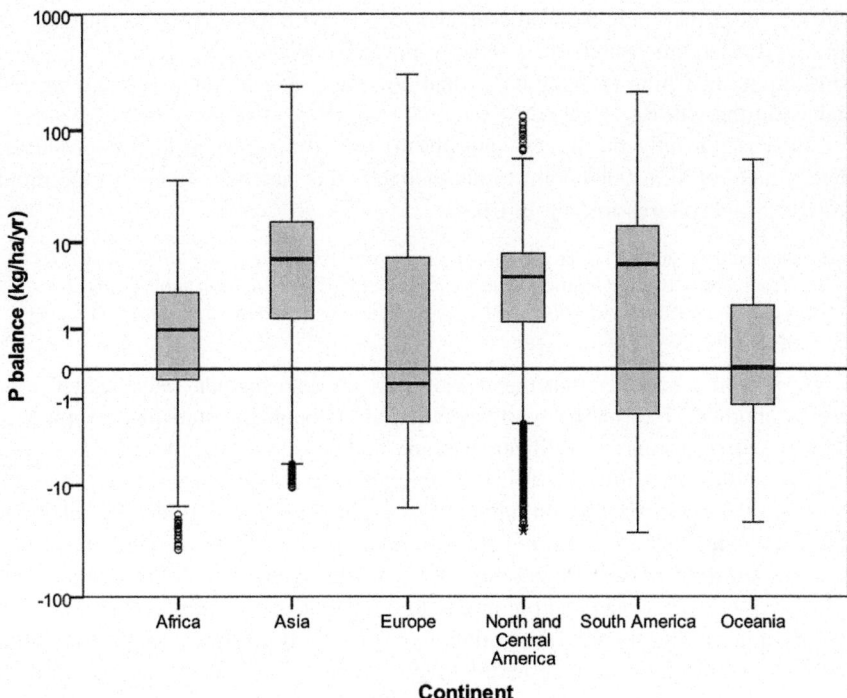

Fig. 14 *Box*-and-*whiskers plot* showing median grid-cell P balances (for all grid cells with >5 % cropland area). Note the logarithmic scale on the *y*-axis. Russia is included in the Europe grouping (MacDonald et al. 2011b)

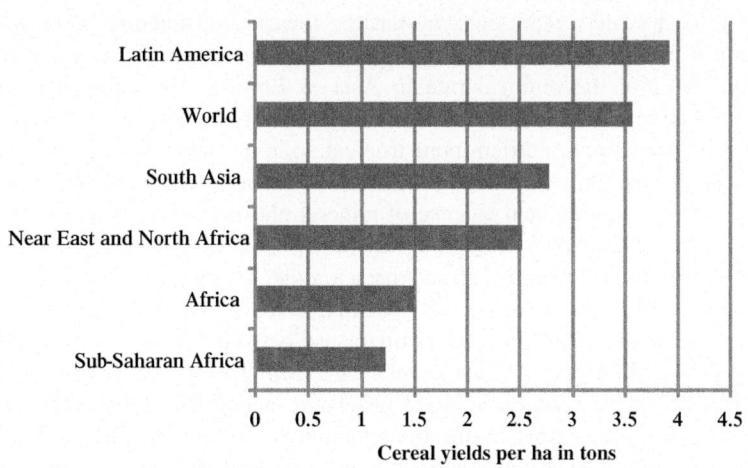

Fig. 15 Cereal yields per hectare by region for 2011/2012 (VFRC 2012, based on FAO data)

whiskers box) that there are also cropland regions with high losses, for instance in northwest Vietnam, which show severe nutrient depletion (Vu et al. 2012). The NPK inputs in China (482 kg ha^{-1}) and Vietnam (354 kg ha^{-1}) are among the highest in the world.

China is not only the largest phosphorus user, but also the largest producer. Here, Zhang et al. (2008) point to another flaw that has been linked to the rapid growth of phosphorus mining in China:

> P resource use efficiency decreased from a mean of 71 % before 1995 to 39 % in 2003, i.e., from every 10 kg P in rock material, only 3.9 kg P was used to produce fertilizer, 5.6 kg of the residues were discarded at the mining site, and 0.5 kg was manufacturing waste. (Zhang et al. 2008)

We should remember that the Asian statistics also include India, which represents another 17 % of the world population. Though the statistics do not distinguish between mineral fertilizer, manure, and sewage-based phosphorus flows, there are indications of overfertilization in other areas of Asia, including India; the average NPK fertilizer input in India is 154 kg ha^{-1}. A recent study (Pathak et al. 2010) indicates that the majority of the Indian states have a positive phosphorus balance, and that manure represents 78 % of the phosphorus input (Tirado and Allsopp 2012). This is in sharp contrast to the situation in China, where 80 % of the phosphorus inputs are from mineral fertilizers (Ma et al. 2011). The high application rates in Chinese agriculture have had negative impacts on a large share of inland lakes, which have been exposed to eutrophication (Tirado and Allsopp 2012). India and China, combined, account for 37 % of the world population and, thus, Asia is by far the largest and most critical region in the pursuit of sustainable phosphorus management.

Africa: Despite having some of the most nutrient-depleted soils, Africa has the lowest fertilizer input per ha of all regions measured and, importantly, with 3.1 kg P ha^{-1} year^{-1}, represents, by far, the lowest crop removal of all regions. Consequently, we find very low cereal yields in Africa (see Fig. 15). This is less than one-third of the removal rate in Asia or Europe, illustrating the difficult situation faced by the continent's smallholder farmers. Given that Africa is faced with highly weathered, nutrient-poor tropical soils in many regions, we begin to understand how important improvements can be from a sustainability perspective.

If we look at the regional balance of mineral phosphorus fertilizer flows based on data from 2000–2005, Cordell et al. (2009) suggest that there are 5.39 Mt P year^{-1} of fertilizers exported from Africa to other countries, but only 0.38 Mt of the mineral fertilizers are used domestically. Here, we may further reflect that there is a substantial difference in fertilizer use between exported crops and subsidized crops. Whereas exported crops (e.g., from Egypt) are grown on highly fertilized soil, subsistence agriculture elsewhere engages in little or no mineral fertilizer use. Further exasperating the situation is the fact that the limited phosphorus that does exist in these soils is being exported from the continent in the form of agricultural production; this should be considered a critical situation. This resource deficit, which is often believed to be "a typical African phenomenon" (Mehlum et al. 2006), may be illustrated by a review of general NPK use across a number of African countries. Though Africa does have phosphorus mines and fertilizer companies, the use of NPK fertilizer is low, for instance, 2 kg ha^{-1} in Uganda, 4 kg ha^{-1} in Angola, 10 kg ha^{-1} in Algeria, 17 kg ha^{-1} in Ethiopia, 34 kg ha^{-1} in Kenya, and 51 kg ha^{-1} in South Africa. Conversely, Egypt has been known to apply and as much as 574 kg ha^{-1} in some areas.

We suggest that the low input of phosphorus is a matter of social equity that most African smallholder farmers tend to face, speaking to the question of fertilizer access. Further, given that the benefits of phosphorus application are not immediately evident (as with nitrogen) and that they provide multi-year benefits, low land ownership in Africa may be seen as another reason for the low phosphorus input. In other words, if farmers do not own the lands that they cultivate, they are less likely to invest in plant health and sustainable soil quality.

South America shows the greatest heterogeneity, or diversity, indicated by the middle box and the spread of the whiskers (i.e., the 5 and 95 % percentile). South America faces negative P balances in many parts of Argentina, and positive balances on large-scale plantations in Brazil and other countries. This is very likely related to NPK fertilizer input, which is 40 kg ha^{-1} in Argentina, 158 kg ha^{-1} in Brazil, and as high as 566 kg ha^{-1} in Chile (which explains the upper whisker boundary of South America). These rates correspond with the high cereal yields (see Fig. 15) that are found in the mostly large-scale farming operations of South America.

Europe: Eastern Europe and Western Europe are not separated in Fig. 14, but they do offer different pictures. Whereas Central Europe and Southern Europe show high surpluses, often as much as 10 kg ha^{-1} year^{-1}, large areas of eastern Europe present negative balances. Here, spatial statistics reveal that—given comparable sizes—the yield and phosphorus recovery in eastern Europe is only 38 % of that of western Europe.

The tremendous differences between the western and eastern European countries may be observed when looking at the range of NPK fertilizer consumption in eastern Europe (15 kg ha^{-1} year^{-1} in the Russian Federation, 21 kg ha^{-1} year^{-1} in Armenia, 48 kg ha^{-1} year^{-1} in Romania, 84 kg ha^{-1} year^{-1} in Albania, 128 kg ha^{-1} year^{-1} in Bulgaria, and 385 kg ha^{-1} year^{-1} in Croatia) and western Europe (84 kg ha^{-1} year^{-1} in Sweden, 121 kg ha^{-1} year^{-1} in Finland, 169 kg ha^{-1} year^{-1} in France, 188 kg ha^{-1} year^{-1} in Germany and 428 kg ha^{-1} year^{-1} in Ireland).

Australia and Oceania: This region shows a very balanced picture. As we review the NPK fertilizer data, we see that the Philippines (134 kg ha^{-1} year^{-1}) and Indonesia (154 kg ha^{-1} year^{-1}) are in the midfield, while Malaysia, which has large palm oil plantations, is at the high end (574 kg ha^{-1} year^{-1}). Conversely, Australia shows low use of NPK fertilizer inputs (40 kg ha^{-1} year^{-1}) on cropland.

North and Central America: In the United States, one may find areas with large phosphorus surpluses, such as in coastal areas, yet negative balances are pervasive in the northern portions of the country. The NPK fertilizer use in the United States averages 113 kg ha^{-1}. However, the level of fertilizer consumption may differ on smaller regional scales, and socioeconomic and political impacts may affect phosphorus use. This may be observed when comparing two Meso-American countries. Nicaragua, for example, uses only 28 kg ha^{-1} NPK fertilizer, while the neighboring Costa Rica, with an NPK fertilizer input of 775 kg ha^{-1}, is ranked third among all countries (if we discard certain extremes, such as Iceland or Qatar).

We should also note that the crop mix may affect the above statistics as we have crops with low- and high-fertilizer-consuming plants.

4.4 Do We Have Low or High Nutrient Efficiency?

Sections 2.3 and 4.3 explore the question of phosphorus use efficiency. In review, phosphorus and potassium fertilizers are much less mobile than nitrogen fertilizer. Thus, the key factor for assessing nutrient efficiency is whether we frame the question in the context of annual or multi-year balancing. Once phosphorus is bound to soil particles, it does not simply disappear or forever become plant unavailable, yet it is in danger of being lost to runoff and erosion. Here, best agricultural management practices are required (Kroger et al. 2012; Sharpley et al. 2004).

We believe that multi-year balancing is not considered in many studies assessing (mineral) fertilizer phosphorus nutrient efficiency. Thus, one may find that unrealistically high losses for fertilizer inputs are reported (Cordell et al. 2009, 2011). Others believe that, given the use of best practices and proper environmental constraints, phosphorus nutrient efficiency may well be as high as 90 % (Syers et al. 2008). We expand on this question in the following section.

4.5 "Losses" and "Sinks" of Phosphorus from a Supply–Demand Chain Perspective

As the terms "sinks" and "losses" are pivotal in the analysis and construction of the MFA-Chart, which is a key representation for identifying means of sustainable phosphorus management, we will define the notion of sustainability in these terms. A definition of losses is also necessary, as phosphorus atoms do not disappear from the earth, but simply take on new inaccessible forms. A loss denotes the "harm or privation resulting from losing or being separated from something …" (Webster's 2002a). In defining sustainability, a loss may mean that phosphorus reserves become excluded from the current and future value chain, for example, due to economically motivated action. As it is impossible to predict which parts of the geopotential of phosphorus may be mined in, say, the next 1,000 years (for example, whether we begin to extract it from sea water), we rather take a present-

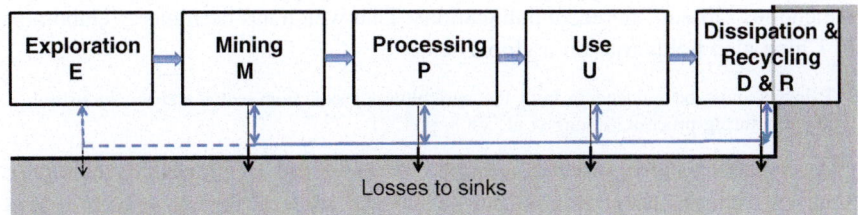

Fig. 16 The five stages of the supply–demand chain

day perspective and, thus, a precautionary view, which does not speculate about currently undeveloped future technologies. But "sink" also includes an evaluative connotation, denoting not only "a pit for the deposit of waste or sewage," but also "a place where vice, corruption or evil collects or gathers" (Webster's 2002b). The latter definition brings to light that, as phosphorus is removed from the economic value chain, the action may cause negative impacts that must be considered in a comprehensive sustainability assessment.

Harms from losses or sinks are, thus, always relative to a particular actor's interests. With respect to the supply–demand chain, we distinguish five stages: *Exploration*, *Mining*, *Processing*, *Use* and *Dissipation* and *Recycling* (see Fig. 16). The major information and material flows move from left to right. From a functional perspective, primary demand is represented in the *Use* stage (phosphorus use for agriculture or technology). The figure presents multiple interactions, such as the emergence of urban mining of municipal waste disposals. The losses or gains from information with respect to recycling are represented by dotted lines.

Identifying and managing uneconomical or unsustainable losses of phosphorus along the supply chain is a prerequisite of sustainable phosphorus management. But we face (at least) two types of losses. One type of loss (within the phosphorus industry) moves along the industrial value chain and includes mining, beneficiation, and downstream processes of producing fertilizer, phosphoric acids for feed additives or yellow phosphorus (P_4). The other loss is linked to phosphorus use, and includes losses from soil, crops, livestock, and industrial flows. These losses may be due to different key actors and the technologies they employ. We include these key actors from 'a Global transdisciplinary roadmap' perspective,[2] which decisively includes the key stakeholders in order to properly design Global Transdisciplinary feedback.

We separate industrial and agrofood losses, as they are of different natures. But even though we deal with them separately, they share common statements about

---

[2] Here, we may refer to (*individual*) consumers, (*groups* such as) households, companies and diverse non-governmental organizations (NGOs) from sectors such as environment, development etc. (as examples of *organizations*), nations (as the major form of *societies*), super-national *organizations* (such as the EU), and the human species (as the *supreme entity* from which the intragenerational and long-term aspect of social responsibility may be defined).

tremendous "losses" in recent publications. That which has been nicely elaborated for China also holds true on a global scale:

> P losses must be reduced in both the agricultural production sector and the industrial manufacturing process. (Zhang et al. 2008, p. 132)

As a starter (and in order to prepare the reader for the previously mentioned factor two uncertainty), we confront the critical reader with two views, both of which speak to the magnitude of the different types of "losses."

> IFDC estimates that between 30 and 50 % of the P_2O_5 equivalents in the mined ore is unrecovered and is contained in waste ponds and piles. (VFRC 2012)

It is interesting that we find the same magnitude of losses in the agro-use phase (the following text refers to 2008 production; Cordell et al. and Greenpeace estimates).

> ... the livestock system loses about 45 % of the phosphorus entering the livestock system itself (which makes 7 Mt P/yr input of 15.6 Mt P/yr to livestock) ... and this represents 29 % loss of the phosphorus entering the agriculture system overall. (Tirado and Allsopp 2012)

The latter data hypothesize an input of a total of 51.1 Mt P year^{-1} (which is 15.1 Mt P^{-1} year^{-1} higher than the above estimate (Liu et al. 2008) which excludes animal feed and results in a tremendous annual "loss" of 21 Mt P year^{-1}). We will discover which of these loss statements may be reasonably considered and which may not.

4.5.1 Phosphorus "Losses" from Industrial Phosphorus Processing

In this section, we look at the losses in the exploration and mining processes. As outlined in Spotlight 4, much of what are referred to as "losses" may neither be physical losses (e.g., by irreversible dissipation which makes phosphorus inaccessible for the foreseeable future) or losses from a business perspective (e.g., by not implementing state-of-the-art mining management principles). And what may be considered a loss today may be viewed a gain in the future. The phosphates that are not mined, but are left on the mines (e.g., by incomplete excavation or by supposed low grades) will be referred to as "ore residues." Please note that in this book, the phrase "ore residues" does not include the potentially risky beneficiation residues or tailings, only the phosphorus that is not processed from the ore.

We should also note that this section (as most parts of this book) deals with 191 Mt PR, *which makes* 25 Mt P *as recorded by USGS* (2012) *for the year 2011*. This 25 Mt P entered processing. One question addressed in the Exploration and Mining sections is how phosphate rock (phosphorus) has been moved or affected (e.g., by separating it from a large ore layer) before processing.

Exploration: We begin with the resources, which we may consider portions of the ore residues that are economically mineable but are not included in the reserves. The largest "losses" are attributable to mine planning, mainly in the decision regarding what parts of the deposits are to be mined and which are to remain untouched. The main factors here are the technological and economic conditions at the time—identified in the feasibility study used to determine the cutoff grade of the ore. Even in the future, the ores beyond this point may rarely be suitable for subsequent extraction because they are most likely located at the margins of the deposit or at considerable depth (Kippenberger 2001). In this context, we may talk about losses due to exploration limits. To the best knowledge of the authors, there are no publications that provide an assessment of this type of "loss" due to the limits of exploration of phosphorus mines.

Estimates of geopotential of phosphorus span some millions of gigatons (Smil 2000). But a comprehensive survey on reserves and resources is missing. There may be various reasons for this; some nations with presumably high phosphorus reserves, such Estonia, may not yet have surveyed and documented possible resources for various reasons. In recent years, USGS (2010) reassessed Morocco's reserves following a report of IFDC (van Kauwenbergh 2010) which updated already existing resources to reserves from 5.7 to 50 Gt PR [which seems to still be considered a conservative estimation (Terrab 2013; quoted from personal communication)]. In similar fashion, Iraq submitted information documenting 5.8 Gt PR to the USGS in 2011 (USGS 2012). But this number was lowered to 0.46 Gt PR in their 2013 reporting (USGS 2013b). The fact that about 90 % of the Iraqi reserves was downgraded and not included in 2013 recording (USGS 2013b) was due to an inaccurate use of the Russian categorization of reserves in the reserves assessment (Al-Bassam et al. 2012), rather than the common US classification (Jasinski 2013). Thus, 5.3 Gt PR were downgraded to resources as phosphate rock production seems to be more expensive (in mining) than the documented reserves at USGS. The ambiguous situation of exactly classifying reserves is also illuminated by the most recent statement of the Geological Survey of Iraq which states "The phosphate rock resources of Iraq are estimated at 9.5 billion metric tons" (Benni 2013). Here, it seems unclear what definition is taken.

Further, given the lengthy static lifetime of at least 350 years for economical phosphorus extraction, currently "… no company nor institution has the interest or the means to invest in exploration which does not contribute to their business plan. In general, companies only spend money at the high risk of exploration if they can bring the deposits quickly into production" (Scholz and Wellmer 2013). We may infer from this that there may be many unidentified reserves in the geopotential. We should note that speculation on what deposits may be economically extracted and processed under current and assumed future market conditions is a core business aspect of mining companies.

Mining: Mining is a multi-step economic and physical activity. Planning includes designing and constructing the mine, mining technology, infrastructure (particularly energy and water supply), extraction, and handling of the ore prior to beneficiation (UNEP and IFA 2001). As described in Spotlight 4, we must distinguish between the *mining ratio*, which describes the percentage of the ore that is excavated and the percentage left in the site (touched or untouched), and the low-

Table 2 IFA and IFDC recording of "ore residues" (for a definition of "ore residues" see text) based on the USGS (2012) with linear extrapolation (assuming the same ore grade in the processed phosphorus and the ore residues)

	IFA		IFDC	
	Operated (Mt P)	"Ore residues" (Mt P and %)	Operated (Mt P)	"Ore residues" (Mt P and %)
Mining	36.3	5.8 (18 %)	39.6	3.8 (9.5 %)
Beneficiation	30.5	5.5 (16 %)	35.8	11.8 (30.2 %)
Processed P	25		25	

grade ore which is put aside before primary beneficiation. The mining ratio does not refer to the overburden, but rather the rock or to the ore (this is sometimes not unambiguously clear in literature) that has been assigned to the ore.

Mining ratios have been assessed by IFA (Prud'homme 2010). Many mining companies obviously have reported very low losses in mining, with mining ratios well above 90 %, whereas others perform poorly, with rates of recovery as low as 45 %. *The global weighted average of the mining ratio is assessed to be* 82 % (Prud'homme 2010), *thus 18 % of the phosphate ore is not excavated.*

The IFDC data (VFRC 2012, p. 12) differ slightly from the IFA data. When using data from the year 2009,[3] IFDC estimates a loss of 9.5 % from 560 Mt ore rock extracted with 1,700 Mt overburden. Of the total of 36.5 Mt P year^{-1}, about 3.5 Mt P year^{-1}, which makes 9.5 %, remains unrecovered. This is roughly half of the IFA estimate (see Table 2).

Given the percentage of loss, one could estimate the total amount of phosphate rock and phosphates that have been affected in the mining process. This, however, would require one to know the amount that enters beneficiation which is currently unknown. We only know the amount (24.8–25.7 Mt P included in phosphate rock granulate) that enters processing or that is used for direct application as fertilizer based on the USGS (2012) and 2013 data. We use 25 Mt P as a rounded-up value of the USGS (2012) data in the following. We may provide estimates of the total amount of phosphate rock after the amount put to beneficiation has been assessed (refer to the formulas in the following sections).

We should note that the reported estimates are below those of Kippenberger (2001) from the German Geological Survey who reports non-used phosphate ores of 36 %. Compared with some other minerals such as aluminum (13 %) or iron (23 %), this is in the higher end of the range (Wagner 1999). The 25 Mt P year^{-1} for 2011 recorded by USGS (2012) include phosphates after primary beneficiation. Based on this, the estimates of a virtually mined amount of phosphate rock provide a broad range of *v*irtually *m*ined phosphate rock (PR_{vm}) that may be transferred to virtually mined phosphorus P_{vm}. We should also note that Kippenberg estimates refer to *tons of ore*. The degree of ore material that is left in the mines (e.g., the

---

[3] USGS reports mine production of 158 Mt in 2009; (Löffler 2013).

below cut-off material) compared with the ore that is processed is unknown. We may assume that the ores not sent for processing are of lower quality. Thus, the "losses" may be overestimated in terms of phosphorus.

(Primary) Beneficiation is most often conducted at the minesite, which results in a higher phosphorus concentration—in particular for igneous phosphate rocks, which have very low phosphorus content. The rock phosphate is subjected to a multi-step process that includes crushing, grinding and flotation, acid washing, magnetic metal extraction, dewatering, and drying. For primary processing recovery, IFA (Prud'homme 2010) presents a statistical analysis including 93 % of world production. These data are attained from private companies and, thus, are not generally available to the public. IFA provides an estimate of 84 % of primary processing recovery. This would calculate to an additional loss at the mining stage of 5.5 Mt P or 16 %.

These data allow for a rough extrapolation of the amount of phosphorus (by means of simplification, we work with phosphorus and not phosphate rock).

(*)   25 Mt P = (25 Mt P 0.84^{-1}) 0.82^{-1} = 36.3 Mt P
(*)   36.3 Mt P (1.0–0.18) (1.00–0.16) = 25 Mt P
(***) 36.3 Mt P−5.5 Mt P−5.8 Mt P = 25 Mt P

This means—if we take the IFA data—that roughly 5.5 Mt P (in phosphate ores) are left at the mines in the process of excavation and 5.8 Mt P is put aside during beneficiation. Thus, about 11 Mt P in ore fractions is put aside. The degree of these fractions is not known. We may assume that they are certainly of lower degrees, but the IFA data refer to P_2O_5 which may be transferred to P. Naturally, one may assume that the concentration of the set-aside from which no phosphorus is recovered and which left on the mines or put to waste deposals is of lower concentration and may not economically processed. This would mean that this part is not lost (from the value chain). Thus, we lower the above estimates and assume that only 50–80 % of the above 11.3 Mt P may be considered as losses. *Thus, a realistic estimation of the lower boundary, referring to the IFA data and based on the above assumptions, may provide an (conservative) estimate of 2.7–4.4 Mt P residues from mining and 2.9–4.6 Mt P from beneficiation.*

The IFDC (VFRC 2012, p. 12) data for beneficiation again differ from the IFA survey and provide a much more unfavorable picture. Of the 38.6 Mt P year^{-1} (contained in P_2O_5 concentrate) which is beneficiated, 30 % are classified as unrecovered by beneficiation (see Table 2).

We presented different estimates of IFA and IFDC in Table 2. There are similarities, but also large differences, which may be due to differences in the definitions of system boundaries (e.g., what is overburden and what is ore in mine spoils), the definition of system elements (what are "other" or "industrial" uses), accessible data, and the author's decision on what information the data are derived, etc. Both surveys are reports by industrial or nonprofit organizations and have not been independently evaluated, as is the case with scientific papers.

Both IFDC (about 31 % if 36.3 Mt P is taken as a reference) and IFA (39 % if 39.3 % is taken as a reference) suggest a total of 35 % ore residues or non-recovered phosphate rock. This means that only two-thirds of the rock phosphate ore becomes a marketable product.

In the following, we use the IFA data for the amount of phosphorus that is affected by mining and beneficiation (i.e., a loss of phosphate rock of 31 %). And we use adjusted estimates (which assume a lower grade for the ore residues of between 50 and 80 %) for the crop residues or losses, or 2.7–4.4 Mt P unrecovered phosphorus in mining and 2.9–4.6 Mt P unrecovered in beneficiation.

Finally, the USGS and the IFA numbers differ as IFA does not include the official China phosphate rock production data (Prud'homme 2013). Thus IFA has smaller data, which may affect, for instance, some estimates in Fig. 21.

Non-processed fertilizer: We should note that according to Prud'homme (2010) in 2007, about 1.9 Mt P_2O_5 was used as direct fertilizer. If we adjust this number linearly to compare directly to 2011 data (USGS 2011), we get an estimate of 1.1 Mt P that was directly used in 2011 as fertilizer without entering wet or thermal processing. If we include the difference between the USGS and the IFA data, we may face even (slightly) higher values here. The thermal route is all dedicated to industrial uses.

Processing: Providing reliable data on phosphorus losses in this node of the supply chain is critical because no comprehensive survey has been performed and the expert estimations often prove to be inconsistent.

Prud'homme (2010) reports that in 2007, 51.7 Mt of P_2O_5 (which makes 22.6 Mt P) was shipped to or processed at plants linked to mines. Here, almost 4 % was directly used, about 5 % was thermally processed (mostly for P_4 production) and the bulk of the remaining 91 % was subject to chemical wet processing, primarily for fertilizer production (see Fig. 17).

According to a mid-term 2011 estimate by IFA (Heffer and Prud'homme 2011), 18.1 Mt P was used for fertilizer. IFDC (VFRC 2012) estimates for 2009 that about 18 % went for uses other than fertilizers (see Fig. 4) including feed and food additives, detergents and technically and industrially used P, including some used as additives (as a strengthening agent) of steel.

If we simply extract 18 % from the 25 Mt P and the 1.1 Mt P for direct use, we get 18.4 Mt P, which become subject to fertilizer processing. The challenge is now to assess how much phosphorus is lost in fertilizer processing.

This estimate is below the IFDC (VFRC 2012a) estimate of 10–15 % that is not recovered in the wet processing of fertilizer which provides 1.8–2.7 Mt P. IFDC estimates that 4–9 Mt P_2O_5 (i.e., 1.7–3.9 Mt P) is lost in stacks or ocean disposals from fertilizer processing in 2009.

We may also look at the survey performed by Villalba et al. (2008) that focuses on the 2004 production streams. This study well distinguishes between fertilizer and non-fertilizer, and wet and thermal phosphoric processing. For wet phosphoric processing, clearly phosphogypsum is identified as a major environmental issue. Villalba et al. (2008) suggest that a range of 2–12 % is lost. The wide span refers to the different properties of the phosphate rock, processing technology, etc.

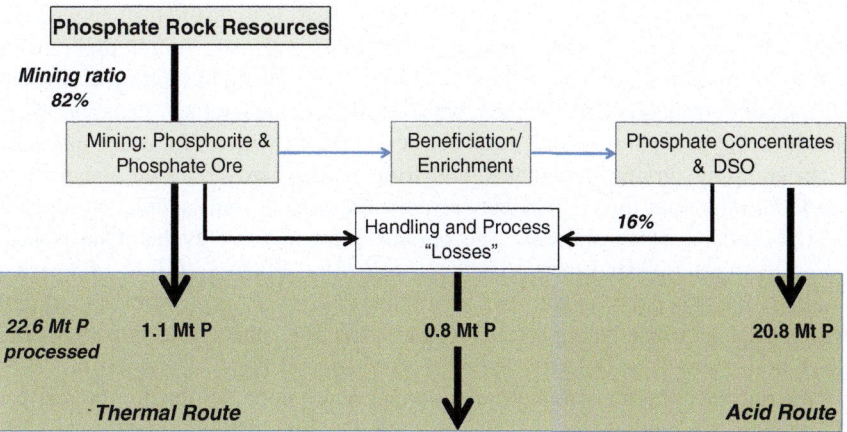

Fig. 17 Mining, beneficiation, and routes taken as described in 2007 data (Prud'homme 2010)

We see that a precise estimate is beyond our capabilities. However, including the non-fertilizer route, a loss range of 2–3 Mt may be seen as a moderate, conservative estimate. Thus, the amount of phosphorus in fertilizers in 2011 was around 16 Mt P.

Virtual flows: Before we deal with agricultural use, we briefly touch on non-chemical industry virtual phosphorus flows. Here, we mean phosphorus that is bonded with mineral resources such as iron, coal, or limestone. In heavy industry, phosphorus causes impurity. Also, in steel production, phosphorus has two faces and—depending on the quality of the steel—is sometimes added and sometimes subtracted and goes to the slag.

Matsubae et al. (2011) provide an interesting analysis on the amount of real (phosphate ore) and virtual phosphorus which is imported (Matsubae et al. 2011). For Japan, virtual phosphorus flows take a large share. Matsubae et al. (2011) state that the consumed agricultural products in Japan make only 12 % of virtual phosphorus consumption. This number is certainly specific to Japan, which has a high proportion of car and metal-based industries. But these data may suggest that the magnitude of the virtual flows may well be of the magnitude of the phosphorus in the food uptake (Jeong et al. 2009). We do not deal with the virtual flows, which may define a new research domain, in detail in this chapter.

4.5.2 Phosphorus Losses from Agricultural Use

The use system is an overly complex, multilayered, partially interconnected system. It includes arable and livestock farming and forestry of very different types. A specific challenge is that the phosphorus flows in the agricultural system sometimes are in a rapid state of change due to changes in the demand/diet side or due to events such as the mad cow disease outbreak that changed the processing of animal bones and carcasses in some countries. In the following, we aim to adjust all data to be representative of the year 2011.

A specific challenge in assessing use system losses is to distinguish between the natural biogeochemical flows, i.e., the flows from weathering, runoff, and erosion, and those from anthropogenic impacts. As we live in the anthroprocene (Crutzen 2002), our current geological epoch, we are experiencing a rapidly changing world in which the phosphorus flows may not be explained without human impacts.

In cropping (grains, vegetables, and fruit) and in terrestrial and aquatic livestock farming, we distinguish between natural/organic and synthetic/processed (e.g., beneficiated) phosphorus. The organic form has two cycles. One is land-based, between soil and terrestrial fauna and flora. Zoomass (10 % of which is anthropomass) is not relevant here, as it makes only 1 % of phytomass (and thus, we may exclude fish-eating terrestrial animals). The other cycle is water-based; phosphorus circulates between sediments and aquatic biota (Chapin et al. 2009). Terrestrial phosphorus is moved to aquatic environments by erosion of soil particles and surface water runoff. This is a natural cycling that nourishes aquatic life. Human activity, in particular plowing and overgrazing by livestock, may change the cycle dramatically. Whereas the cycling of phosphorus may repeat 100 times in a natural biotic environment before it is transported via rivers to the sea system, intensive cropping and grazing without best management practices may reduce the cycling to just a few times. Thus, the anthropogenic regional and global change (e.g., climate change effects on runoff) may affect the speed by which phosphorus is removed from terrestrial systems. From a human-environment systems (HES) perspective, the rate at which phosphorus is lost depends on the functionality of the land use. Natural land use losses are much less than losses associated with grazing land and cropland.

If we consider losses from crop production, we are interested in how much applied phosphorus is taken up by the crop, recognizing that some of the applied phosphorus is lost by erosion, runoff, or leaching (possible on sandy soils), while much is converted to less soluble forms in the soil. Naturally, both the organic and the mineral fertilizer inputs must be considered. While there is more or less reliable data on phosphorus fertilizer use, the recording of organic inputs such as manure, slurry, peat, seaweed, leaves, compost, human excreta, irrigation is not a common subject of statistical surveys. Consequently, estimates of manure or organic fertilizer are more difficult to obtain, though better standards have been developed (Wang et al. 2011). We find similar uncertainties in the diverse system of animal production with respect to household waste, crop residues, etc.

4.6 Runoff and Erosion

Surface runoff and erosion cause two large sets of losses of phosphorus in agriculture. Runoff occurs if the soil is saturated or cannot absorb rain or melting ice. Erosion includes soil and nutrient removal by wind and water. As much of the phosphorus runoff is of particulate phosphorus that is bound to soil particles, the two processes overlap. Wind combined with droughts may become a major source of erosion. Both runoff and erosion depend on the geographical setting (e.g., soil type, slope) of the cropland and grassland agrotechnology (e.g., tilling technologies, drainages) and intensity of use, e.g., overgrazing (Haygarth and Jarvis 1999; Sharpley et al. 2003). Both runoff and erosion of soil and fertilizer are most vulnerable to "incidental losses" (Haygarth and Jarvis 1999), i.e., discrete events such as heavy rains or storms. These *event-specific losses* surprisingly have "received relatively little study" (Hart et al. 2004) though they make up a major share of phosphorus losses.

> ... event-specific losses often make the dominant contribution (50–98 %) to P in runoffs from field plots ... (Hart et al. 2004)

We must also be aware that climate change increases the vulnerability of cropland and grassland by extreme event-caused erosion (Nearing et al. 2004).

We illuminate the difficulties in providing reliable estimations with the case of wind erosion. For instance, Liu et al. (2008) refer to Schlesinger (1991) who provided an estimate of 4.6 Mt P year^{-1} of atmospheric phosphorus deposition (with a mean residence time of about 80 h). But this does not tell us about the wind erosion transfer to oceans. Ruttenberg states that "atmospheric deposition is relatively unimportant" (Ruttenberg 2003). In his phosphorus flow model, 4.3 Mt P year^{-1} is transferred from land to the atmosphere and 3.1 Mt P year^{-1} return to soil, which would provide an estimated loss of 1.2 Mt P year^{-1} by wind erosion to the sea.

With respect to erosion, we may also look at the data provided by Graham and Duce (1979), which, however, was developed in a time when world phosphorus mining was about 13 Mt P year^{-1}. Graham and Duce estimate that the atmospheric burden is 0.28 Mt P year^{-1} (corresponding to a lifetime of 80 days and a wind erosion of 1.2 Mt P year^{-1}). In their balance, 1.4 Mt P year^{-1} is moved from the continent to the sea and 3.2 Mt P year^{-1} is transferred to other soil. But there are some flows back to the land (0.33 Mt P year^{-1}) from seawater; a net balance of 1.0 Mt P year^{-1} is the result. But here, we must acknowledge that "50 % of this transport is due to the flux of dust from the Sahara desert to the North Atlantic at 15° and 25° N" (Graham and Duce 1979).

Another important flow affecting erosion and runoff is the annual input by weathering. Carpenter and Bennett (2011) distinguish between a natural, non-anthropogenic weathering of 10–15 Mt P year^{-1} (which may differ between interglacial and warm stages) and human-induced weathering of around 5 Mt P year^{-1}. This brings the weathering input to 15–20 Mt P year^{-1}.

The transfer from soil to the water system is a very critical one. Smil provided an estimate of erosion and runoff of 18–22 Mt P year^{-1} for particulate and 2–3 Mt P of dissolved P (Smil 2000). Ruttenberg (2003) is very close to this, estimating 18.3–20.2 Mt P year^{-1} for particulate and 1.0–1.8 Mt P year^{-1} for dissolved P. These data have been challenged by Hart et al., who refer to studies that conclude that runoff from organic (poultry) fertilizer included 67 % dissolved phosphorus, and runoff from mineral fertilizer was more than 95 % phosphorus. Because of the chemical dynamics of phosphorus, this data may not be transferred to the phosphorus transported in rivers. Here, the phosphorus transported in global river sediments to the ocean (Beusen et al. 2005) amounts to 9 Mt P year^{-1}. We suggest that the Liu et al. estimate of 36 Mt P year^{-1} of runoff and erosion from pastures and cropland seems to be too high and will not be referred to in the following. We follow Carpenter and Bennett (2011) with an estimate of 22 Mt P year^{-1} for the losses of phosphorus from soil sea via freshwater.

According to a rough balance, in 2002, USGS accounted for 17.4 Mt P year^{-1} after beneficiation and about 10–15 Mt P year^{-1} input resulting from manure. This totals 25–30 Mt from anthropogenic sources compared with 22 Mt P year^{-1} from weathering. This provides considerable stock building in terrestrial systems.

Of course, we are interested in how much is lost from the 12 % of cultivated land and the 22 % of grassland (Leff et al. 2004). The review of Hart et al. (2004) illuminates the difficulties in providing a reliable estimate. This starts with difficulties in distinguishing between particulate phosphorus and dissolved phosphorus, as phosphorus has specific physicochemical characteristics:

> Phosphorus can occur in a continuum of sized down to near-molecular dimensions, and thus the definition of particulate and dissolved forms of P is rather arbitrary, defined by analytical convenience … (Hart et al. 2004)

The surveys on phosphorus losses differ by scale, soil type, slope, weather conditions, tillage and drainage systems, crops, and types of fertilizer. Thus, it is not surprising that the conclusion is that these runoffs are highly site-specific. Hart et al. review 29 studies on phosphorus losses from different land uses related to fertilizer applications. The losses vary from 0.03 to 42 % with a mean of 17 %. But the studies by no means can be considered representative. The difference between estimates of soil erosion of 10 t ha^{-1} year^{-1} on US cropland compared with China at 40 t ha^{-1} year^{-1} (Pimentel et al. 2010) shows the large variance. Cordell et al. (2011) provide an estimate of 8 Mt P of *erosion losses* from agricultural soils and pastures. Again, it seems clear that the Liu et al. (2008) estimation of 19.3 Mt P year^{-1} from cropland and 17.2 Mt P year^{-1} from pastures seems to be an overestimation.

When providing an assessment of the losses, we face uncertainties and a lack of understanding. Given an input of 16 Mt P year^{-1} in 2011 from mineral P, between 15 and 20 Mt P year^{-1} from manure, sewage, crop residues, dry collection system etc. we may assume that around 35 Mt P year^{-1} is put to soil.

Rockström et al. provide (without considering the losses) this estimate:

> Some 20 million tonnes of phosphorus is mined every year and around 8.5 million–9.5 million tonnes of it finds its way into the oceans. (Rockström et al. 2009)

Transferred to 25 Mt P year^{-1} of mined phosphorus in 2011, this provides between 10 and 12 Mt P year^{-1} that is transferred from mined phosphate to the sea. If we wish to provide an estimate for the total inflow to oceans, we may refer to the Carpenter and Bennett (2011) and assume that most of the weathering input of 15–20 Mt will be transferred to the ocean. Thus in 2011, an estimate of 25–30 Mt P year^{-1} may be a reasonable estimate for the phosphorus input to the sea, more than double the non-anthropogenic input.

4.7 Changing Lifestyles Increase Anthropogenic Phosphorus Flows

Economic growth in China, India, and many other countries of the world is directly linked to a change in dietary habits, including an increase in the consumption of meat, fish, and dairy. "World meat production is projected to double by 2050, most of which is expected in developing countries" (FAO 2012a). The impact of this dietary change on phosphorus consumption may be inferred from the data currently available. Measured in calories, 1 kg of meat includes 1,500 calories, whereas, 1 kg of cereals provides about 3,000 calories. Moreover, it takes 3 kg of grain to produce 1 kg of meat, even if we assume that part of the feed is taken from rangeland and organic waste (Nellemann et al. 2009). This translates to a factor 6 increase in nutrient efficiency for food uptake if you refer to energy.

Today, we find the following data for cereal production:

> Of the 2.4 billion tonnes of cereals currently produced, roughly 1.1 billion tonnes are destined for food use, around 800 million tonnes (35 percent of world consumption) are used as animal feed, and the remaining 500 million tonnes are diverted to industrial usage, seed, or wasted. (FAO 2012b)

For 2050, UNEP assumes that "at least 1.45 billion t [Mt] cereals are used as animal feed" (Nellemann et al. 2009). This means that—all things being equal and taking the uncertainties of the numbers into account—several hundred (around 600–750) Mt of cereals must be produced in addition to the current consumption. Assuming a population growth by 23 % through 2050 (and the factor 2 calorie efficiency differences between cereal and meat), this would mean that just about 10 % (i.e., 235 Mt) more cereals as UNEP assumed must be produced annually to account for dietary change to more meat production.

The latter estimate is very much in line with an estimate by Hu (2011), FAO (2012b) on the impacts of the migration of Chinese farmers into cities and townships (75 % in 2050 compared with 47 % in 2010). Given that 300 million rural persons move to cities, Hu estimates an increase in phosphorus demand by about 20 % (0.36 Mt P year^{-1}) due to diet change, and a further 10 % increase (0.18 Mt P year^{-1}) due to sewage loss and increased biofuel consumption. Given the current wastewater treatment technology in China, urban lifestyle change, for instance, is linked to the recycling of sewage at 30 % compared with 94 % which was the recycling rate related to traditional sewerage treatment. Naturally, these are rough estimates under certain assumptions. But this reveals that dietary and lifestyle changes may induce additional phosphorus use and provides some information about the magnitude of the increase.

Fig. 18 There are historic trade-offs to the use of dung as fertilizer or fuel (picture taken in Tibet, R. W. Scholz)

4.8 Insufficient Phosphorus Recycling from Manure and Crop Residues

Possibilities exist for improvements in some anthropogenic flows of phosphorus related to agricultural food production, i.e., through the reclamation of phosphorus in *manure, crop residues* and *sewage*. All three flows are central from a soil management perspective and, thus, are significant factors in future food production. All three flows must be viewed from a historical, sociotechnological perspective and, thus, viewed in the context of the evolution of human-environment systems. This holds true in particular since gatherers and hunters are nearly extinct and the world has faced almost simultaneous agrotechnology developments across our agrarian, industrial, and post-industrial societies. As a result, history shows how new trade-offs have emerged (see Fig. 18).

4.8.1 Learning from History

Sustainable farming requires sophisticated amelioration and fertilization of the soil. About 7,000 years ago, Asian Neolithic farmers learned to improve soil fertility of otherwise non-arable land by dunging the cropland with the excreta of grazing cattle, sheep, and goats (McNeill and Winiwarter 2004; Bellwood 2005). Though livestock manure may improve soil quality by increasing organic matter or expanding the water-holding capabilities of soils, it does not possess the optimal composition of nutrients for maximum crop production in a given field. Various types of composting performed by early agrosocieties are known to have improved the efficacy of manure.

Fig. 19 A flotilla of manure boats on Soochow creek, Shanghai, prepare to transport excreta to cultivated fields (taken from King 1911, p. 195)

Another important step of *dislocation* has been linked to urbanization. About 5,000 years ago, when the first cities emerged (Benevolo 1980), nutrients were drawn from the fields to the cities. The importance of human excrement for soil fertilization may be observed in a 1649 decree in Tokyo that banned toilets that emptied into canals or creeks (Smil 2004). Here, we may have met an early trade-off inherent in manure recycling, which is that the Tokyo decree may have improved the ecosystem health in canals, but may have increased the risk to human health, as manure may contain a variety of pathogens. The latter rebound effect, in particular, holds true if, for example, the cycling system for human sewage is too short, and the product is applied directly to vegetable fields while pathogens still remain.

Clearly, the specialization and industrialization of agriculture may have been a significant factor in the reduced use of manure as a crop fertilizer. If large-scale swine or cattle production became clustered, transportation costs may have become a barrier for the economical application of manure to cropland. And in times of increasing rural residential settlement, odor became another factor preventing widespread manure, and sewage use.

With respect to sewage, the anthropologist, King (1911/2004), reported how sophisticated the amelioration of sewage had become in highly populated areas such as the Hankow-Wuchang-Hanyang area in China, where 1.8 million people lived in a radius of four miles (see Fig. 19). Here, history shows that utilizing manure and sewage requires treatment before reuse and transportation.

Finally, if reviewing the history of crop residues, three distinct stages and loss mechanisms emerge. The first is between *soil and farm*, where the widespread burning of straw and stalks resulted in losses of nitrogen, phosphorus, and other nutrients (which has implications for long-term nutrient management and climate change). The second may be between *farm and table*, and the third is *after-table* losses.

4.8.2 Phosphorus Recycling from Manure

While manure represents a major portion of inefficiently used and wasted phosphorus, providing reliable data on a global scale remains difficult. Cordell et al. (2009) provide a figure which suggests that the current share of manure being used for fertilization amounts to about 3 Mt P year^{-1}. A recent thorough analysis of MacDonald et al. (2011a) estimated the agricultural input of manure to be 9.6 Mt P year^{-1}, or 40 % of total manure phosphorus excreted by livestock in 2000. But the lack of data outside of those cited indicate the uncertainty and bias that may be tied to data on manure.

Nevertheless, the practice of manure recycling differs from country to country, and even within regions of a single country. The practice may also differ among cropping systems. Sattari et al. (2012, SI p.1) provides an estimate of 15–24 Mt P year^{-1} for annual manure production. This study points to the different uses of manure in industrialized countries, where about half is supposedly used for grassland and half for cropland.

The comprehensive report *Lifestock's long shadow* (FAO LEAD 2006) deliberates on many scales the need and potential for better nutrient management. For instance, a cow excretes 18–20 times a much phosphorus load as a human (Novotny et al. 1989). And the energetic potential of manure as well as the pollution potential may be seen by the estimate that "Methane released from animal manure may total up to 18 tonnes per year" (FAO LEAD 2006, p. 113).

In developing countries, 95 % of the available manure is assumed to be applied on cropland (Sattari et al. 2012, SI p.1). This assumption may be questioned as manure is competing with other uses such as fuel (see Fig. 18). A very high estimate of 24.3 Mt P year^{-1} is provided by Potter et al. (2010) while other studies provide estimates of 21.1 Mt P year^{-1} (Sheldrick and Lingard 2004).

In order to provide a more comprehensive picture on the potential, but also the limits of phosphorus use, the most available and widely elaborated statistics from the United States are reviewed (Ruddy et al. 2006; MacDonald et al. 2009; EPA 2012). If we look at the total phosphorus nutrient input on the land surface in 1997, 50.4 % of total farmland phosphorus input was mineral fertilizer, 48.5 % was manure phosphorus, and 1.1 % mineral fertilizer was applied to non-farmland. These figures, compared with the MacDonald et al. (2009) report cited below, suggest that significant amounts of phosphorus from manure production remain on the grazing land, are contained in lagoons, were incinerated or ended up in streams and rivers from surface water runoff. There were high farm-fertilizer inputs in the upper Midwest areas, along the east coast and in irrigated areas of the west. A 2007 estimate of animal manure provides a phosphorus total of 2.04 Mt P year^{-1} just in the United States, which is higher than the 1.72 Mt P year^{-1} in 1997 and the 1.62 Mt P year^{-1} in 1987.

Given animal manure production increases—nearly 50 % of total phosphorus input to the land was contained in manure and "Manure was spread as fertilizer on 15.8 million acres" (D, see MacDonald et al. 2009)—the increasing need to recycle nutrients and manage waste is clear. Maize (the largest single crop accounting for 25 % of the total acreage in the United States) accounted for over half of acreage fertilized with manure (9.1 million acres), with hay and grasses accounting for 4.2 million acres and soybeans accounting for slightly less than 1 million acres. Maize being a major beneficiary of manure is due to a number of beef feedlots and swine operations in the Midwest located close to large maize farms.

Due to high transportation costs, manure markets tend to be highly localized. And obviously its value as fertilizer is not unanimously acknowledged, as about 60 % of swine and broiler manure that is not used on the farm is given away for free in the United States. Finally, the odor associated with manure spread on croplands is offensive to nearby neighbors and peri-urban areas, creating friction and complaints about air quality and elevating health concerns as a result of recent food contamination occurrences.

The 2007 USDA ERS estimate of 2 Mt P year^{-1} of 2012 (Frear 2012) allows for a first rough estimate about the magnitude of phosphorus in livestock production. The world population is 22 times larger than the population in the United States, which is known as the world's leading meat-producing country. However, meat production and protein consumption from livestock is rapidly changing in most parts of the world. Between 1967 and 2007, the production (per person) of pig meet increased by 152 % and poultry meat by 369 %, whereas beef and buffalo meat decreased slightly to 93 %. If we acknowledge that the world population has more than doubled, one may see that manure has become more important. Even acknowledging that the phosphorus contents of manure depends on feed, species (see Table 5), and many other factors, we may suppose that the 9.6 Mt P year^{-1} of MacDonald et al. (2011a) for the global phosphorus in manure is assuming that there are more than four times as much phosphorus in animal manure in the United States (with a large share of beef with relatively low phosphorus content) than for the average world citizen. Given that the world protein consumption is increasingly shifting to livestock-based dietary demands, the global phosphorus in manure tends to have 10–15 Mt P year^{-1} and more rather than below 10 Mt P year^{-1}.

In Europe and other developed countries, the application of manure developed a negative image due to the odor and human health concerns, despite advanced environmental regulations in most developed countries. The regulations focus on preventing direct application of manure to salad or other directly consumable products without extensive processing. Regardless of the current situation, the nutrient content of and the possibilities for nutrient recycling from manure must be assessed. Proper strategies for improving manure quality in combination with crop residues, other waste, or mineral fertilizers are possibilities. While the potential of nutrient recovery from manure is not assessed on a global scale, it appears that in developed countries, a large share is used in incineration (e.g., in the cement industry), or dumped in waste fields. With the recent growth in the renewable energy sector, manure has become of interest as feedstock for energy production. The major processes to be distinguished are (a) the thermal combustion path, and (b) biological anaerobic digestion.

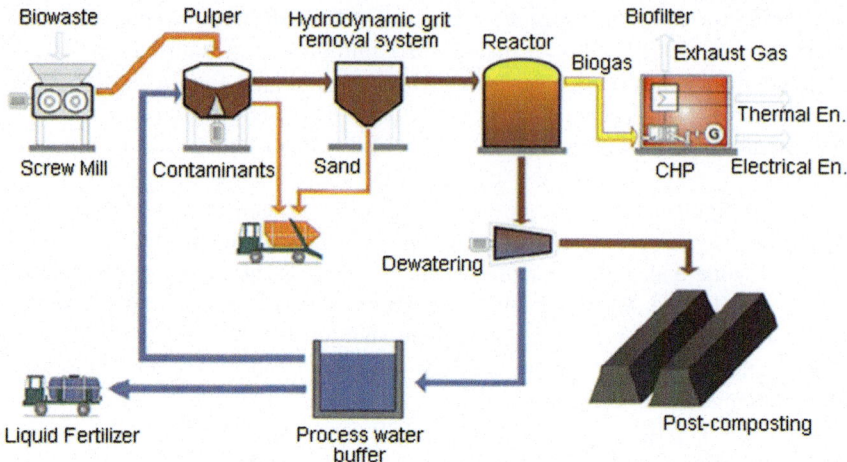

Fig. 20 The process schematic of the anaerobic digestion of biowaste (BTA International 2013)

Combustion will burn most nitrogen, but the ash residues retain the phosphorus and potash. The challenge here is to process the nutrients in the ash in a way that they become plant-available. The interest in manure as fuel is evident (see Fig. 18), as the heating value of horse manure, for instance, is on the same level as wood, with about 20 MJ kg^{-1} dry matter (Edström et al. 2011).

Anaerobic digestion (see Fig. 20) has the advantage that the nutrients N, P, and K are (mostly) retained, and both the effluent of the digestion process and the solid biowaste may be processed and used after energy production. The operation of an anaerobic digester requires technical knowledge, but is applicable at different scales, e.g., from the farm or on a municipal/community scale.

4.8.3 Phosphorus Recycling from Crop Residues

> Crop residues incorporate more than half of the world's agricultural phytomass … Consequently, it would not be inappropriate to define agriculture as an endeavor producing mostly inedible phytomass … [But] No nation keeps statistics on the production of crop residues. (Smil 1999)

The role of *crop residues* in agricultural systems prior to biofuels was primarily seen as a contributor to maintenance of the soil structure and soil organic matter content. The management of residues is central to conservation and no-till practices common in intensified agriculture. In other parts of the world, crop residues are extremely important sources of fuel and fodder. We may also need to acknowledge that the classifications of crop residues and plant-based animal feed are "fuzzy."

Table 3 Phosphorus contents in different animal and crop products (Lamprecht et al. 2011, most data refer to Swiss or German reference values)

Product	Phosphorus content (%)
Bone shred (CH, FS)	9.0–11.0
Bone meal (CH, FS)	7.7
Meat meal (CH, FS)	5.1
Fresh bones (CH, FS)	3.0
Sewage sludge (DS)	2.5
Animal waste from meat processing (CH 2002, FS)	1.1
Soy beans, dried seeds (91.5 % DS)	0.650
Cheese: Tilsit cheese (FS)	0.500
Cereals: feed wheat, grains, FS (87 % DS)	0.380
Compost (DS)	0.301
Beef, FS (24.7 % DS)	0.237
Pork, lean meat without fat (FS)	0.204
Waste (Weinfelden and St. Gallen, DS, 2003; water content 22 %)	0.092
Cow milk, raw (FS)	0.092
Potatoes, raw, freshly harvested (FS)	0.050
Vegetables, Swiss production, 1998 (FS)	0.039
Cola drinks (FS)	0.014
Apple, unpeeled, raw (FS)	0.010

DS dry substance, *FS* fresh substance, *CH* Values from Switzerland

Crop residues gained additional attention with the advent of biofuel production. Smil's initial estimate of 3.75 Gt (Smil 1999) was affirmed by independent bioenergy researchers who estimate that residuals for cereal crops are between 2.80 and 3.76 Gt for 27 food crops (Lal 2005). However, in terms of phosphorus content, cereals and residuals differ significantly. For instance, soybean contains about twice the phosphorus found in cereal (wheat grain), and 15–20 times more than potatoes (see Table 3).

The estimates in Table 4 also reveal the potential of phosphorus recycling from crop residues. According to Smil (1999), the potential of phosphorus in annual crop residues is about 4 Mt P and is thus a significant share of the 14 Mt P of phosphorus in organic fertilizers, which is about the same magnitude as the roughly 16 Mt P of mineral phosphorus fertilizers that were used around the year 2000 (USGS 2000).

The burning of crop residues is higher (around 25 %) in developing countries than in the developed world, primarily due to slash/burn agriculture, pest control and the use of residues as fuel for cooking. Slash and burn agriculture to bring forested land into production or to prepare land set-aside (or fallow) for cropping is most critical from a nitrogen conservation and greenhouse gas perspective. We note here that there is an impending trade-off with bioethanol use (Kim and Dale 2004), which would require a differentiated assessment (see Sect. 5.2.9).

Table 4 Estimates of macronutrient content in crops, crop residues, and inorganic fertilizer in terms of MT year^{-1} (taken from Smil 1999)

Outputs and inputs	Nitrogen	Phosphorus	Potassium
Crop residues	25	4	40
Harvested crops	50	10	20
Total crops phytomass	75	14	60
Inorganic fertilizer	80	14	19

Table 5 Average N, P, and K content of different types of manure in % of dry matter from different livestock (Schnug et al. 2011)

Manure source	Nitrogen	Phosphorus	Potassium
Cow slurry	9.7	0.8	5.9
Cow manure	2.8	0.9	2.6
Swine slurry	8.7	2.4	6.3
Swine manure	3	2.8	4.6
Poultry manure	4.7	4.7	2.6

4.9 Phosphorus Recycling from Animal Carcass and By-products

The part of the animal carcass (e.g., meat, bowels, blood, bones, skin, hooves, feathers) that is used for food, feed, composting, anaerobic digestion, or other purposes differs among cultures and times. In developed countries, carcass waste may cover a considerable amount of the phosphorus cycle. For instance, in Switzerland, there are about 7,000 t P year^{-1} in sewage compared with 2,800 t P year^{-1} in by-products from meat processing, animal and pet carcasses (Lamprecht et al. 2011). Further, there are 3,500 t P year^{-1} in waste and separate collection. Due to the high concentration of phosphorus in bones, in Switzerland about 50 % of the phosphorus was in livestock carcass and by-products of meat processing (Lamprecht et al. 2011).

The animal carcass-related flows are higher concentrations and provide a very good option for recycling. The losses on a global scale are difficult to quantify but deserve increasing attention from a sustainable phosphorus management perspective—though severe trade-offs with health protection must be acknowledged (Scholz 2011a; Sharrock et al. 2009).

4.10 Phosphorus Recycling from Sewage

Although there is a wide range of recycling options along all nodes of the supply–demand chain (see Fig. 16), the popular focus today is very much on sewage sludge (also referred to as "sludge," "compost," or "biosolids"). Options for recycling from sewage, manure, and solid industrial waste are addressed comprehensively in Chap. 6 of this book (Smith 1995). As this is the case, a brief illustration is presented which shows that there is no panacea for recycling sewage, and that the technological options may be site-dependent and will have a historic tendency to change.

Natural sewage recycling has become the object of environmental concerns due to the pervasiveness of toxic elements and compounds and their potential negative effects on humans, livestock, and ecosystem health—including soil fertility (Davis 1996). The dry weight of sewage sludge after multiple (primary, secondary and even tertiary) treatments in the EU is still 90 g per person per day (Scholz et al. 1990). If one looks at the current streams of sewage use, three main streams for recovering nutrients may be roughly and ideally identified along with one stream centered on the use of manure for energy.

Stream one is the recycling of human excreta and manure to the nutrient chain in a *direct or moderately processed way*. This is certainly the most natural and historically common stream and may include waste treatment processes such as drying, fermentation, and other composting-type processes. Here, it is noteworthy that even historic sewage recycling in Japan or Korea did not use direct application, but rather, treated the sewage to mitigate certain unwanted effects. Sewage fields, in which high concentrations of sewage were applied to soils, were dependent on the hope that soil microorganisms would remove associated toxicants that could endanger human health—the effects of which were measured in residential gardens where waste disposal biomonitoring took place (Polprasert 2007). But the issue of reuse (in the sense of organic reuse) may function properly if the wastewater is properly treated (Quazi and Islam 2008); this treatment is also common and considered to be effective in aquaculture and the potential to detect struvite. This has already been stated by Ulex (1845).

Stream two is the recovery of phosphorus *by crystallization from sewage water*. Struvite is produced by adding magnesium and ammonium to phosphorus. The mineral, which also forms naturally—a fact that was first scientifically documented in 1845 (Johnston and Richards 2003)—is, in general, a low-pollutant fertilizer. It may be produced in a water-soluble form with high plant availability (Baur 2010), it is available on a commercial scale (Chen et al. 2012) and it can be economically attractive in various sociocultural contexts (Kabbe 2013; Herrmann 2012, September, 4). Struvite processing is of interest from an operational wastewater treatment point of view. When applied it may provide process stability (avoidance of incrustation and abrasion) for sewage plants, reducing maintenance costs, lower sludge disposal costs by less sludge production, and less chemicals consumption (e.g., polymers for flocculation). Lower energy consumption and incomes by selling struvite are further positive components,

Stream three is fertilizer produced from *mono-incinerator-based sewage sludge*. A challenge here is to economically extract and process phosphorus in a plant-available form for a variety of soil types. Sewage sludge ash contains high amounts of either iron or aluminum and is, thus, not a material appropriate under traditional P fertilizer processing methods. Nevertheless, some thermal and perhaps wet processes may be developed to extract phosphorus from ash (van Otterdijk and Meybeck 2011).

The *fourth stream* is using the sludge as an *energy source* that is partially involved in stream three. Here, the recycling of phosphorus has not yet been sufficiently investigated.

In principle, there is a sustainability competition among the four described streams of sewage recycling though streams 2 and 3 do not directly compete because the catching of dissolved phosphorus in the aqueous phase may reduce the amount of particulate phosphorus extracted. To determine which operation is environmentally and economically viable is very much dependent on local conditions and constraints.

4.11 Food Waste

Simply stated, large amounts of food are wasted after agricultural production. The food waste is estimated to be 95–115 kg annually per capita in Europe compared with low-waste regions such as Sub-Saharan Africa at 6 kg per capita per year and South/Southeast Asia at 11 kg per capita per year (van Otterdijk and Meybeck 2011).

The primary reasons for food losses in the developing countries are multiple inefficiencies and underdeveloped infrastructure in storage- and transportation-related, given difficult climatic conditions and a lack of temperature control, proper storage, packaging, unpaved roads, and processing technologies. In the developed world, food losses occur at food industry, retailer, and consumer levels. Consumers in the developed countries tend to consume only goods perfect in appearance and, thus, waste perfectly edible food. They also do not plan their purchases carefully, in part because most are more readily able to afford the food waste. The industrial and retailer management of food requires closer analysis to identify both causes and effects. Nonetheless, there is believed to be about 1 Mt P that is wasted annually through food loss in the developed countries, which might have been diverted to use as animal feed, composted or avoided altogether. The total estimate is that "one-third of food produced for human consumption is lost or wasted globally, which amounts to about 1.3 billion tons." (Scholz and Wellmer 2013, SI 7).

4.12 How May We Define Sustainable Phosphorus Use?

Food security and supply security With today's agrosystem and food demand, food security means avoiding the scarcity of phosphorus. Taking an anthropogenic perspective, one must consider a time range that is relevant for *human individuals* and the *human species*. Since the human body requires food and phosphorus, scarcity must be avoided on a daily and an evolutionary timescale of the magnitude of 100,000 years. Accordingly, with the specific dissipative nature of phosphorus, in our (geological) age, a large amount is estimated to dissipate (from surface waters and rivers to the sea) just from agricultural systems. This phosphorus is not easily accessible, as the sedimentation of phosphorus from seawater occurs over an extremely long period. While extraction of phosphorus from seawater is technically possible, the required scale of mining is neither technically or economically feasible (MacDonald et al. 2011a). Thus, from a precautionary and socially responsible view, the currently known phosphorus reserves and resources must be optimally managed to provide for food security and to prevent phosphorus scarcity in the long term.

Efficiency If we consider *efficiency*, much insight may be gained from simply answering the question, "how much phosphorus is mobilized to produce the food which we consume?" If we take a (simple) functional perspective, phosphorus uptake is the "target variable." Based on human uptake/excretion of 1.4 g P per day, we get an annual uptake of 3.4 Mt P for the current global population. Compared with this, we have an input on the magnitude of 15 Mt P in mineral fertilizers annually, and—with a very low estimate—at least 10 Mt P in manure (Liu et al. 2008). These numbers are supplemented by phosphorus that enters the agrosystem due to weathering. Liu et al. (2008) provide an estimate of 17 Mt P from pastures, which is partly accounted for in manure. Then, we account for 1 Mt P from feed additives, and about 1 Mt from food waste as well as 1 Mt from human excreta. Further, we must incorporate the input from weathering (which is also accounted for in the pasture based phosphorus input) and some other sources such as fertilization by slurry. This amounts *to a magnitude of 40–50 Mt, which we are mobilizing for a human uptake of 3 Mt P in food.* Naturally, the annual input from fertilizer and geogenic phosphorus is lower, but remains around 30 Mt P.

In a recent analysis organized by the European Union (EU), 15 nations provided slightly more friendly data, stating a phosphorus consumption average of 4.7 kg P annually per consumer, of which 1.2 kg P are consumed and only 0.77 recycled (Dumas et al. 2011).

Avoiding pollution When defining sustainability with respect to *environmental quality*, referring to the essence of the Brundtland definition (Rockström et al. 2009), the risk or vulnerability of fulfilling human needs in relation to available ecosystem functions is the reference. Here, we must ponder in what ways the pollution of aquatic systems and the potential overload of phosphorus by fertilizer may become critical, and what *environmental responsibility* should be taken.

Finally, we mention again the stark differences in P fertilizer use between developing countries and the developed world. This is directly related to extreme differences in agricultural yields, which subsequently may cause a critical state in the food supplies of many developing countries, with large shares of crop systems dependent on soils that require the highest input of P fertilizers.

5 CLoSD Chain Management

5.1 The Vision of Closing Anthropogenic Material Flows

We know that humans will eventually triple the phosphorus flows. And phosphorus may become a pollutant.

> At the planetary scale, the additional amounts of nitrogen and phosphorus activated by humans are now so large that they significantly perturb the global cycles of these two important elements. (Carpenter and Bennett 2011)

Rockström et al. define ten times the natural cycle as *planetary boundaries*. But as the global use of phosphorus is uneven, the environmental impacts are uneven. Thus, we must reflect that phosphorus is spatially not evenly distributed. Ecosystems that are of different scales—including marine systems—may be highly vulnerable with respect to phosphorus impacts. We argue that the "average planetary boundary" may be a questionable concept for many pollutants such a phosphorus. For assessing the planetary boundary of phosphorus load rather a regionalized view seems adequate that assesses unwanted environmental impacts in aquatic and other ecosystems, perhaps from a pattern of contaminated area perspective.

Carpenter and Bennett elaborate that we must distinguish between planetary boundaries of (average) seawater and freshwater and that:

> ... planetary boundaries for eutrophication of freshwaters by P have already been surpassed. (Udo de Haes et al. 1997)

A different perspective is provided from a resources management perspective. High ore phosphate rock reserves are finite, non-renewable on the human scale and phosphorus use shows a very low efficiency that should be increased. Thus, it seems desirable if not necessary to close the anthropogenic fertilizer loop. A first focus in a sustainable transitioning would be the reduction in losses of phosphorus by dissipation in the supply–demand chain. The vision here would be the closing of the anthropogenic material flows, or—to express it in other terms—to approach the issue with *Closed Supply-Demand Chain* (CloSD Chain) management of anthropogenic phosphorus.

The Material Flow Analysis (MFA) is a simple, easily understandable method that may help to represent the main losses, sinks, obsolete stock building and options for increasing efficiency by changing consumption, technology development, and recycling. This chapter provides a blueprint for this effort on a global scale.

MFA is a quantitative accounting tool for representing the flows and stocks of materials and energy. Its starting point is mass balances of inputs by extraction, etc. We organize the MFA according to the supply chain (see Fig. 16). We should note that this chapter focuses on the chemical element phosphorus. We also talk about substance flow analysis (SFA) (Brunner and Rechberger 2003), whereas MFA focuses on goods (Binder et al. 2004). As the presented MFA includes (some non-quantified) information about the chemical processes underlying fertilizer and food production, we continue to use the term MFA.

A challenge in the research of MFA is to move *from flows to actors* (Matsubae et al. 2011). As the first step, we organize the MFA along the supply–demand chain and thus may identify key stakeholder groups who may take responsibility in *Closed Supply-Demand Chain* (CloSD Chain) management.

Key Message

MFA may serve as a tool to identify key stakeholders along the supply–demand chain for *Closed Supply-Demand Chain* (CloSD Chain) management.

5.2 A Blueprint of Global Phosphorus Flows

5.2.1 System Boundaries

Figure 21 is the pillar of this section. The bold-lined box represents the *system boundary* of the "supply–demand chain system." The *Bedrock phosphate* and the *Atmosphere* (considered as a constant pool) are external systems. Reserves are partly external and become internal if they become economically accounted values in the supply–demand chain.

The shaded bottom box, *Losses to sinks*, includes (currently) uneconomic stocks in "bedrock" (e.g., the stock "Losses of mining," bottom left), *Sediments*, *Landfills & cesspit*, *Ash dumps*, etc. In principle, the S-D chain system should also include the virtual flows, which are presented as a bottom flow. As these are not addressed in this chapter, they are represented separately. The presented MFA is called a *Blueprint of the Global Phosphorus Material Flows*, as it only provides an initial rough outline, which asks for elaboration and validation of many data.

A classical MFA includes *stocks*, *processes*, and *flows*. The *boxes* present the *stocks and (inner) processes*. The *arrows* represent the *flows* that are linked to material metabolisms. We present the non-quantified flows, as the changes and innovation in fertilizer processing and products (goods) are considered an important issue. Most data are linearly transformed to the 2011 (USGS 2012) input data (referenced to the 25 Mt year^{-1}) if survey data are available for certain processes or flows for 2011. For this reason, among many others (years with extreme weather conditions affecting runoff or erosion, such as different classifications of national statistics, uncertain estimates), one is only allowed to consider the many data as rough estimates within factor 2 precision.

The reader who critically examines the data of Fig. 21 will face some general problems with the consistency of the data of the global phosphorus flows. For instance, the main (top down) reference for the amount of global phosphorus flows are 25 Mt P which enter the processing stage. This is referring to the USGS data for 2011 that have been published in the USGS (2012) Mineral Commodity Summaries (191 Mt PR = 24.8 Mt P year^{-1}). The data in the USGS 2013 report for 2011 differ from that of those published in 2012 and suggests 25.7 Mt P year^{-1}. There are many causes for this inconsistency including the fiscal year in some countries (and thus the basic statistic recording) does not end on December 31 but on March 31. The diverting time frames spoil the coherence of most annual global statistics referring to economic data. Further, in general, there are also competing sources from different statistics with different types of classification system, e.g., what is considered manure may differ from country to country. Many statistics are officially published by various governmental authorities due to different interests. And these interest and thus also statistics may differ from those of market research institutes (which may correct "mis"-classifications). Also, tax declaration may serve as a distractor, e.g., if fertilizer-related use receives special treatment. Thus, it may seem that the principle of mass conservation seems to be violated, e.g., if we look at the input (i.e., 25 Mt PR) or the sum of outputs of processing.

The *virtual flows* at the bottom of Fig. 21 include the phosphate bonded to iron in metals; iron may build an important part of the anthropogenic flows (FAO 2010a). Here, research has just begun (see Sect. 4.5.1).

All figures are Mt if we consider stocks, or Mt P year^{-1} if we consider flows. The latter are presented without units. For most of the figures, there is at least one reference cited as upper quotes. These may be found in literature. The non-referenced figures are derived and discussed in 4.

Key Messages

The system boundaries include the economically accounted flows of goods with respect to phosphorus.

Some data are of high uncertainty and only allow for a factor 2 certainty.

5.2.2 System Inputs

The *system boundaries* (inside the box) are the operations along the phosphorus supply–demand chain and the waste-recycling activities.

The first natural *input* in this system is that from the weathering of rocks to the soil system. There are three main input processes: (a) *natural weathering*; (b) *human-induced weathering*; and (c) *mining*. We have discussed (a) and (b) in Sect. 4.5. Here, soil surface management and climate change are important factors.

We may also look at natural (via seabirds) and anthropogenic (via fisheries) biotic terrestrial inputs. The fishery is relevant. In 2009, 145 Mt of fish were caught; 118 Mt used for food and the remainder for feed (Ruttenberg 2003). About 55 Mt of fish are from fish farms (whose input–output relationship is not represented). The phosphorus concentration of fish is about 0.25 %, which translates to an input of 0.36 Mt P year^{-1}, which is compatible with the 0.31 Mt P year^{-1} suggested by (Ruttenberg 2003). We work with 0.34 Mt P year^{-1}, of which 0.24 is part of food and the rest (not explicitly presented) is part of feed. By means of simplicity, we do not include these flows in Fig. 21.

1 Sustainable Phosphorus Management

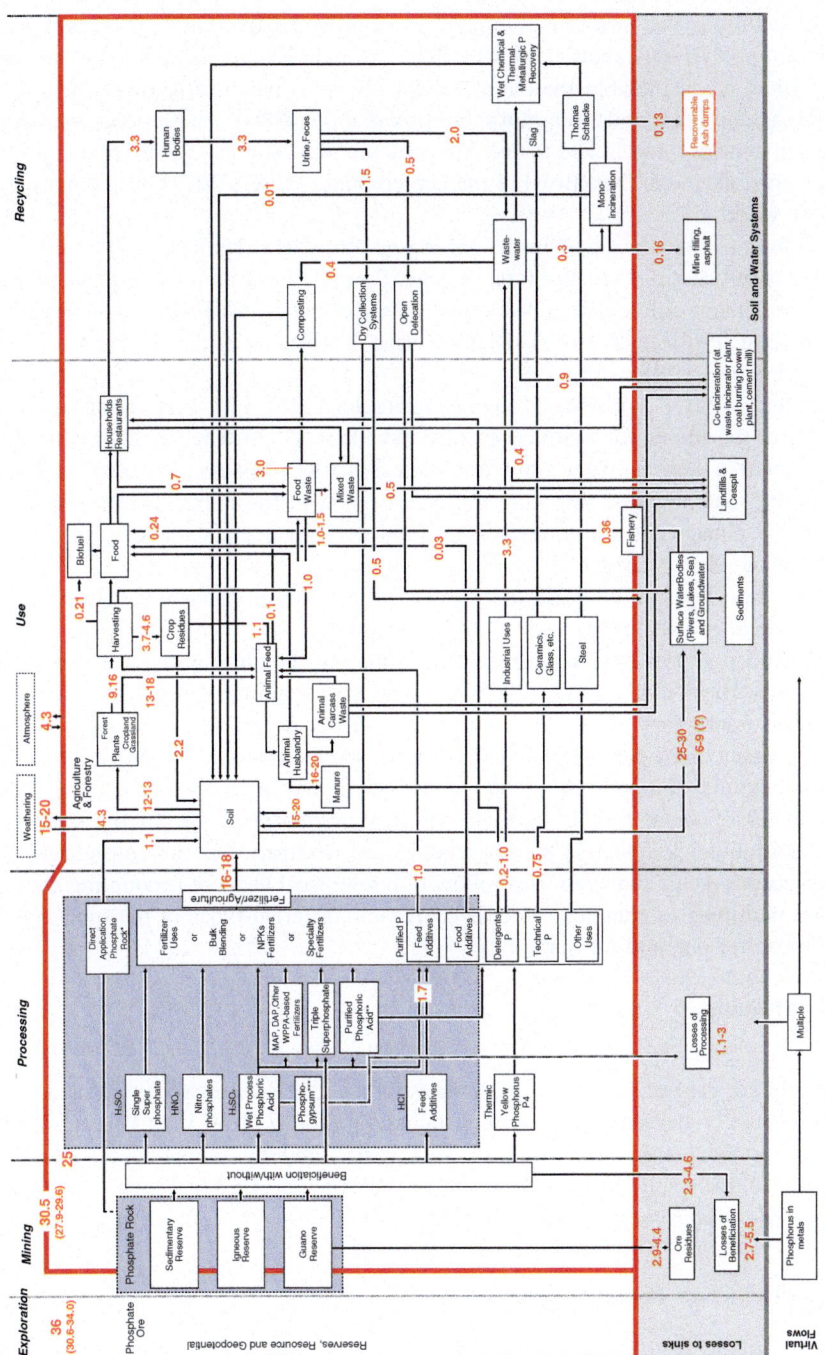

Fig. 21 A blueprint of global phosphorus flows 2011 (USGS 2012) along the steps of the supply-demand chain (non-incorporating virtual flows) referring to an equivalent of 25 Mt P year^{-1} (Fig. 21 is electronically accessible by http://dx.doi.org/10.1007/978-94-007-7250-2_1)

In addition, the *atmosphere* may be mentioned here. Geologists assume a constant pool of only about 0.28 Mt P. The "residence time" is asserted to be only 80 h; thus, a considerable amount of 4.3 Mt P year^{-1} may be assumed to be linked to the annual land (soil) to atmosphere flux (Cunfer 2004). These geological data are—in general—not linked to specific contemporary human activities such as the huge erosions in the Dust Bowl of the United States (USGS 2012) or in other parts of the world.

Mining (c) transferred about 25 Mt P year^{-1} in 2011 (Heffer and Prud'homme 2011) to processing. Of this total, 1.1 Mt P year^{-1} was directly applied to soil after or without differentiated beneficiation, and about 16 Mt P year^{-1} was applied as mineral fertilizer. Thus, 17–18 Mt P year^{-1} (Villalba et al. 2008) of mineral fertilizers entered the soil.

In Sect. 4.5.1, we provide a rough estimate that 2.9–4.4 Mt P of ore phosphorus is left in the mines for various reasons. Whether or which parts of this may be considered a loss in what time frame is difficult to assess. We may assume, however, that much of this may not be assigned to the currently economically mineable phosphates, as otherwise it would not be understandable as to why it would not be excavated.

We may argue similarly for the 2.3–4.6 Mt P from ore phosphate that has been put aside after excavation in primary beneficiation.

We do not provide numbers for the different chemical process engineering pathways. But Villalba et al. (2008) provide some estimates about the share that different tracks may take. Here, we start with the rough estimate that wet-processed ammonium phosphate (MAP, DAP) makes around 47 %. Nitric or nitro-phosphate may amount to 28 %. Single superphosphate provides 19 %. The remainder of about 6 % is taken by triple superphosphate. In all the data, the reader should acknowledge the uncertainty and fuzziness that are caused by various reasons—e.g., that countries follow different fiscal years of accounting data or that a multitude of chemical process chains and different types of phosphate rock are linked to one and the same box.

Key Messages

- There are varying changing phosphorus inputs to the (terrestrial) food and non-food supply–demand chain system including *natural* (10–15 Mt P year^{-1}) and *anthropogenic weathering* (up to 5 Mt P year^{-1}) *mining* (more than 30–34 Mt P year^{-1}), *fishery* (0.36 Mt P year^{-1}), in which about 4.3 Mt P year^{-1} is *transported through the atmosphere*.

5.2.3 The Crop–Forest System

With respect to the crop systems, we are interested in: (1) how much phosphorus is taken up by cultivated crops; (2) how great is the phosphorus use efficiency (NUE), and in particular how much phosphorus is taken up by the plant from the applied

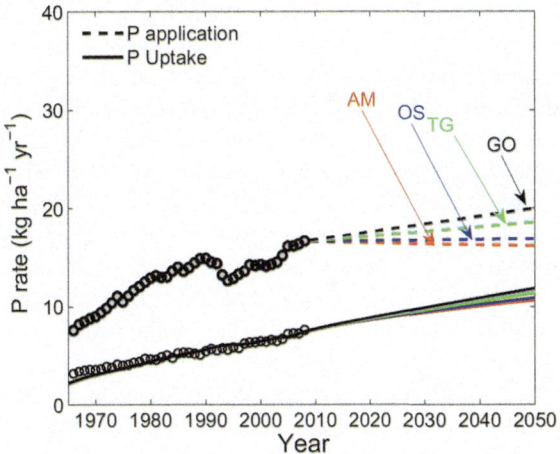

Fig. 22 Annual application of mineral fertilizer and manure per hectare (kg P ha year^{-1}) between 1965 and 2007 and phosphorus uptake in cropland and simulation for future phosphorus uptake for the MEA scenarios [Sattari et al. (2012); *AM* Adapting Mosaic, *OS* Order by Strength, *TG* Techno garden, *GO* Global Orchestration, see (MacDonald et al. 2009; Smith et al. 2008)]

mineral and organic (in particular the manure) fertilizers; (3) what role do crop residues play; and (4) how much is leached or lithified (i.e., transferred to the stable, not to the plant bioavailable pool) in the sense that it will not be available to plants in the foreseeable future.

Also, managed forests receive fertilizers (Tanner et al. 1992), and fertilizers have a tremendous effect on trunk growth (The World Bank 2012). Despite this fact, there is no data available on the global inputs to forest systems other than the inputs to palm oil crops and similar plantation operations. Thus, this part of biomass management is not included here, though it may well be of the magnitude of 1 Mt P year^{-1} and above, given that the average consumption of NPK fertilizer in Malaysia is 770 kg NPK ha^{-1} (Frossard et al. 2000; Dumas et al. 2011; The World Bank 2012).

With respect to the phosphorus uptake (1) of crops, we face a difficult task because of the heterogeneous soil conditions (Liu et al. 2008). In addition, literature provides diverging estimates. Liu et al. (2012) provide an estimate of annual crop uptake in 2005 of 12.7 Mt P year^{-1}. Figure 6 identifies the average world phosphorus nutrient use efficiency (NUE) on a per-hectare basis. According to Sattari et al. (Sims et al. 1998), the total uptake of phosphorus in crops in 2007 was 11.6 Mt P year^{-1}. Thus, given that food production has increased, we work with an estimate of 12–13 Mt P year^{-1} uptake in 2011.

The Sattari et al. (2012) paper also provides insight into (2). The phosphorus nutrient efficiency (P-NUE) increased between 1965 and 2012 only slightly, from 42 to 46 % (see Fig. 22). However, the input from natural and anthropogenic weathering is not included here. Nevertheless, this long-term, equilibrium-like efficiency term allows one to check some input and output data.

A crop uptake of 13 Mt P year^{-1} and an efficiency of 46 % provide an input of 28.3 Mt P year^{-1} and a loss of phosphorus from cropland of 15.3 Mt P year^{-1}. This loss may be assigned to runoff and erosion, transfer to the stable, non-bioavailable pool of the topsoil or leached to the subsoil. This would also imply that about half of the total phosphorus runoff and erosion of 25–30 Mt P year^{-1} that stems from the 10.3 % of the earth's surface grassland erosion is not included in the 15.3 Mt P year^{-1}.

The crop residues (3) play an important part, as the question is how much of the crop may be eaten. Here, we estimate that roughly 50 % of the crop is not removed from the fields (see Sect. 5.2.5). We should note that the current trend of breeding and plant modification to increase the edible part of the plant may cause a rebound effect with respect to fertilizer needs.

Finally, (4) the leaching of phosphorus or the losses in drainage has not been studied for a lengthy period. However, there are losses in sandy or high organic matter soils, or soils with high phosphorus fertilizer overshoot (Oenema et al. 2005). There is also an in–out aspect to groundwater, but it is much lower and less critical with respect to human hazards than that of nitrogen (Alcamo et al. 2006).

Key Messages

- There is a crop uptake of 12–13 Mt P year^{-1}.

- We are facing a long-term global plant nutrient efficiency of about 45 % with losses by erosion, runoff, leaching, crop residues (which are partially reused), etc. of 15.3 Mt P year^{-1}.

5.2.4 Animal Production

Given that animal bones contain up to 10 % phosphorus concentration, a great deal is accumulated in animals (see Table 3). But the phosphorus content in animals depends on feed and differs among species. Poultry manure has about five times more phosphorus in dry matter than cow manure.

We only focus on annual flows in this section. Based on the comprehensive discussion in Sect. 4.8.2, we assume that about 15–20 Mt P year^{-1} of manure is produced and used as a soil amendment.

There are many value-laden controversies regarding manure that touch on energy issues (e.g., for transporting manure, producing organic fertilizers), soil quality, health issues, etc. (Schipanski and Bennett 2012). Thus, it is impossible to provide a reliable estimate of non-soil manure use.

The estimate of phosphorus in livestock feed may be estimated by the output, i.e., the amount of phosphorus in excreta/manure plus the livestock products including meat, milk, eggs, etc. Phosphorus in livestock products is difficult to access due to the high variance of phosphorus content in numerous animal feeds. According to an estimate of 12 countries by Schipanski and Bennet (USGS 2000), about 6 % of the phosphorus in manure becomes animal products. This would provide an estimate of 0.9–1.2 Mt P year^{-1} as animal product. The waste of animal carcasses and the wasted by-products of meat processing are difficult to assess on a global scale, but are increasing in industrial nations (see Sect. 4.9).

Key Messages

- There are about 16–21 Mt P year^{-1} in animal feed from grassland, feed additives and other animal feed.
- There are increasing losses due to animal carcass waste and by-products of meat processing in developing countries.

5.2.5 From the Farm to the Table

In 1999, Smil provided an estimate of food residues of 3.75 Gt, including about 4 Mt P year^{-1}. From 1991 to 2011, the world population increased by 17 %, from 6 to 7 billion, and the fertilizer increased by 35 %. This may be roughly estimated from the total amount of phosphate rock production if we assume that the share of fertilizer in total phosphate consumption did not change over years. The phosphate rock production increased from 145 Mt PR year^{-1} in 1998 (Liu et al. 2008) to 191 Mt PR year^{-1} in 2011. We thus may roughly estimate that crop residues increased by 10–20 %, and we take 4.6 Mt P year^{-1} as a rough estimate for the phosphorus content of the crop residues. We add (Kim and Dale 2005) that 50 % of crop residue is not removed from fields. We may assume that some of the residues that are used as animal feed will end in manure, but we also acknowledge that some portion is burnt.

A critical issue is the trade-off with energy production. Here, the perspective is changing. Rather than calories, we refer to joules: "The functional unit is defined as 1 ha of arable land producing biomass for biofuels to compare the environmental performance of the different cropping systems" (Kim and Dale 2004). And crop waste becomes the status of alternative energy. The high expectations from the energy domain are characterized by the following statement:

> There are about 73 Tg of dry wasted crops in the world that could potentially produce 49.1 GL year^{-1} of bioethanol. (Mihelcic et al. 2011)

Key Messages

- In a time of transition to alternative energy, crop residues have become a hot topic for the energy market. The rebound effects of this option for agriculture must be pointed out.
- The phosphorus balance of biofuel production requires special attention as the recycling streams of the different types of biofuel are not yet well assessed.
- Crop residues and waste of food products in retailing also requires special attention.

5.2.6 After the Table

The *daily intake of phosphorus per person* differs among diets and shows extraordinary variances depending on age, weight, gender, nation, etc. Thus, the uptake in the Democratic Republic of Congo is estimated to be 490 mg P d^{-1} compared with 2,000 mg P d^{-1} in Israel (Walther and Schmid 2008)—or consumer groups in Switzerland, for example, who tend to eat processed food with phosphates as additives, may show an intake of 0.35 g P d^{-1} (Liu et al. 2008; Mihelcic et al. 2011).

We assume an average intake of 1,250 mg P d^{-1}, which makes about 3.4 Mt P year^{-1} (Mihelcic et al. 2011).

If we follow the mass balance for the human population after intake, some 0.05 Mt P year^{-1} (1–2 %) accumulates in the body, but a larger share is excreted by urine and a smaller share via feces. This greatly depends on the digestibility of the food in the diet. Estimates in Sweden assume 68 % in urine, yet those for China reflect only 20–60 %. Often, a rough 50:50 split is assumed for a global estimate (2010).

If we take a simplified look at the flows in the world, we distinguish between *Open Defecation*, *Dry Collection Systems*, which are prevalent in rural areas, and different types of centralized, connected *wastewater treatment treatments*, which are dominant in urban systems.

According to the WHO and UNICEF (UN 2011), about 15 % of the world population represents open defecation, which makes about 0.5 Mt P year^{-1}.

The UN statistics on *population connected to wastewater collection* systems is incomplete and varies greatly among countries. There are percentages below 5 % for countries such as Yemen, Maldives, and Kenya. China is recorded with 32 %, Brazil with 26 % and Croatia with 27 %. And some European countries such as Belarus, Germany, or the United Kingdom rate above 95 % (Ott and Rechberger 2012). There are many countries without data, such as India and the United States. And the unweighted average of 101 recorded countries comes to 50.1 %, which provides an estimate—including an average loss of collection of 10 % or more—of about 35 % connected to wastewater treatment systems. This provides an estimate of 1.2 Mt P year^{-1} treated in various types of wastewater treatment (WWT) plants.

Naturally, the performance of sludge extraction differs greatly among these systems. Estimating the current recycling and recycling potential from different types of WWT plants, septic tanks, cesspits, pit drainage, or other systems is difficult. Estimates for the EU 15 countries, which show very high standards of sanitation and make 5 % of the world population, show that 79 % of the population is connected with WWT plants and that about 70 % of the phosphorus of the influent is contained in the sludge, which would provide a 55 % extraction of phosphorus by WWT in the countries with the highest standards of WWT systems (van Otterdijk and Meybeck 2011). This includes households, not including the losses of wastewater by leakage before reaching the WWT plant, which represent about 5–10 % in the EU 15 sample.

1 Sustainable Phosphorus Management

Fig. 23 Urine separation and separated dry collection and processing are used by the Guatemalan Mopan Mayas in the region of Peten (*Photo* R.W. Scholz)

Finally, we define a category called *dry collection systems* that includes 1.6–1.9 Mt P year^{-1}, with different types of cesspits, septic tanks, and decentralized management of excreta. There are a large number of culturally driven management systems. As urine is known to include far fewer pathogens and less organic content, urine separation is common and was so even in some ancient cultures (see Fig. 23). The nutrients in urine are directly applied. The feces is processed and composted separately and is assigned to the dry collection path. We work with the rough estimate that two-thirds, i.e., between 1.1 and 1.3 Mt P year^{-1}, of the phosphorus on that track is reused.

Finally, we take a brief look here at the extraction of phosphorus in wastewater treatment plants (see Chap. 6 of this book). There are, in principle, three ways of utilizing treated wastewater. The most common is the direct reuse of the sludge in agriculture, which is represented by the *Composting* box of Fig. 23. The second is the incineration of ash, which became popular in various developed countries such as Japan or many European countries. The third is the extraction of phosphorus by biological or chemical precipitation. We do not discuss the different option of chemical or thermal recycling in this section (see Sect. 4.9), but we do provide an estimate of 0.2–0.3 Mt P year^{-1} that is treated, with about half in asphalt and cement, half ends in dumps and a minor part already used in agriculture.

The food waste by consumers (including restaurants) at the end of the supply chain has been estimated to amount to 1 Mt P year^{-1} in the developed countries (1–1.5 billion people waste 100 kg food waste per year including 0.06–1 % phosphorus). Conversely, the food loss of the 5.5 billion people in the South is estimated to be 10 kg per person, with lower phosphorus concentrations and with a lower phosphate content (perhaps of 0.3 % P kg^{-1} food waste is of marginal magnitude of 0.25 Mt P $year^{-1}$ (Emsley 2000b).

Key Messages

- Related to the total anthropogenically caused flows, sewage is a relatively small fraction of about 3.3 Mt P $year^{-1}$.
- The amount of phosphorus in sewage differs by factor 2 and depends on the diet.
- Phosphorus recycling has to be adapted to the wastewater system.
- The recycling of sewage on a global level is on a very low scale. There are different options for phosphorus recycling, all of which have strengths and flaws. There are examples of economically beneficial recycling procedures.
- The recycling of sewage may become a paradigm and an object of demonstration in how CloSD-Loop Management may look.

5.2.7 Industrial Use

We will distinguish among a wide range of technical uses (including military use) and virtual flows of phosphorus in heavy industry.

The technical part makes about 0.7 Mt P $year^{-1}$ and is based on white phosphorus. "The enigma of phosphorus lies in chemistry" (Webster's 1913). It shows specific chemical characteristics such as the clustering (e.g., P4) or affinity to oxygen. But it is a key element at the boundary between organic and inorganic chemistry and thus may have a huge potential for technological application.

The virtual flows (see Sects. 2.5 and 4.5.1) may have a large potential for resource management. The amount of phosphorus that is included in many mineral and metal processing may become of interest in the future. Thomas slag may be viewed as an historic issue:

> A by-product from the manufacture of steel by the basic process is used as a fertilizer. It is rich in lime and contains 14 to 20 percentage of phosphoric acid. Called also Thomas slag. (Babenko 2012)

But the issue remains of interest (Wetzel 1983). With Thomas slag, as with other ashes and wastes, we are seeking proper technologies that allow one to economically extract the valuable, i.e., economically scarce, materials.

Key Messages

- About 0.7 Mt P $year^{-1}$ is used for a wide range of technical processes.
- The flows of phosphorus in heavy industry and exploration of technology options should become a subject of research.

5.2.8 Detergents

The amount of phosphorus used in laundry and dishwater detergents varies historically and regionally. During the 1980s, in the United States, 2 Mt year^{-1} phosphorus was used for detergents (Wetzel 1983). The detergent industry "was reluctant for quite a long time to look for and tried to minimize its (i.e., phosphorus) role in eutrophication process" (Knud-Hansen 1994). A conflict trade-off emerged between aquatic pollution of STPP, in particular for regions with low profile sewage treatment without phosphorus extraction and the higher toxicity of alternatives (see Sect. 3.2). After the phosphate bans in a few states in the US States in the midst of the 1980s, the share of phosphorus in detergents by weight decreased in the United States and other developed countries. There had been a worldwide discussion at that time and seven nations out of the EU-25 banned detergents from laundry detergents. Other countries relied on the voluntary action of industry. This has been insufficient, to resolve eutrophication problems, for instance in particular in the Baltic Sea and the Danube River (EU 2012a). These cases may also be taken as an example for the necessity of high performance wastewater treatment plants with phosphorus extraction. Factually, in 2007, only 66 % of the detergents of EU-25 were classified as "phosphate free." "Where STPP is used as a builder in household detergents it contributes to up to 50 % of soluble (bioavailable) phosphorus in municipal wastewater..." (EU 2002). Thus a EU regulation (EU 2012b) will set limits for the total phosphorus content in both laundry and dishwater consumer detergents. One should note that the above data are under discussion, also as detergent formulation and phosphorus use in detergent is subject to change in Europe as in other countries of the world.

A reliable estimate of the amount of phosphate use in detergents is difficult. One reason is the high volatility of production in the STPP market. Other reasons are that industry data are not public because it is subject to anti-trust and commercial disclosure restrictions and the only information available are therefore (heterogeneous) estimates. These estimates are provided by marketing studies or industry experts and some customs data which only covers some cross-frontier tonnages.

In the past, the share of STPP in detergents and cleaners was much larger than its use in drinking water, water treatment, metal treatment, food and beverages fire safety, phytochemicals, chemical industry, ceramics etc. This may have changed in the last years (Shinh 2012). Shinh reports that the amount of P in detergents and cleaners has decreased in the last years from about 0.41 Mt P in 2006 to 0.19 Mt P in 2011 worldwide. However, we should acknowledge that China became a key market player in STPP. According to Chinese customs-based market reports in 2011, China exported 0.31 Mt STPP which includes about 0.08 Mt P (Zheng 2013). The total elemental P production (also including a minor share of non-detergent use) in China in 2011 is supposed to be between 0.7 and 0.8 Mt P (Schipper July 9, 2013). This suggests that the Shinh estimate seems too low and rather a lower bound estimation.

A differentiated analysis of regional STPP demand based on twelve business areas resulted in an estimate of 0.71 Mt P year^{-1} in 2011 just for detergents and cleaners (Mew 2013). How much the STPP production may have decreased may be taken from the Global Phosphate Forum homepage, the industry association of companies which are manufacturing phosphates for detergents. At one time, the "World detergent phosphate production is estimated to be 4.7 million tonnes STPP per year" (GPF 2013). But the recent estimate indicates production of only "1.0–1.7 million tonnes/year (as STPP)" (GPF 2013). In terms of phosphorus consumption, this would mean a decrease from 1.19 Mt P year^{-1} to 0.25–0.43 Mt P year^{-1}. The data presented in Sect. 2.1 of 1.7 Mt P for total STPP production in Fig. 4 (Prud'homme 2010) would—linearly extrapolated—provide around 2 Mt P year^{-1} for the year 2011 but keeps unspecified how much phosphorus much is used for detergents and how much for other purposes. Therefore, we are receiving a picture that goes beyond the factor 2 uncertainty in the estimation of phosphorus in detergents. This may ask for clarification, also given that a significant amount of detergents are used in megacities of the developing world where they may cause pollution of waterbodies.

Key Message and Data

It is difficult to reliably access the amount of phosphorus in detergents. There is evidence that it has decreased significantly from far above 2 Mt year^{-1} to around or even below 1 Mt year^{-1} worldwide. But the number of produced tons is highly volatile, most data are not public and the published data show high inconsistency. The uncertainty here may go even beyond the proposed factor 2 uncertainty. Thus, based on the presented literature, we assume that in 2011 the amount of phosphorus in detergents may be between 0.2 and 1.0 Mt P year^{-1}.

5.2.9 Biofuel

Utilizing biofuel and mastery of fire 250,000 years ago played a key role in human evolution and wood has been. But the generation of biofuel is shifting from wood to crops. Since 1900, the share of crops in total human biomass extraction increased from 21 to 35 %. This is countered by declines of wood from 15 to 11 % (Alexandratos and Bruinsma 2012). Thus biofuel now sometimes is named agrofuel. The extensive and increasing use of bioethanol and biodiesel is a factor of agricultural land expansion (Heffer 2013). And there is exceptional high demand on fertilizer, for instance for oil plantations, which may be seen from the fact that in—for instance in the year 2008—Malaysia used 1036 kg fertilizer per hectare of arable land and palm oil plantation are a major consumer (The World Bank 2012).

The biofuel feed stocks are expected to increase. According to a 2011 IFA estimate, biofuel feedstocks received 0.21 Mt P year^{-1} (Heffer 2013). This makes around 3.0 % of world phosphorus fertilizer applications. We should note that but most of the phosphorus found in the feedstocks ended up in oilcakes and slurry which is recycled and thus not lost. Thus, the net impact (after deduction of the phosphorus ending up in co-products) would be much smaller, below 1 % (Heffer 2013).

Key Messages

About 0.5 Mt P year^{-1} was used for biofuel. Due to the high recycling of bioethanol and biofuel by-products (for feeding and livestock), there are few losses and it reduces demand on feed grain and oilseed, and the annual net impact of biofuel on the increase in phosphorus consumption is smaller.

5.3 Actor-Based MFA for Changing Flows

This book aspires to contribute to sustainable phosphorus management and thus seeks to go beyond a mere description of the phosphorus flows from a material flow or resource science perspective. Despite this, most parts of Chaps. 4 and 5 remain in line with classical technology or natural science-based descriptions of flows of phosphorus. This is certainly due to the manifoldness, diversity, scale-dependency and complexity of phosphorus flows. Following the principle that we must *start with a thorough understanding of how the environment works* (Scholz 2011a; Scholz and Binder 2004), the MFA is a simple tool which also may serve from joint problem representation among scientists and practitioners in transdisciplinary processes.

But most flows are affected by human actors. Thus, we must shift our attention from "flows to actors." This means that we must link the material flows with human actions and decisions. This brings us to the modeling of coupled human-environment systems—what may be understood as the cutting edge of complexity research. As phosphorus has been and is an important public good which becomes a commodity, much of the flows may be understood from a supply–demand chain perspective, which focuses the drivers of transactions and material metabolisms and both incorporates and goes beyond pure value chain thinking (see Sect. 5.4).

A first step in developing a comprehensive theory of coupled human-environment systems (Binder et al. 2004) is to integrate material flux analysis with agent analysis (Merton 1938). This means that we must identify each stock, process and flows, the key actors and key persons concerned, and in particular their drivers and the constraints of their behavior. This is a challenging task. We may easily see when looking at Fig. 22 that this may be easy if we look at the mining and beneficiation node. Here, mining companies, their technology providers, and the phosphate processing industry may be identified. But already, when differentiating between private and government-owned companies in different political and economic systems (free market vs. centrally planned economy), we may learn that just looking at "unspecified" actors is insufficient.

What decisions an individual or a company makes depends on the *societal framing*, which is primarily given by the economic system (e.g., free market vs. planning society), the political and legal system, the culture and the available knowledge (USGS 2013a). We may also consider whether a state-owned mining company is seen as an institution, which is conceived as a special organization established by the state to guarantee the reproduction of society; institutions such as traffic or water departments are established to guarantee that certain public services or goods become available. Here, mining companies, just as utilities, are at the interface of companies and institutions. The USGS is a typical representation of an institution and follows six goals which "emphasize the critical role of the USGS in providing long-term research, monitoring, and assessments for the Nation and the world and describe measures that must be undertaken to ensure geologic expertise and knowledge for the future" (Cordell et al. 2011).

We will not delve further into the subject here, but only point out that it is very helpful to work with the concept of a hierarchy of human systems. Scholz (2011a) distinguished among the *individual, group, organization* (e.g., companies and NGOs), *institutions, societies* (the primary subdivision of human society, which is currently given by nations), *supra-national institutions* (such as the EU) and the *human species*. Each of these human systems has generic drivers (goals, motives, preference functions), and there are specific social sciences that may define how these systems function. Psychology may explain how individuals (consumers) function, business science explains how companies work, or administration science may illuminate rationales of administrations.

Two issues are important if we wish to apply the hierarchical view on the phosphorus MFA. One is that the hierarchical levels interact such as each human system interacts with its natural and social environment. The second is that we must distinguish between the specific and the generic. All companies have the primary goal to thrive in the market and generate profit. But how this may be done depends on the mission of the company. Likewise, people may follow different motives and values (explained by psychology), or societies may pursue different national goals (explained among others by political philosophy). A challenge of the future work of Global TraPs will be to utilize this knowledge in transdisciplinary processes.

5.4 Supply–Demand Analysis for Improving Technologies

Each market and technology innovation has a *push* (supply) and a *pull* (demand) functions. When we aspire to *CloSD-Loop phosphorus management*, we must identify incentives, means, etc. in the current trend of increased use, what losses may be averted and how more efficient use may lead to that reduction, and we must identify efficient recycling and better environmental performance.

Before we deal with supply–demand chain analysis, let us clarify how the different methods are related. *Actor-based MFA* is a method to represents the global phosphorus management system. *The SD chain analysis* is a means of "faceting" the phosphorus management system. Such faceting is necessary to cope with the complexity of the system (Scholz and Tietje 2002). And such complexity is a challenge in the case of global phosphorus management. Thus, we utilize MFA and SD analysis as the method for structuring, faceting, and representing the case.

The CLoSD-Loop management of phosphorus takes the role of a goal and a vision. And both MFA and SD analysis function as specific tools that are used in the global transdisciplinary process.

The SD analysis may help to understand and to identify the specific pushes and pulls for the different human actors identified in Sect. 5.4. SD analysis goes beyond value chain analysis. Supply analysis is a core concept of Operations Research and may be used for sustainably transitioning the phosphorus chain.

> Supply Chain Management [is the] design, planning, execution, control, and monitoring of supply chain activities with the objective of creating net value, building a competitive infrastructure, leveraging worldwide logistics, synchronizing supply with demand and measuring performance globally.

SD dynamics takes a global view and embeds options for recovery and reuse of phosphorus in an economic frame, which is part of the goal system of any human actor. We use SD instead of supply analysis to emphasize that the interaction of the human systems makes the market. Against the background of this book, there are five aspects of SD that are central to global phosphorus management.

1. *Supply security of phosphorus.* This key for food security requires the monitoring of changing demand (we require 17–21 Mt P year^{-1} in the next decades), securing worldwide infrastructure logistics from a geopolitical view, and synchronizing supply and demand dynamics that follow different timescales. Due to the relative abundance and the relatively low costs, the demand side may provide the requested amount of mineral fertilizer in the next decades. As announced Morocco is planning to increase its annual phosphate rock production from 30 to 55 Mt PR year^{-1} by 2018 (Jasinski 2011b) and also increase its fertilizer production (Jasinski 2011b). A challenging question here is how regulatory processes and conducive policies to support a viable and sustainable mining industry may be fostered.
2. *Framing markets for increasing efficiency and inducing recycling.* We have identified a set of losses and residues that may require better use (if external or future costs are incorporated). Here, political framing that promotes the development of technologies for recycling or efficient use is essential. The EU (Kanton Zürich 2007) statement "to make use of best practice in the field of resource efficiency …, for example phosphorus, with a view to achieving virtually 100 % reuse by 2020 and optimizing their use and recycling;… should receive direct funding from the EU" may be seen as one example. Likewise, the activities of the Canton Zürich (as an example for an activity of a highly developed country) to recycle phosphorus from 100 % of all sewage reflects back to a governmental decree (Kraljic 1983).
3. *Coping with turbulence in the phosphorus market.* Anticipating and preparing for as well as mitigating financial market turbulence or national political imponderabilities requires sophisticated planning, purchasing, and fallback planning (Binder et al. 2004) including options such as stock building and other means.
4. *Mitigating collateral negative impacts of (current) phosphorus use.* The environmental impacts, the finiteness of phosphate reserves, and the differential

access to phosphorus (political instability due to undersupply) may cause unwanted feedbacks. Sustainable phosphorus management requires meaningful management here.
5. *Changing systems.* Sustainable phosphorus management will require fundamental changes along all parts of the SD chain. Mining and beneficiation may become more (eco-)efficient, more efficient fertilizers may be produced, industrial use of phosphorus may become more efficient, farming systems may use the right or new types of fertilizers in a meaningful way, and recycling may work at the tailings of mining, gypsum, manure, crop residues, food waste, sewage, etc. These changes may induce an evolution of technology and sustainable phosphorus use.

6 The Global TraPs Project: Goals, Methodology, Organization, and Products

This section introduces and defines the terms *transdisciplinarity* and *transdisciplinary processes* and explains, in greater detail, the Global TraPs project.

6.1 What is Transdisciplinarity?

Transdisciplinarity may be conceived as a third mode of science, complementing *disciplinarity*, and *interdisciplinarity*. Whereas interdisciplinarity is defined as the integration of concepts and methods from different disciplines, transdisciplinarity additionally integrates different epistemics (i.e., ways of knowing) from science/theory and practice/stakeholders. Transdisciplinarity begins with the assumption that scientists and practitioners are experts in different aspects of knowledge, where both sides may benefit from a mutual learning process. Thus, co-leadership between science and practice, based on equal footing on all levels of the project (i.e., the umbrella project, the nodes and the case studies), is required to assure that the interests and capacities of theory and practice are equally acknowledged. Spotlight 2 describes in detail the principles of transdisciplinarity.

Transdisciplinary processes (td-processes) target the generation of knowledge for the sustainable transition of complex, societally relevant real-world problems. Td-processes include: (1) joint problem definition; (2) joint problem representation; and (3) joint preparation for sustainable transitions (see Scholz 2000). Section 1 of this book, for instance, may be considered an interdisciplinary review of the anthropogenic phosphorus fluxes along the supply–demand chain. The identification of the key actors that are responsible for the flows may lead to an actor-based Material Flow Analysis (Eilittä 2011; Scholz et al. 2013), which may become a key element of the multi-stakeholder discourse (see Spotlight 2,

Fig. 29). This approach links key actors/stakeholders to the flows of the MFA and thus opens a management perspective.

In transdisciplinary processes, the stakeholders are not only identified but are incorporated into the process of problem definition. A pivotal part of any transdisciplinary process is the formulation and consenting of the *guiding question*. This was a major outcome of the second Global TraPs workshop (see Table 6):

> Guiding question of the Global TraPs Project: *What new knowledge, technologies and policy options are needed to ensure that future phosphorus use is sustainable, improves food security and environmental quality and provides benefits for the poor?* (http://www.globaltraps.ch/)

In general, *transdisciplinary processes provide an improved problem understanding and robust orientations on policy options or business decisions for the practitioners*. Transdisciplinary processes serve for capacity building of all participants and facilitate consensus formation, for instance, about what the most important flows may be, and which options for changing them should be explored. In transdisciplinary processes, scientists benefit by obtaining in-depth insight into the dynamics of complex systems and mechanisms of sustainable transitions. The mutual learning between science and practice is the basic principle of a transdisciplinary process. Scientists and practitioners work on equal footing; the co-leadership of the transdisciplinary project is a key property. In principle, we distinguish among three types of agents: (a) a legitimized decision-maker from practice; (b) a representative from a university or public science institution; and (c) those concerned with or affected by the problem addressed or by the decision made by the legitimized decision-maker.

The dynamics of an ideal transdisciplinary process are presented in Fig. 24. Here, a legitimized decision-maker and members of the science community decide to collaborate about, for instance, sustainable phosphorus use (see Fig. 24) and incorporate stakeholders. In the case of the Global TraPs project, the project was initiated by researchers from ETH Zürich with IFDC assuming co-leadership on behalf of practice (Scholz 2011a).

In a follow-up step, key stakeholders were identified and joined the Global TraPs project (key stakeholder groups may be seen in Fig. 24), which was planned as a five-year project. The core phase of the Global TraPs project was planned for that same five years (from 2011 to 2015). This phase is expected to end by providing what we refer to as "socially robust orientations" on sustainable phosphorus use. As a result, it is expected that science will be enriched in its understanding of sustainable phosphorus management and IFDC and other decision-makers will have attained the capacities to improve sustainable decisions on phosphorus use.

By *socially robust knowledge* (often referred to as sociotechnologically robust knowledge), we refer to orientations on sustainable phosphorus use which: (1) meet science's state-of-the-art knowledge; (2) are based on the integration of knowledge and values from practice; (3) receive acceptance from the practitioners; and (4) acknowledge not only the uncertainties but also the unknowns of scientific knowledge (1996). As the reader may agree upon review of this book,

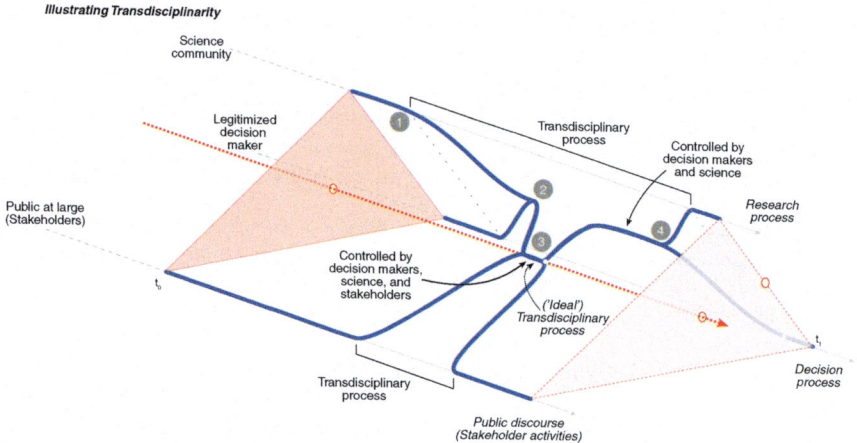

Fig. 24 The dynamics of an ideal transdisciplinary project (Scholz 2011a, p. 375)

acknowledging the unknown is important with respect to the many aspects of sustainable phosphorus management. For example, we do not have exact knowledge of world reserves, nor do we know the exact technologies that may evolve for efficient mining complexity. Also, we do not yet know the most efficient methods of processing manure or recycling sewage.

The following definition refers to the discourse theory (Habermas 1996) postulated by the German sociologist Jürgen Habermas (Regh 1996):

> According to Habermas, conflict resolution on the basis of reasoned agreement involves at least three idealizing assumptions: members must assume they mean the same thing by the same words and expressions; they must consider themselves as rationally accountable; and they must suppose that, when they do arrive at a mutually acceptable resolution, the supporting arguments sufficiently justify a (defeasible) confidence that any claims to truth, justice, and so forth that underlie their consensus will not subsequently prove false or mistaken. (Scholz et al. 2006)

Naturally, we must acknowledge that this is an idealized notion that refers to a European model of discourse and democracy. The history of more than twenty years of transdisciplinary projects (Scholz 2011a), as a means of sustainability learning, has shown that transdisciplinary processes are possible in many cultures, but not all, and that certain constraints must be fulfilled on the side of the participants.

Collaboration in a transdisciplinary project requires *certain rules*. A crucial issue is that all agents remain in their roles and positions. This holds true in particular for scientists who are tasked to provide a problem representation that is "as close to reality as possible," which may be utilized by all key actors, independent of their interests and values.

In order to allow for learning, a td-process must provide a *"protected discourse arena,"* which allows for thought experiments and "unproven ideas." As there are key actors from industry and business participating in the Global TraPs, special care is taken that only issues of precompetitive character are addressed (Scholz et al. 2013).

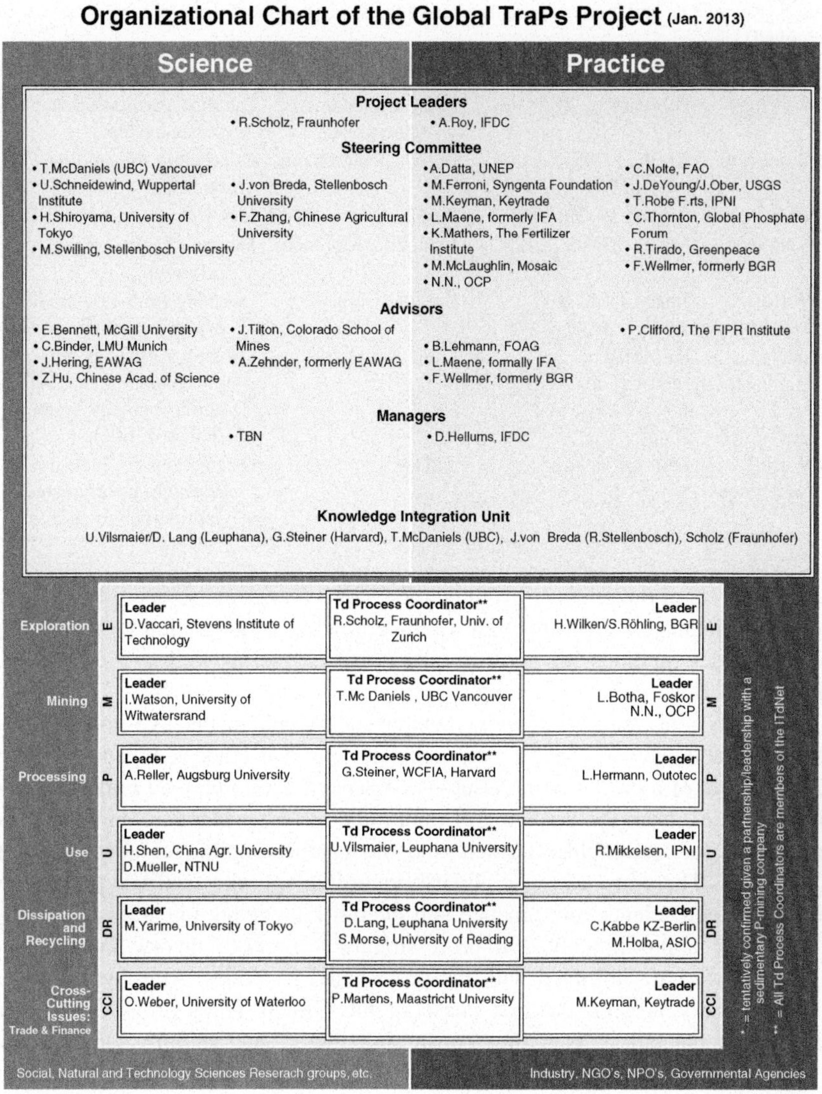

1 Sustainable Phosphorus Management

Organizational Chart of the Global TraPs Project (Jan. 2013)

Science		Practice	
colspan across — see below			

Project Leaders
- R.Scholz, Fraunhofer
- A.Roy, IFDC

Steering Committee

• T.McDaniels (UBC) Vancouver • U.Schneidewind, Wuppertal Institute • H.Shiroyama, University of Tokyo • M.Swilling, Stellenbosch University	• J.von Breda, Stellenbosch University • F.Zhang, Chinese Agricultural University	• A.Datta, UNEP • M.Ferroni, Syngenta Foundation • M.Keyman, Keytrade • L.Maene, formerly IFA • K.Mathers, The Fertilizer Institute • M.McLaughlin, Mosaic • N.N., OCP	• C.Nolte, FAO • J.DeYoung/J.Ober, USGS • T.Robe F.rts, IPNI • C.Thornton, Global Phosphate Forum • R.Tirado, Greenpeace • F.Wellmer, formerly BGR

Advisors

• E.Bennett, McGill University • C.Binder, LMU Munich • J.Hering, EAWAG • Z.Hu, Chinese Acad. of Science	• J.Tilton, Colorado School of Mines • A.Zehnder, formerly EAWAG	• B.Lehmann, FOAG • L.Maene, formally IFA • F.Wellmer, formerly BGR	• P.Clifford, The FIPR Institute

Managers
- TBN
- D.Hellums, IFDC

Knowledge Integration Unit
U.Vilsmaier/D. Lang (Leuphana), G.Steiner (Harvard), T.McDaniels (UBC), J.von Breda (R.Stellenbosch), Scholz (Fraunhofer)

	Leader	Td Process Coordinator**	Leader
Exploration (E)	D.Vaccari, Stevens Institute of Technology	R.Scholz, Fraunhofer, Univ. of Zurich	H.Wilken/S.Röhling, BGR
Mining (M)	I.Watson, University of Witwatersrand	T.Mc Daniels, UBC Vancouver	L.Botha, Foskor N.N., OCP
Processing (P)	A.Reller, Augsburg University	G.Steiner, WCFIA, Harvard	L.Hermann, Outotec
Use (U)	H.Shen, China Agr. University D.Mueller, NTNU	U.Vilsmaier, Leuphana University	R.Mikkelsen, IPNI
Dissipation and Recycling (DR)	M.Yarime, University of Tokyo	D.Lang, Leuphana University S.Morse, University of Reading	C.Kabbe KZ-Berlin M.Holba, ASIO
Cross-Cutting Issues: Trade & Finance (CCI)	O.Weber, University of Waterloo	P.Martens, Maastricht University	M.Keyman, Keytrade

Social, Natural and Technology Sciences Reserach groups, etc. | Industry, NGO's, NPO's, Governmental Agencies

(Umbrella Project / Node Level; * = tentatively confirmed given a partnership/leadership with a sedimentary P-mining company; ** = All Td Process Coordinators are members of the ITdNet)

Fig. 25 Organizational chart of the global TraPs project (Jan 2013)

6.2 How is the Global TraPs Project Organized?

The Global TraPs project is organized on three levels (see Fig. 25). Level 1 is referred to as the "Umbrella Project," which is determined by the Project Leaders who take responsibility for the overall project. The strategic decisions are made by the Steering Committee, which shows equal representation—such as in all levels

Table 6 Key events and focus-guiding themes of the seven key meetings of Global TraPs[1] to be determined)

Event	Date	Location	Focus/guiding theme
1st Workshop	February 6, 2011	Muscle Shoals, AL/ Phoenix AZ	Building partnership and co-leadership
2nd Workshop	April 31–May 1 m 2011	Zürich, Switzerland	Consenting the guiding question
3rd Workshop	August 19–30, 2011	Zürich, Switzerland	Identifying critical questions
4th Workshop	March 15–16, 2012	El Jadida, Morocco	Defining Cases–setting priorities
1st World Conference	June 18–21, 2013	Beijing, China	Learning from case studies—exploring policy options
Workshop	Fall 2014	Latin America	Looking for new paradigms—understanding the soil and agrotechnology system (tentative title)
2nd World Conference	Fall 2015	TBD	Orientations for industry, research, governments, and the public at large

and subprojects of the project—from members of practice and science. The project leaders and managers are supported by the Knowledge Integration Group (KIG), which provides methods for transdisciplinary discourse, as well as the representation, evaluation, and transitioning of complex structures.

Level 2 is represented by *Nodes*, which correspond to the five nodes of the supply–demand chain in Fig. 16. Additionally, there is a Trade & Finance Node—as an example of a cross-cutting issue—as financial actors play an important role along all nodes, from the financing of mines and fertilizer plants to micro-finance mechanisms for smallholder farmers and the venture capital for new recycling technologies. There are transdisciplinarity coordinators who facilitate the collaboration between science and practice and overall knowledge integration.

There are approximately 200 professionals from both theory and practice institutions who are affiliated with the project and its nodes. As is typical for a transdisciplinary project, the first phase of the Global TraPs project, i.e., joint problem representation, is a time-consuming issue. The guiding question was determined in the 2nd Workshop by means of defining the critical questions. These questions served to identify knowledge gaps, environmental impacts, social equity, technology options, policy means, etc. Each of the six node groups authored a chapter of the book, *Sustainable Phosphorus Management: a Transdisciplinary Roadmap* (Scholz and Wellmer 2013).

The discussion on peak phosphate theory and the validity of data from Moroccan mines were disputed topics in the first two workshops. With the assistance of key experts from industry, science, and public institutions such as USGS and IFDC, a comprehensive view on geological data, resource economics, mathematical modeling, price dynamics, etc. was developed. In addition, some mining companies provided insight to the methods they use to access reserves and resources. This revealed (in particular, for two large mines) that the data recorded by USGS may be rather conservative estimates. A comprehensive scientific paper emerged from these activities (Scholz and Tietje 2002).

Based on the portfolio of critical questions, a set of approximately 15 case studies were identified, and many are currently in progress. One cluster of Global TraPs members—including representatives of the University of Freiberg, BGR Hannover and USGS, Washington—is addressing procedures that may improve the homogeneity of the data provided by the mining companies to the Mineral Commodity Survey (see also Fig. 27). Another case study, in the case of Manila Bay and Laguna, surveys the hypothesis that phosphorus may be environmentally uncritical in the case of high technology WWTP (Knud-Hansen 1994), which is also valid for megacities in the developing world. Also included in this study is a partnership among key stakeholders, ranging from Greenpeace to detergent producers, that is being sought, just as on other levels of the Global TraPs project. Naturally, in each of these case studies, the principles of transdisciplinarity will be strictly applied. Finally, we mention that questions of social equity (some linked to the socioeconomic divide between developing and developed countries) are an important issue in Global TraPs as well. The project, Smallholder Farmers Access to Phosphorus ([SMAP], financed by Syngenta Foundation), began in January 2013 with two case studies in Vietnam and Kenya. This project focuses on smallholder farmer access to (avoiding underuse) and proper use of phosphorus.

The follow-up steps of the Global TraPs project may be reviewed in Table 6.

6.3 Knowledge Integration and Mutual Learning as Components of the Global TraPs Project

Transdisciplinarity—as a new way of engaging in and utilizing science—requires defined methods. This holds true, in particular, for knowledge integration (i.e., capacity building) and for consensus building, two main functions of transdisciplinary processes. Transdisciplinary processes must establish five types of knowledge integration (see Spotlight 2). Here, we highlight different types of knowledge integration that are utilized within the Global TraPs project.

1. *Interdisciplinarity*: integrating knowledge from the natural, engineering, and social sciences. If we wish to improve farmers' use of phosphorus, knowledge of plant nutrition must be combined with agrotechnological knowledge and knowledge about the willingness or the preparedness of farmers to utilize the technology. From a method perspective, Formative Scenario Analysis (FSA), Quantitative System Analysis (SA), or System Dynamics (SD) genuinely establish interdisciplinarity (Scholz 1987; Kahneman 2011).
2. *Integrated systems analysis*: From a sustainable transitioning perspective, different aspects of phosphorus use are interrelated. Here, the definition of the system boundaries is an important prerequisite, as looking at the whole of a

system also requires defining the endogenous and exogenous variables. Besides FSA, SA, or SD in particular, methods of evaluation must consider the holistic perspective taken in sustainability.
3. *Integrating different modes of thought*: The key to transdisciplinarity is that the different modes of thought, knowing, and epistemics are related. Here, the experiential knowledge on the side of practitioners must be integrated with the analytic knowledge of science, which is linked to academic rigor (Scholz and Stauffacher 2007).
4. *Integrating interests and worldviews from different stakeholders*: What you see, what you like, and what you prefer depends on your perspective and interests. This suggests that a transdisciplinary process must acknowledge the different (partial) knowledge of the participants and the different preferences. Usually, there is a conflict of interest related to any sustainable transformation. If for instance, the phosphorus content in manure is reduced, this affects the interest of the inorganic phosphorus feed additives industry. There are methods of "analytic mediation," which may measure consent and dissent of the stakeholders (Godeman 2008; Scholz 2011a).
5. *Relating different cultures*: The most challenging issue in transdisciplinarity is the relating of different cultures (Scholz 2011a, b), which may also be referred to as worldviews or cosmologies. We conceive *culture* as the total pattern of human behavior and action embodied in values, thought, religion, language, and learning transmitted as institutional knowledge to succeeding generations. The way fertilizer is used and agriculture is practiced differs among cultures. The same holds true for the corporate responsibility that companies may take with respect to the environment. These issues must be acknowledged within the Global TraPs project.

6.4 Mutual Learning Sessions and Dialogue Sessions as Instruments of Transdisciplinary Processes

In this section, *Mutual Learning Sessions* (MLS) and *Dialogue Sessions* (DS) are seen as methods or techniques under which *capacity building* and *consensus formation* among key stakeholders may be developed for sustainable transitioning of phosphorus use. When entering a learning process for proper knowledge integration, scientists from different disciplines and practitioners (ranging from phosphorus traders to members of environmental NGOs) should *acknowledge the otherness of the other* and must *meet on equal footing* (Thompson Klein et al. 2001).

6.4.1 Why does Sustainable Phosphorus Management Require Transdisciplinary Processes that Include MLS and DS?

Sustainably transforming the P cycle is a remarkably challenging task. The phosphorus cycle is complex in its natural and anthropogenic matrix on local, regional, and global levels. Today, we face an overly complex, rapidly changing, anthropogenic-shaped system of material flows and agricultural food and technological transformations related to phosphates. These transformations involve the many activities, needs, interests, cultural settings, and other aspects of various agents. Both the deficiency and abundance of phosphorus may have negative effects on humans, as well as agro- and ecosystems.

We argue that the complexity and multidimensionality of sustainable transitioning on different scales require the *integration of knowledge* (epistemics), insights, and interests of all key stakeholders involved in the P supply–demand chain. The MLS and DS should help to: (1) sufficiently understand the (dynamics of the) P cycle; (2) identify and appraise critical or negative aspects of current phosphorus use; and (3) properly identify, develop and establish options and policy means for changing/closing the (anthropogenic) nutrient loop. Here, a *transdisciplinary multi-stakeholder* approach is required to provide *interperspectivity* on different scales.

There are three prerequisites, which have been agreed upon by all participants of the Global TraPs project.

First, the exchange of ideas and knowledge integration takes place in a "protected discourse arena." All participants are expected to bring their own personal ideas. None will be allowed to cite or make reference to that which has been stated or distributed during the discourse without explicit agreement of those who have provided the information. Second, all discourse issues, including day-to-day politicization or policies related to programs of political parties, are excluded. Third, all discussions run in a precompetitive setting.

Knowledge integration and mutual learning are key elements of transdisciplinary processes. Transdisciplinary processes serve four functions: (a) *capacity building* of and among all key stakeholders (e.g., regarding how sustainable phosphorus management could be conducted both on a small and a large scale); (b) *consensus building* (e.g., what may be priority fields of action and which issues could be postponed); (c) *mediation* (which includes reflecting on the negative effects that may result for certain stakeholders caused by a change in the practice of phosphorus use); and (d) *legitimization* of practical solutions that are socially robust and have been developed in extended multi-stakeholder discourses, making them more likely to find stronger political support and behavioral acceptance.

The two types of sessions, MLS and DS, were first employed at the Zürich 2000 conference on *Transdisciplinarity: Joint problem solving among science, technology, and society* (Scholz and Tietje 2002). The First Global TraPs World Conference in Beijing (June 2013) utilized both forms of discourse in order to develop a jointly shared view on how the current options and obstacles of sustainable phosphorus management should be addressed.

6.4.2 Mutual Learning Sessions

The basic concept of this type of session—as an element of the transdisciplinary process—is that different stakeholders/key agents jointly deliberate on how to approach a complex, real-world case of (un-) sustainable phosphorus management. The MLS at the First Global TraPs World Conference offered a full day of discourse among key stakeholders on a complex real-world case that represents/embodies a challenging problem/barrier to sustainable phosphorous management. In terms of the theory of science, the type of problem dealt with in mutual learning sessions is referred to as an *ill-defined* (or *wickedly-defined*) *problem*. Such problems typically neither encompass a straightforward solution, nor is it clear whether technical, social, or economic barriers are the most important to be overcome.

Ideally, MLS aspire to merge *experiential wisdom* from practitioners with *academic rigor* from scientists. This means, in particular, that practitioners and farmers meet on equal footing. Practitioners may be key agents from the case or decision-makers or actors who have experience with similar cases.

The First Global TraPs World Conference dealt with cases from the Beijing area with settings and innovation in Chinese agriculture phosphorus or fertilizer management. Participants of these MLS had a direct case encounter by visiting sites in the vicinity of the conference location. There were also opportunities to directly interact with case agents.

6.4.3 Dialogue Sessions

DS offer a protected discourse arena for difficult, contested, and sometimes taboo topics. Dialogue sessions (DS) will primarily explore and discuss policy strategies. In general, a DS session begins with triangulation. This means that contested issues will be portrayed/described from different disciplines or perspectives/interests. Usually, a dialogue session includes a mixture of triangulated input lectures which open a controversial space of hypotheses and propositions. Based on this input, additional propositions may be provided, which are moderated in group discussions. DS may structure the problem space.

6.5 Transdisciplinary Case Studies

A case "is unique; one among others … and always related to something general. Cases are empirical units, theoretical constructs, and subject to evaluation, because scientific and practical interests are tied to them" (Müller 2002). According to this

1 Sustainable Phosphorus Management

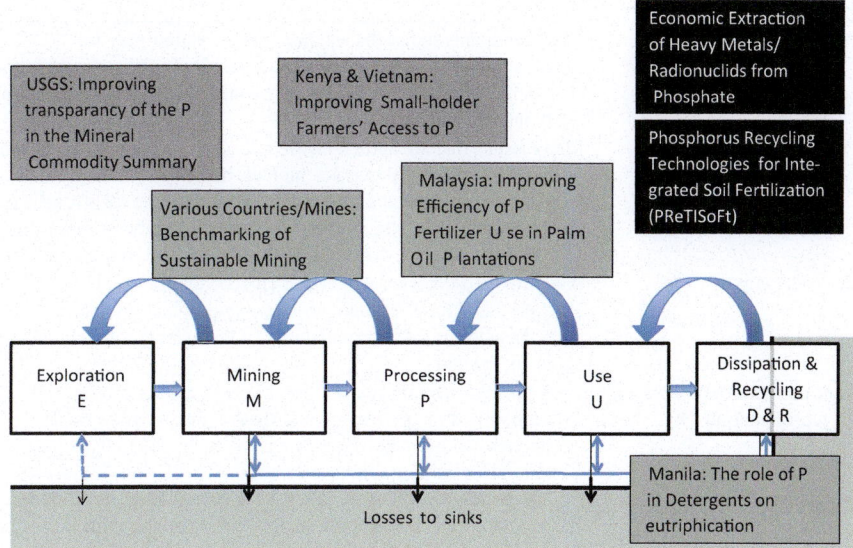

Fig. 26 Some of the planned and realized case studies (*gray-shaded boxes*) and topics for case study research (*black boxes*) of the Global TraPs project

statement, phosphorus issues may stand as a case for phosphorus, just as the Global TraPs project may stand as a case for transdisciplinary processes.

But case studies in the Global TraPs project function as the means of answering the identified critical questions and thus are tools for learning. For each of the nodes, critical questions have been identified and transdisciplinary case studies have been launched and are in progress. Exemplary case studies are listed in Fig. 26. The characteristics of transdisciplinarity (such as co-leadership), as they are stated in Table 7, are addressed in each of these studies. Thus, USGS (Director John H. deYoung) and Prof. Jens Gutzmer (University of Freiberg) are the practice and science leaders of the *Transparency of USGS Data* case study, which examines how the mining companies report reserves and how systems may be harmonized.

7 The Challenge of Increasing Efficiency, Avoiding Environmental Pollution, and Providing Accessibility

Phosphorus is essential for food security and is an important element for many industrial, medical, and technological processes. Due to its importance, humans have almost *tripled the phosphorus flows*, including the virtual anthropogenic phosphorus flows in heavy industry.

Table 7 Principles of mutual learning sessions

Protected discourse arena	All participants agree that nothing that is stated may be communicated without explicit agreement of the person who has made the statement. This promotes learning and allows for thought experiments
Pre-competitive issues	Many issues that are addressed are of economic interest. MLS and transdisciplinary processes include industry-to-industry and industry-to-science dialogues that deal with the early stages of development in which competitors may collaborate in a pre-competitive process of mutual learning or research partnership
Co-leadership	A legitimized decision-maker and (independent) scientist(s) build a partnership. They take co-leadership, which includes responsibility and accountability of the *orientations* developed in the project
Joint problem definition and representation	Joint problem definition asks that the case agents present the problems they are facing, and that scientists adapt their interests to the real case setting. MLS should ratify a (case-related) specific guiding question that is of generic interest, and which answer is of interest for sustainable transitioning. As practice and sciences (such as different disciplines in science) use different languages, a joint representation of the case and its problems is an important issue. Here, graphical representation is an important and common tool
Differentiation of roles	Acknowledging different roles means that the otherness of the interest, perspectivity and role (functionality) of the different actors are acknowledged. If the differentiation of roles (e.g., among fertilizer producers, traders and farmers) is acknowledged, it is possible to identify behavioral changes that affect such issues as the efficiency and environmental impacts related to a specific role. This is a prerequisite of actor-based MFA
Acknowledging constraints	The efficacy and efficiency of learning and the generation of knowledge depend on the time, motivation (money), prerequisites (what knowledge do the participants have), intensity of preparation and many other aspects of the process and the participants (e.g., their diversity). Given that MLS are limited in time, as much time as possible should be spent in preparation
Orientations instead of recommendations	The development of (socially robust) orientations is a goal of the MLS. Here, insight into causal chain logics ("if you do A, then Z is more likely than for you to do B") may lead to an identification of the "do-s & don'ts." We also use the term "orientation" to avoid a doctrinaire flavor that may be linked to (lists of) recommendations

The first two years of the Global TraPs project provided what we refer to as *robust* orientations *on sustainable phosphorus management*. Recognizing phosphorus' crucial role in food production, Global TraPs summarizes the weaknesses in global phosphorus flows when we focus on three aspects: efficiency, pollution, and social responsibility. This chapter focuses on phosphorus use. Naturally phosphorus flows in agriculture are interlinked with other (essential) nutrients. Here, one future task will be to elaborate in what way the use of phosphorus is interconnected with those of other nutrients, in particular nitrogen (Sutton et al. 2013) and what role phosphorus plays in the nexus with energy, or water and the flows of other materials or chemical elements.

7.1 Efficiency as an Indicator of Unsustainable Phosphorus Use

Currently, phosphorus use shows extraordinarily low use efficiency across the entirety of the supply–demand chain. If we just look at before processing, about 30–50 % of phosphate rock assigned to mine ore is not used and that much of it is excluded from value chain before processing. Given a conservative estimate that means that from around 36 Mt PR which may become subject of mining 25 Mt PR enters the processing or direct application stage. Please note that all these figures are fuzzy. There is uncertainty in the data (e.g., runoffs depend on whether) and knowledge (for instance the losses in processing are rather based on expert judgments than on empirical data). But, of course, also classification (e.g., what is included in the reserves and what not) may differ between the deposits.[4]

The apparently low efficiency may also be illustrated by another input–output relationship. There is an input on the magnitude of 20–22 Mt P from mining for food production. Only a little more than 3 Mt P are eaten. These numbers become even less favorable if the non-used phosphorus in mining and primary beneficiation/processing to concentrate ore and the use of geogenic, weathered phosphorus is included. With a rule of thumb calculation (referring to the flows of Fig. 21), we may see 30–50 % of unrecovered phosphate from deposits before processing. Here, we should acknowledge that just the mining ratios that were surveyed show a high range between 95 % for many mines and 50 % for some. A moderate estimate of the extracted and not used phosphate rock during extraction and beneficiation increased the 25 Mt P annually used in 2011 according to 36 USGS data up to 36 Mt P year^{-1} (see Sect. 4.5.1). What are factual losses before beneficiation asks for a closer look and may be considered as a knowledge gap such as the ore grades of the phosphate in operating mines.

If we account the anthropogenic flows, we also have to take a look at the increased weathering of phosphorus. Here, estimates go up to 5 Mt P year^{-1} which may be partly promoted by agriculture (see Sect. 4.6). Also in this place, a more substantiated estimate is missing. There are certainly losses in livestock and

---

[4] One option of representing the uncertainty is by probability distributions. The figures presented may be considered as means of—partly empirically and partly subjectively reasoned—probability distributions. Thus, as the distributions are not independent, the additivity for the means must not be given.

manure management that may amount to the magnitude of 15 Mt P year^{-1} (see Sect. 4.6). In order to reflect on double accounting, this includes phosphorus feed additives and of course also mineral fertilizer inputs.

Of course, we also have to account for the stock building of phosphorus in soil by fertilization which is judged to have a magnitude of presumably 2–3 Mt P year^{-1} or even more (see Sect. 4.2). A coarse figuring may amount to 50 Mt P year^{-1} that may become conceivably subject of CLoSD (Closed-Loop Supply–Demand Chain) phosphorus management. What is considered as loss and what definitions of efficiency may be used to improve the sustainability of the supply chain is a matter of future discussion and research.

When reflecting on whether efficiency is meaningful concept, we further should acknowledge the data from the regional statistics. For instance, Africa shows the highest efficiency but the lowest yield. This certainly asks for properly combining efficacy and efficiency and may illustrate that efficiency is neither a necessary nor sufficient condition for sustainability, but rather a useful means.

We also have to question how efficiency may be improved. There seems to be poor efficiency in processing, recycling, and the general use of phosphorus, at least in the nutrient chain. We have identified unnecessary sinks along all steps of the supply chain. It is obvious that increasing the efficiency of phosphorus use not only offers multiple business and technology options but also may contribute to the sustainable use of this non-substitutable element of life, in particular from an environmental and long-term supply security perspective. We conclude that a multi-stakeholder discourse including industry-to-industry and industry-to-science dialogues among all key stakeholders also may help to set priorities for avoiding unwanted losses of phosphorus.

7.2 Avoiding Pollution from Phosphorus

Phosphorus is a basic nutrient required for plant and animal life, but following the rule that *the dose matters*, phosphorus overuse may lead to pollution, while underuse contributes to land degradation. There is irrefutable evidence that excessive amounts of phosphorus can adversely shift freshwater and marine ecosystems due to algae blooms, eutrophication, hypoxia, or anoxia in freshwater systems and the world's oceans. The collective anthropogenic input by agriculture, sewage, industry, detergents, etc. will require thorough monitoring and assessment in many areas around the globe. This monitoring may start with regional or national phosphorus balances. But these large-scale views must be supplemented with impact assessments of anthropogenic phosphorus flows on the small-scale, local environment. In relation to underuse, large areas of land in sub-Saharan Africa are being depleted of nutrients (including phosphorus) resulting in decreasing soil fertility and productivity and culminating in land degradation. In turn, land degradation is intricately linked to food insecurity and poverty. Figure 27 shows that this is possible. The Lake Constance is the third largest

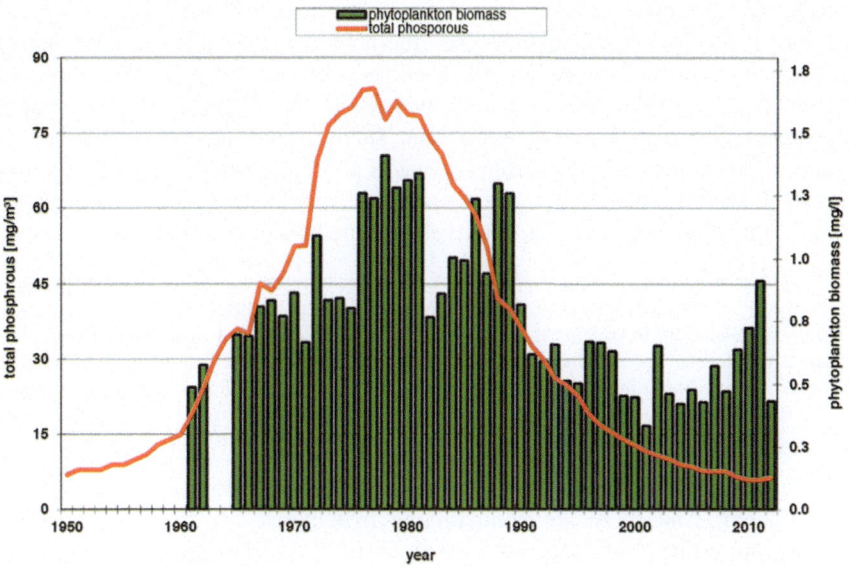

Fig. 27 Development of phosphorus and algal biomass concentration in Lake Constance (There is no clear explanation for the 2011–2012 peak of biomass, changes in the phytoplankton could be observed. By what these changes are caused in under investigation) (Föllmi 1996)

freshwater lake in Europe. Naturally, the lake is phosphorus limited, was heavily polluted by phosphorus, and showed a maximum of 87 µg/l (LUBW 2013; Müller 2002). Due to a joined action of Austrian, German, and Swiss policies for balanced fertilization, phosphorus extraction in sewage plants, limiting phosphorus content in laundry detergents, etc. the phosphorus content declined to 13 µg/l (Fig. 27). The case shows that aquatic systems may be protected if concerted action is taken.

Phosphorus deposits, whether igneous or sedimentary, share space with a number of other elements. Due to its evolution, sedimentary phosphate is linked with critical contaminants such as cadmium, thorium, and uranium. Thus, phosphorus use may become, through possible elemental contamination of compounded elements, potentially hazardous. There are concerns that these compounded elements, if errantly included in fertilizer manufacturing processes, may cause long-term contamination of soils. Here, it is clear that a long-term perspective must be taken, and that the issue of irreversibility of potentially large-scale contamination or high costs of remediation requires thorough research and the development of technologies to economically separate the contaminants from the mineral. As these compounded elements may also be viewed from a co-mining of other elements perspective, there may be multiple incentives for developing such technologies.

A third consideration is currently based on theoretical arguments. We are living in the anthropocene, an age in which most geological and ecological processes are affected by anthropocentric activity. We have learned from sulfur (i.e., acid rain), chlorine, and bromide compounds (i.e., the hole in the ozone) and from human carbon emissions (i.e., emissions-related climate change) that the human impact on biogeochemical cycles can cause unwanted secondary feedback loops that can be irreversible in the short and medium term. There is evidence that—in geological and evolutionary terms—the abrupt tripling of phosphorus flows may cause severe and imponderable changes in the resilience of the world's ecosystem, biodiversity, soil fertility, and other—perhaps not yet identified—areas of impact.

> It is clear from the above, that phosphorus is beneficial to the biosphere in many ways; however, examples from the present and the geological past also show that phosphorus may be deleterious to large parts of the biosphere when released and applied above natural threshold levels. Through the excessive use of phosphorus, an intricate network of feedback mechanisms comes into play, a play that may evolve into a drama, if we look at the present-day anthropogenic release rates of phosphorus and the already visible impact on ecosystems. (Emsley 2000b)

The speed of the change rate of the global phosphorus flows is the rationale of the concern. And there are contradicting hypotheses whether this is critical (Rockström et al. 2009; Carpenter and Bennett 2011). Here, science and society are challenged to develop the knowledge that allows for proper anticipation of the unintended consequences of increased phosphorus flows:

> The overuse of phosphorus contributed to environmental damage because this simply took the brake off this limiting factor in certain vulnerable locales. (Emsley 2000b, p. 301).

7.3 Securing Access to Phosphorus

Phosphorus is essential for food production. Thus, sufficient access to phosphorus is vital for any society to survive, and ultimately thrive. As has been the case for pollution, we look at this issue on different scales.

Undernourishment is rampant, and securing basic food supplies is a global issue. There are about one billion undernourished people in the world. And many of them are *smallholder farmers*, whose soils are extremely deficient in nutrients and often particularly in phosphorus, requiring fertilizer to supplement the deficiency. However, most smallholders do not have the financial means to buy fertilizers, nor have knowledge of their proper use. But beyond this fact, there are a number of other social equity issues. We argue that these issues require suitable policies on both the global and local level. And we argue that these policy means should be elaborated through a transdisciplinary discourse that includes key stakeholders along the supply–demand chain.

The issue of access offers an interesting perspective, as phosphorus may be conceived as a human right; it belongs in the domain of public goods that humans require to survive—simply because they *are* human. Against this backdrop, sustainable phosphorus management is a valuable tool for the *public good*. But we should note that this holds for all essential elements and phosphorus seems to be rather with the use of white phosphorus as a weapon (Mojabi et al. 2010), an issue that has not been not focused in the chapter.

Another level can be defined if the scale was to be *access to phosphorus on a nation level*. As other critical or essential elements, phosphorus is unevenly deposited throughout the world. Although 1,600 phosphorus mines are currently identified (Jasinski 2010), the large, commercially viable, high-grade ore reserves are located in only a handful of countries. Here, the geopolitical dimension is to be considered. However, we wish to note that not only are the recorded reserves important, but so too are the functionality of mines, efficiency of fertilizer production plants, etc. Further, the diversification of, and access to, phosphorus may be increased by recycling and other means such as buffer stocks. Many developed countries may benefit from intense (over) fertilization over some decades which has provided a level of phosphorus stocks that may allow for nutrient-efficient high-yield agriculture. A critical question which asks for further research (Dumas et al. 2011) is how much nutrient inputs soils with poor nutrients or soils which have just recently have become agricultural land (such as tropical forest soils) must receive that they provide high yield with high efficiency.

It is clear, when investigating geological data that there is no forthcoming physical scarcity with respect to phosphate rock in the short- or mid-term future. The *scarcity* of phosphorus is primarily an *economic issue*. Though phosphate rock is finite, we will have access to future phosphorus reserves, albeit for a higher (but feasible) processing cost. The prospective long-term use of phosphorus, however, requires the precautious use of phosphate reserves. Thus, given that one billion people of the world are undernourished, the short-term social dimension of sustainability, i.e., intragenerational equity, is an issue. This has been expressed by one tenet of the Global TraPs project "providing access to phosphorus for the poor."

Finally, we must reflect on the long-term supply security of phosphorus for food production. The domestication of species and sedentary farming began around 7,000 years ago; systematic phosphorus fertilization dates back 500 years, and the chemical processing of fertilizers about 150 years. No one may predict the kind of agrotechnology that will be employed 1,000 years from now. But we may confidently predict that humans will continue to require phosphorus, and that phosphate rock would remain a vital source. The issue is not availability—phosphorus atoms do not disappear, they simply take another form. But due to its dissipative characteristics, phosphorus may get lost and we have to reflect how it may disseminate along the value chain and become the subject of recycling. The issue is (economic) accessibility of phosphorus on different scales of space and time. Thus, social responsibility, as it has been defined, must dictate and provide sufficient access for future civilizations.

7.4 A Transdisciplinary Roadmap Toward a Demand-Based Peak Phosphorus

Against the aforementioned conclusions, the current increase in phosphorus in mineral fertilizer is critical. Figure 28 showed the historical trajectories. There was a demand peak following the dissolution of the Soviet Union. But there seems to be an even more pronounced increase 10 years after, indicating that there seem to be many parts of the world where inefficient use is taking place. The use in these areas dominates the reduction in Europe, North America, and some parts of South America, where phosphorus use stabilized or peaked from the demand side because soil phosphorus nutrient capital and improved nutrient use management practices have built up over time. Given the current knowledge in the Global TraPs project, recycling technologies and smarter agrotechnology may offset or equalize any increased demand in some parts of the world such as Sub-Saharan Africa associated with a transition to intensive agriculture. Thus, sustainable phosphorus management may work toward a demand peak on a global scale, but may face different trends in different parts of the world.

What form that sustainable phosphorus management will take is the subject of the Global TraPs project. There are practice representatives from far more than 100 key stakeholders along the supply–demand chain, including industry, trade, finance, farmer organizations, NGOs, international organizations, and scientists from a broad range of disciplines involved. Global TraPs focuses on phosphorus on a global scale. The knowledge gained in this book is and will be related with that developed by other initiatives on sustainable nutrient use such as GPNM (Global Partnership of Nutrient Management) and with other national or international multi-stakeholder or governmental initiatives related to sustainable phosphorus use and management such as the PCPRJ (The Promotion Council of Phosphors Recycling of Japan) or EPP (European Phosphorus Platform), just to mention a few. The present book focuses global transdisciplinary processes. Chapters 2–5 present pathways on a roadmap that points to transitioning to sustainable phosphorus use. As is typical in constructing a new map of a complex terrain, one may not find all roads recorded. And, of course, some highways are missing that, once identified, may facilitate quicker access to solutions, and the implementation of more desirable methods of phosphorus use. Despite these unknowns, we are confident that the present volume, which includes presentations of 52 practitioners and researchers, will serve as a reference for the better understanding and promotion of sustainable phosphorus use.

Acknowledgments We wish to thank Fridolin Brand, Patrick Heffer, Christian Kabbe, Kazuyu Matsubae, Daniel B. Müller, Michael Mew, Gregoire Meylan, Michel Prud'homme, Terry Roberts, Desirée Ruppen, Sheida Sattari, Willem Schipper, Andy Spoerri, Christopher Thornton, and Andrea E. Ulrich for their valuable input to various sections or the whole text of a previous version and Clyde Beaver and Donna Venable for the editing of this chapter.

References

Adger WN (2006) Vulnerability. Global Environ Change 16(3):268–281
Äikäs O (1989) Phosphate resources in early Proterozoic supracrustal rocks, Finland with reference to the Baltic Shield. In: Northholt AJG, Sheldon RP, Davidson DF (eds) Phosphate deposits of the world, vol 2., Phosphate rock and resources. Cambridge University Press, Cambridge, pp 429–437
Al-Bassam K, Fernette G, Jasinski SM (2012) Phosphate deposits of Iraq. Paper presented at the PHOSPHATES 2012, El-Jadidam, Morocco, 20–23 March 2012
Alcamo JD, Van Vuuren D, Cramer W (2006) Changes in ecosystem services and their drivers across scenarios. In: Carpenter SR et al (eds) Ecosystem and human well-being: scenarios. Island Press, Washington, pp 279–354
Alexandratos N, Bruinsma L (2012) World agriculture towards 2030/2050. The Revision. FAO, Rome
Alford WAL, Parkes JW (1953) Murray, James—a pioneer in the making of superphosphate. Chem. Ind 33:852–855
Allen HL (1987) Forest fertilizers. J Forest 85(2):37–46
Andrews JB (1910) Industrial diseases and occupational standards Proc Nat Conference of Charities and Corrections 27th session
Aven T (2011) On some recent definitions and analysis frameworks for risk, vulnerability, and resilience. Risk Anal 31(4):515–522. doi:10.1111/j.1539-6924.2010.01528.x
Babenko AA (2012) Removal of phosphorus in the final stages of oxidative refining of phosphorus hot metals. Steel Translation 41(12):985–987. doi:10.3103/s0967091211120035
Battista OA (1947) Got a match? The Irish Monthly 75:446–449
Baur R (2010) Operating North America's first full scale nutrient recovery facility. In: Federation WE (ed) Proceedings of the water environment federation, residuals and biosolids, pp 492–506
Bellwood P (2005) First farmers. The origins of agricultural societies. Blackwell, Malden
Benevolo L (1980) The history of the city. The MIT Press, Cambridge
Benni T (2013) Phosphate deposits of Iraq. Paper presented at the UNFC Workshop, Santiago de Chile
Beusen AHW, Dekkers ALM, Bouwman AF, Ludwig W, Harrison J (2005) Estimation of global river transport of sediments and associated particulate C, N, and P. Global Biogeochem Cycles 19(4). doi:Gb4s0510.1029/2005gb002453
Binder CR, Hofer C, Wiek A, Scholz RW (2004) Transition towards improved regional wood flows by integrating material flux analysis and agent analysis: the case of Appenzell Ausserrhoden, Switzerland. Ecol Econ 49(1):1–17. doi:10.1016/j.ecolecon.2003.10.021
Body JJ (2006) Breast cancer: Bisphosphonate therapy for metastatic bone disease. Clin Cancer Res 12(20):6258S–6263S. doi10.1158/1078-0432.ccr-06-0840
Bohle H-G (2001) Vulnerability and criticality: perspectives from social geography. IHDP Update 2 (1): http://www.ihdp.uni-bonn.de/html/publications/update/IHDPUpdate01_02.html
Bouwman AF, Beusen AHW, Billen G (2009) Human alteration of the global nitrogen and phosphorus soil balances for the period 1970-2050. Global Biogeochem Cycles 23. doi:Gb0a0410.1029/2009gb003576
Bouwman L, Goldewijk KK, Van Der Hoek KW, Beusen AH, Van Vuuren DP, Willems J, Rufino MC, Stehfest E (2012) Exploring global changes in nitrogen and phosphorus cycles in agriculture induced by livestock production over the 1900–2050 period. Proceedings of the National Academies of Sciences USA
Bowers JA (1995) Criticality in resource constrained networks. J Oper Res Soc 46(1):80–91. doi:10.1057/jors.1995.9
Brand CJ (1937) The fertilizer industry. Ann Am Acad Polit Soc Sci 193:22–33
Brandt AR (2010) Review of mathematical models of future oil supply: historical overview and synthesizing critique. Energy 35(9):3958–3974. doi:10.1016/j.energy.2010.04.045

Brundtland GH (1987) Our common future. World Commission on Environment and Development, Oxford

Brunner PH, Rechberger H (2003) Practical handbook of material analysis. Lewis, Boca Raton

BTA International (2013) Verfahrensschema einstufiges Vergährungsverfahren. http://www.wtert.eu/default.asp?Menue=13&ShowDok=17

Carpenter SR, Bennett EM (2011) Reconsideration of the planetary boundary for phosphorus. Environ Res Lett 6(1). doi:01400910.1088/1748-9326/6/1/014009

Chapin FS, Kofinas GP, Folke C (2009) Principles of ecosystem stewardship: resilience-based natural resource management in a changing world. Springer, New York

Chemicals Unit of DG Enterprise (2004) Draft proposal relating to cadmium in fertilizers. European Commission, Brussels

Chen H, Yan SH, Ye ZL, Meng HJ, Zhu YG (2012) Utilization of urban sewage sludge: Chinese perspectives. Environ Sci Pollut Res 19(5):1454–1463. doi:10.1007/s11356-012-0760-0

Childs PE (2000) Phosphorus: from urine to fire. 3. Food from old bones: the fertilizer connection. Chemistry in Action (60)

CNCIC (2008) China fertilizer export taxes, China Fertilizer Market Week, No. 8 Volume 34. China National Chemical Information Center. http://www.sinofi.com/english/show_news.asp?id=88. Accessed 21 Aug 2012

Cordell D, Drangert JO, White S (2009) The story of phosphorus: global food security and food for thought. Glob Environ Change-Human Policy Dimensions 19(2):292–305. doi:10.1016/j.gloenvcha.2008.10.009

Cordell D, Rosemarin A, Schroder JJ, Smit AL (2011a) Towards global phosphorus security: a systems framework for phosphorus recovery and reuse options. Chemosphere 84(6):747–758. doi:10.1016/j.chemosphere.2011.02.032

Cordell D, White S, Lindström T (2011) Peak phosphorus: the crunch time for humanity? The Sustainability Review, 4 April 2011

Crutzen PJ (2002) The "anthropocene". Journal de Physique IV France 12(10):11–15. doi:10.1051/jp4:20020447

Cuesta J (2013) A world free of poverty... but of hunger and malnutrition? Eur J Dev Res 25(1):1–4. doi:10.1057/ejdr.2012.43

Cunfer G (2004) The dust bowl. EH.Net Encyclopaedia, edited by Robert Whaples. August 19, 2004. http://eh.net/encyclopedia/article/Cunfer.DustBowl. Accessed 10 May 2010

DACH (2008) Referenzwerte für Nährstoffzufuhr (3rd. edn.) http://www.sge-ssn.ch/de/ich-und-du/rund-um-lebensmittel/Referenzwerte-fuer-die-Naehrstoffzufuhr/

Daly M (1984) The deposed capital. Cork University Press, Cork

Datta NC (2005) The story of phosphorus. Universities Press, Andhra Pradesh

Davidson EA (2012) Representative concentration pathways and mitigation scenarios for nitrous oxide. Environ Res Lett 7(2):024005. doi:02400510.1088/1748-9326/7/2/024005

Davis RD (1996) The impact of EU and UK environmental pressures on the future of sludge treatment and disposal. CIWEM Water Environ J 10:65–69

de la Vega G (1609/1990) Comentarios reales historia general del Perú Santander de Quilichao/Madrid

de Vries B (ed) (1998) Umm el-Jimal. A frontier town and its landscape in northern Jordan: Fieldwork 1972–1981. Journal of Roman Archaeology. Supplementary series no. 26. Journal of Roman Archaeology. Supplementary series no. 26, Portsmouth, RI

Déry P, Anderson B (2007) Peak phosphorus. Energy Bulletin (Retrieved September 22, 2011) Deutsche-Rohstoffagentur/BGR (2011)

Diaz RJ, Rosenberg R (2008) Spreading dead zones and consequences for marine ecosystems. Science 321(5891):926–929. doi:10.1126/science.1156401

Dumas M, Frossard E, Scholz RW (2011) Modeling biogeochemical processes of phosphorus for global food supply. Chemosphere 84:798–805

Dumas MJ, Boussingault JB (1844) The chemical and physiological balance of organic nature: an essay. Saxton, Peirce & Co, Boston

Editorial (2010) How to feed a hungry world. Nature 466(7306):531–532

Edström M, Schüßler I, Luostarinen S (2011) Combustion of manure: manure as fuel in a heating plant. Baltic MANURE WP6 Energy potentials

EFSA Panel on Dietic Products Nutrition and Allergies (2005) Opinion of the Scientific Panel on Dietetic Products, Nutrition and Allergies on a request from the Commission related to the Tolerable Upper Intake Level of Phosphorus (Request N° EFSA-Q-2003-018). The EFSA Journal 233:1–19

Eilittä M (2011) The Global TraPs Project. Transdisciplinary Processes for Sustainable Phosphorus Management (2010–2015). Multi-stakeholder forum to guide and optimize P use. ETH-NSSI and IFDC, Zürich and Muscle Shoals

Elser J (2014) Health dimensions of phosphorus. In: Scholz RW, Roy AH, Brand FS, Hellums DT, Ulrich AE (eds) Sustainable phosphorus management: a global transdisciplinary roadmap. Springer, Berlin, pp 229–231

Emsley J (2000a) The 13th element: the sordid tale of murder, fire, and phosphorus. Wiley, New York

Emsley J (2000b) The shocking history of phosphorus. The biography of the devil's element. MacMillan, London

EPA (2012) Estimated animal agriculture nitrogen and phosphorus from manure. http://water.epa.gov/scitech/swguidance/standards/criteria/nutrients/datasetagloads.cfm. Accessed December 20, 2012

Erdmann L, Graedel TE (2011) Criticality of non-fuel minerals: a review of major approaches and analyses. Environ Sci Technol 45(18):7620–7630

Escueta SC, Tapay NE (2010) Soil and nutrient loss on swidden farms, and farmers' perception in Bazal-Baubo watershed, Aurora Province, Philippines. Asia Life Sci 19(2):395–418

EU (2002) Phosphates and alternative detergent builders—final report. vol WRc Ref: UC 4011. Brussels

EU (2012a) EP supports ban of phosphates in consumer detergents. Press release. IP/11/1542, 14/12/2011. Brussels

EU (2012b) Regulation (EU) No 259/2012 of the European Parliament and of the Council of 14 March 2012 amending Regulation (EC) No 648/2004 as regards the use of phosphates and other phosphorus compounds in consumer laundry detergents and consumer automatic dishwasher detergents Text with EEA relevance. Brussels

Evans M (2012) Phosphate resources: future for 2012 and beyond. Paper presented at the AFA International Fertilizer Forum and Exhibition, Sharm El-Sheikh

Evenson RE, Gollin D (2003) Assessing the impact of the Green Revolution, 1960 to 2000. Science 300(5620):758–762. doi:10.1126/science.1078710

Fan S, Menon P, Brzeska J (2013) What policy changes will reverse persistent malnutrition in Asia? Eur J Dev Res 25(1):28–35. doi:10.1057/ejdr.2012.47

FAO (2008) The state of food and agriculture. Food and Agriculture Organization of the United Nations, Rome

FAO (2009) 1.02 billion people hungry. FAO. Accessed 15 Aug 2013

FAO (2010a) Fishery and aquaculture statistics, commodities. FAO. ftp://fao.org/FI/CDrom/CD_yearbook_2010/navigation/index_content_commodities_e.htm

FAO (2010b) The state of food security in the world: addressing food insecurity in protracted crises. Rome

FAO (2011) Save and grow. A policy maker's guide to the sustainable intensification of smallholder crop production. FAO, Rome

FAO (2012a) Meat and meat products. FAO Agriculture and Consumer Protection Department. http://www.fao.org/ag/againfo/themes/en/meat/home.html

FAO (2012b) Statistical Yearbook 2012. FAO, Rome

FAO LEAD (2006) Livestock's long shadow. Environmental issues and options. FAO The Livestock, Environment and development Initiative, Rome

Färber E (1921) Die geschichtliche Entwicklung der Chemie (Fragments translated by Smithsonian institution United States National Museum)

Finnish Environment Institute (2000) Cadmium in fertilizer, risks to human health and the environment. Finnish Ministry of Agriculture and Forestry, Helsinki
FIPR (2013) Phosphate primer: phosphogypsum and the EPA Ban. Florida Industrial and Phosphate Research Institute. http://www1.fipr.state.fl.us/PhosphatePrimer/0/684AE64864D115FE85256F88007AC781. Accessed 12 Aug 2013
Föllmi KB (1996) The phosphorpus cycle, photogenesis and marine phosphate-rich deposits. Earth-Sci Rev 40:55–124
Foster JB (1999) Marx's theory of metabolic rift: classical foundations for environmental sociology. Am J Sociol 105(2):366–405
Frear C (2012) Farm based anaerobic digestion and nutrient recovery. In: Bio cycle conference, Portland, 17–18 April 2012
Fresenius W, Kettrup A, Scholz RW (1995) Grössenordnung Faktor 2: Zur Genauigkeit von Messungen bei Altlasten. Altlasten Spektrum 4:217–218
Frossard E, Condron LM, Oberson A, Sinaj S, Fardeau JC (2000) Processes governing phosphorus availability in temperate soils. J Environ Qual 29(1):15–23
Gantner O, Schipper W, Weigand JJ (2014) Technological use of phosphorus: the non-fertilizer, non-feed and non-detergent domain. In: Scholz RW, Roy AH, Brand FS, Hellums DH, Ulrich AE (eds) Sustainable phosphorus management—A sustainable roadmap. Springer, Berlin
Ghosh J (2010) The unnatural coupling: food and global finance. J Agrar Chang 10(1):72–86
Global Phosphate Forum (2012) Phosphates in detergents. http://www.hosphate-forum.org/index.php?option=com_content&task=view&id=18. Accessed 11 Nov 2012
Godeman J (2008) Knowledge integration: a key challenge for transdisciplinary cooperation. Sustain High Educ Res 14(6):625–641
Gong P, Liang L, Zhang Q (2011) China must reduce fertilizer use too. Nature 473(7347):284–285
González-Andradea F, Sánchez-Qa D, Martĺnez-Jarretab B, Borja J (2002) Acute exposure to white phosphorus: a topical problem in Ecuador (South America). Leg Med 4(3):187–192
Gossel TA, Bricker JD (1994) Principles of clinical toxicology, 3rd edn. Raven Press, New York
GPF (2013) Phosphates in detergents. Retrieved 10 Aug 2013. Global Phosphate Forum. http://www.phosphate-forum.org/index.php?option=com_content&task=view&id=18
Graedel TE, Barr R, Chandler C, Chase T, Choi J, Christoffersen L, Friedlander E, Henly C, Jun C, Nassar NT, Schechner D, Warren S, Yang MY, Zhu C (2012) Methodology of metal criticality determination. Environ Sci Technol 46(2):1063–1070. doi:10.1021/es203534z
Graham WF, Duce RA (1979) Atmospheric pathways of the phosphorus cycle. Geochim Cosmochim Acta 43(8):1195–1208. doi:10.1016/0016-7037(79)90112-1
Habermas J (1996) Contributions to a discourse theory of law and democracy. Translated by William Regh. MIT Press, Cambridge
Halden RU (2010) Plastics and health risks. In: Fielding JE, Brownson RC, Green LW (eds) Annual review of public health, vol 31. Annual Review of Public Health. Annual Reviews, Palo Alto, pp 179–194. doi:10.1146/annurev.publhealth.012809.103714
Hammond JP, Broadley MR, White PJ (2004) Genetic responses to phosphorus deficiency. Ann Bot 94(3):323–332. doi:10.1093/aob/mch156
Hart MR, Quin BF, Nguyen ML (2004) Phosphorus runoff from agricultural land and direct fertilizer effects: a review. J Environ Qual 33(6):1954–1972
Haygarth PM, Jarvis SC (1999) Transfer of phosphorus from agricultural soils. In: Sparks DL (ed) Advances in Agronomy, vol 66. pp 195–249. doi:10.1016/s0065-2113(08)60428-9
Heffer P (2013) Personal communication, phosphate fertilizer for biofuel, 24 July 2013. Paris
Heffer P, Prud'homme M (2011) Fertilizer Outlook 2011–2015. Paper presented at the 79th IFA Annual Conference, 23–25 May 2011, Montreal (Canada)
Hein JR (ed) (2004) Handbook of exploration and environmental geochemistry. Life Cycle of the Phosphoria Formation—From Deposition to the Post-Mining Environment, vol 8. Elsevier, Amsterdam

Hellstein JW, Marek CL (2004) Bis-phossy jaw, phossy jaw, and the 21st century: Bisphosphonate-associated complications of the jaws. J Oral Maxillofac Surg 62(12):1563–1565. doi:10.1016/j.joms.2004.09.004

Herrmann L (2012) Personal communication, 4 Sept 2012

Herrmann L, Schipper W, Langeveld K, Reller A (2014) Processing: what improvements for what product? In: Scholz RW, Roy AH, Brand FS, Hellums DT, Ulrich AE (eds) Sustainable phosphorus management: a global transdisciplinary roadmap. Springer, Berlin

Hesketh N, Brookes PC (2000) Development of an indicator for risk of phosphorus leaching. J Environ Qual 29(1):105–110

Hilton J, Dawson CJ (2012) Enhancing management of and value from phosphate resources. Waste Resour Manage 165(4):179–189

Holling CS (1973) Resilience and stability of ecological systems. Annu Rev Ecol Syst 4:1–23

Hu Z (2011) 300 million small holders off to cities/townships. What are potential impacts for the P-demand? In: 1st global TraP workshop, Tempe, AZ, 2 Feb 2012

Hubbert MK (1956) Nuclear energy and the fossil fuels. Paper presented at the Meeting of the Southern District, Division of ProductionAmerican Petroleum Institute., San Antonio, TX

IFAD (2011) Rural poverty report. New realities, new challenges: new opportunities for tomorrow's generation. Rome

IPNI (2013) Nutrient source specifics. Single superphosphates. International Plant Nutrition Institute. http://www.ipni.net/publication/nss.nsf/0/5540C741907C7657852579AF007689EC/%24FILE/NSS-21SSP.pdf. Accessed 10 Aug 2013

Jackson BJ (1892) Eben Norton Hosford. Proceedings of the American academy of arts and sciences. pp 340–346

Jansa J, Frossard E, Stamp P, Kreuzer M, Scholz RW (2010) Future food production as interplay of natural resources, technology, and human society a problem yet to solve. J Ind Ecol 14(6):874–877. doi:10.1111/j.1530-9290.2010.00302.x

Jasinski SM (2009) Phosphate rock. In: US Geological Survey (ed) Mineral commodity summaries. USGS, pp 120–121

Jasinski SM (2010) Phosphate rock. In: US Geological Survey (ed) Mineral commodity summaries. USGS, St. Louis, pp 118–119

Jasinski SM (2011a) Phosphate rock. In: US Geological Survey (ed) Mineral commodity summaries. USGS, Reston, pp 118–119

Jasinski SM (2011b) Phosphate rock [advanced release]. In: USGS (ed) Minerals Yearbook. USGS, Washington

Jasinski SM (2012) Phosphate rock. In: US Geological Survey (ed) Mineral commodity summaries. USGS, Mineral commodity summaries, pp 118–119

Jasinski SM (2013) Personal communication by e-mail. 5 Feb 2013

Jasinski SM, Lee WH, Causey JD (2004) Handbook of exploration and environmental geochemistry. In: Hein JR (ed) Life cycle of the Phosphoria formation—from deposition to the post-mining environment, vol 8. Elsevier, Amsterdam, pp 45–71

Jeong Y-S, Matsubae-Yokoyama K, Nagasaka T (2009) Recovery of manganese and phosphorus from Dephosphorization slag with wet magnetic separation. Tohoku University, Graduate School of Environmental Studies

Johnson J, Harper EM, Lifset R, Graedel TE (2007) Dining at the periodic table: metals concentrations as they relate to recycling. Environ Sci Technol 41(5):1759–1765. doi:10.1021/es060736h|ISSN.0013-936X

Johnston AE, Richards IR (2003) Effectiveness of different precipitated phosphates as phosphorus sources for plants. Soil Use Manage 19(1):45–49. doi:10.1079/sum2002162

Kabbe C (2013) Sustainable sewage sludge management fostering phosphorus recovery. Bluefacts 4:36–41

Kahneman D (2011) Thinking, fast and slow. Farrar, Straus and Giroux, New York

Kanton Zürich (2007) Regierungsratsbeschluss (RRB) 572/2007. Zürich

Kauffman JB, Cummings DL, Ward DE, Babbitt R (1995) Fire in the Brazilian Amazon. 1. Biomass, nutrient pools, and losses in slashed primary forests. Oecologia 104(4):397–408. doi:10.1007/bf00341336

Kaufmann D, Kraay A, Mastruzzi M (2009) Governance matters VIII. Aggregate and individual governance indicators, 1996–2008. Policy research working paper

Kaufmann D, Kraay A, Mastruzzi M (2011) The Worldwide Governance Indicators (WGI) project. The World Bank Group. http://info.worldbank.org/governance/wgi/index.asp. Accessed 1 Feb 2012

Kelly R (2012) The hunger grains. The fight is on. Time to scrap EU biofuel mandates. Oxfam Briefing Paper 161:1–32

Kenkel P (2012) Managing Fertilizer Price Risk. vol AGEC-262. Oklahoma State University, Oklahoma

Kim S, Dale BE (2004) Global potential bioethanol production from wasted crops and crop residues. Biomass Bioenergy 26(4):361–375. doi:10.1016/j.biombioe.2003.08.002

Kim S, Dale BE (2005) Life cycle assessment of various cropping systems utilized for producing biofuels: bioethanol and biodiesel. Biomass Bioenergy 29(6):426–439. doi:10.1016/j.biombioe.2005.06.004

King FH (1911/2004) Farmers of forty centuries: organic farming in China, Korea and Japan. Dover Publications, Mineola

Kippenberger C (2001) Materials flow and energy required for the production of selected mineral commodities, vol SH 13. Wirtschaftsgeologie, Berichte zur Rohstoffwirtschaft. Bundesanstalt für Geowissenschaften und Rohstoffe und Staatliche Geologische Dienste in der Bundesrepublik Deutschland, Hannover

Knud-Hansen C (1994) Historical perspective of the phosphate detergent conflict. Paper presented at the Natural Resources and Environmental Policy Seminar, Boulder

Köhler J (2006) Detergent phosphates: an EU policy assessment. J Bus Chem 3(2)

Krafft F (1969) From elemental light to chemical element. Angew Chem Int Ed 8:660–671

Kraljic P (1983) Purchasing must become supply management. Harvard Business Review

Kroger R, Perez M, Walker S, Sharpley A (2012) Review of best management practice reduction efficiencies in the Lower Mississippi Alluvial Valley. J Soil Water Conserv 67(6):556–563. doi:10.2489/jswc.67.6.556

Krohns S, Lunkenheimer P, Meissner S, Reller A, Gleich B, Rathgeber A, Gaugler T, Buhl HU, Sinclair DC, Loidl A (2011) The route to resource-efficient novel materials. Nat Mater 5(9):800–901

Lal R (2005) World crop residues production and implications of its use as a biofuel. Environ Int 31(4):575–584. doi:10.1016/j.envint.2004.09.005

Lamprecht H, Lang DJ, Binder CR, Scholz RW (2011) Animal bone disposal during the BSE crisis in Switzerland—an example of a "disposal dilemma". Gaia 20(2):112–121

Lavoisier AL (1776) Essays physical and chemical. (trans: Thomas Henry). Johnson, London

Laws D, Scholz RW, Shiroyama H, Susskind L, Suzuki T, Weber O (2004) Expert views on sustainability and technology implementation. Int J Sustai Dev World Ecol 11(3):247–261

Lee GF, Jones RA (1986) Detergent phosphate bans and eutrophication. Environ Sci Technol 20(4):330–331. doi:10.1021/es00146a003

Leff B, Ramankutty N, Foley JA (2004) Geographic distribution of major crops across the world. Global Biogeochemical Cycles 18(1). doi:Gb100910.1029/2003gb002108

Leibniz GW (1710) Historia inventionis Phosphori. Miscellenea Berolnensia ad incrementum scientarium. Sumptibus J. Ch. Papenii, Berlin

Lelle MA, Gold MA (1994) Agroforestry systems for temperate climates. Forest Conserv History 38(3):118–126

Leubner C (2013) The seed biology place. http://www.seedbiology.de. Accessed 12 July 2013

Lifset R, Graedel TE (2002) Industrial ecology: goals and definitions. In: Ayres RU, Ayres LW (eds) A handbook of industrial ecology. Edward Elgar Publishing Inc, Cheltenham, pp 3–15

Litke DW (1999) Review of phosphorus control measures in the United States and their effects on water quality. Water-Resources Investigations Report 99–4007. USGS, Denver

Liu Y, Villalba G, Ayres RU, Schroder H (2008) Global phosphorus flows and environment impacts from a consumption perspective. J Ind Ecol 12(2):229–247

Löffler H (2013) Personal communication on biomass increase in Lake Constance after 2002, 1 Aug 2013. Langenargen

LUBW (2013) Total phosphorus and phytoplacton biomass in the Lake Constance, updating of Figure 3 of Müller H., Lake Constance—a model for integrated lake restoration with international cooperation, Water Science and Technology, vol 46 (6–7). Landesanstalt für Umwelt, Messungen und Naturschutz Baden-Württemberg, Langenargen, pp 93–98

Ma WQ, Ma L, Li JH, Wang FH, Sisak I, Zhang FS (2011) Phosphorus flows and use efficiencies in production and consumption of wheat, rice, and maize in China. Chemosphere 84(6):814–821. doi:10.1016/j.chemosphere.2011.04.055

MacDonald GK, Bennett EM, Potter PA, Ramankutty N (2011a) Agronomic phosphorus imbalances across the world's croplands. Proc Nat Acad Sci USA 108(7):3086–3091

MacDonald GK, Bennett EM, Potter PA, Ramankutty N (2011b) Agronomic phosphorus imbalances across the world's croplands. Supporting information. Proc Nat Acad Sci USA 108(7):1–9

MacDonald JM, Ribaudo MO, Livingston MJ, Beckman J, Huang W (2009) Manure use for fertilizer and for energy. Report to Congress. USDA

Marx RE (2008) Uncovering the cause of "Phossy Jaw" Circa 1858 to 1906: Oral and maxilofacial surgery closed case files-case closed. J Oral Maxillofac Surg 66(11):2356–2363. doi:10.1016/j.joms.2007.11.006

Matsubae K, Kajiyama J, Hiraki T, Nagasaka T (2011) Virtual phosphorus ore requirement of Japanese economy. Chemosphere 84(6):767–772

Matsubae-Yokoyama K, Kubo H, Nakajima K, Nagasaka T (2009) A material flow analysis of phosphorus in Japan. The iron and steel industry as a major phosphorus source. J Ind Ecol 13(5):687–705

McDonough W, Braungart M, Anastas PT, Zimmerman JB (2003) Applying the principles of green engineering to cradle-to-cradle design. Environ Sci Technol 37(23):434A–441A. doi:10.1021/es0326322

McNeill JR, Winiwarter A (2004) Breaking the sod: humankind, history, and soil. Science 304(5677):1627–1629. doi:10.1126/science.1099893

Mehlum H, Moene K, Torvik R (2006) Institutions and the resource curse. Econ J 116(508):1–20. doi:10.1111/j.1468-0297.2006.01045.x

Merton RK (1938) Science, technology and society in the seventeenth century England. Osiris 4:360–632

Metzger MJ, Schroter D, Leemans R, Cramer W (2008) A spatially explicit and quantitative vulnerability assessment of ecosystem service change in Europe. Reg Environ Change 8(3):91–107. doi:10.1007/s10113-008-0044-x

Mew M (2013) Personal communication, STPP production for detergents, 27 July 2013

Mihelcic JR, Fry LM, Shaw R (2011) Global potential of phosphorus recovery from human urine and feces. Chemosphere 84(6):832–839. doi:10.1016/j.chemosphere.2011.02.046

Minsch J, Feindt P-H, Meister H-P, Schneidewind U, Schulz T (1998) Institutionelle Reformen für eine Politik der Nachhaltigkeit. Springer, Berlin

Mitchell D (2008) A note on rising food prices. World Bank Policy Research Working Paper Series, vol 4682. World Bank—Development Economics Group (DEC), Washington

Mojabi SM, Feizi F, Navazi A, Ghourchi M (2010) Environmental impact of white phosphorus weapons on urban areas. 2010 International conference on environmental engineering and applications (ICEEA), pp 112–116. doi:10.1109/iceea.2010.5596102

Molina M, Aburto F, Calderon R, Cazanga M, Escudey M (2009) Trace element composition of selected fertilizers used in Chile: phosphorus fertilizers as a source of long-term soil contamination. Soil Sediment Contam 18(4):497–511. doi:10.1080/15320380902962320

Moss DA (1994) Kindling a flame under federalism—progressive reformers, corporate elites, and the phosphorus match campaign of 1909–1912. Bus Hist Rev 68(2):244–275. doi:10.2307/3117443

Moyle PR, Piper DZ (2004) Western phosphate field—depositional and economic deposit models. In: Hein JR (ed) Life cycle of the phosphoria formation: from deposition to post-mining environment. Handbook of Exploration and Environmental Geochemistry, vol 8. Elsevier, Amsterdam, pp 575–598

Müller H (2002) Lake Constance—a model for integrated lake restoration with international cooperation. Water Sci Technol 46(6–7):93–98

National Research Council (2008) Minerals, Critical Minerals, and the U.S. Economy. National Academies Press, Washington

Nearing MA, Pruski FF, O'Neal MR (2004) Expected climate change impacts on soil erosion rates: a review. J Soil Water Conserv 59(1):43–50

Nellemann C, MacDevette M, Manders T, Eickhout B, Svihus B, Prins AG, Kaltenborn BP (2009) The environmental food crisis—The environment's role in averting future food crises. A UNEP rapid response assessment. United Nations Environment Programme, GRID-Arendal, Nairobi

Nicholson FA, Jones KC, Johnston AE (1994) Effect of phosphate fertilizers and atmospheric deposition on long-term changes in the cadmium content of soils and crops. Environ Sci Technol 28(12):2170–2175. doi:10.1021/es00061a027

Novotny V, Imhoff KR, Otthoff M, Krenkel PA (1989) Handbook of urban drainage and wastewater. Wiley, New York

Nziguheba G, Smolders E (2008) Inputs of trace elements in agricultural soils via phosphate fertilizers in European countries. Sci Total Environ 390(1):53–57. doi:10.1016/j.scitotenv.2007.09.031

OED (2012) Online Etymology Dictionary. Douglas Harper. http://www.etymonline.com/index.phpp?allowed_in_frame=0&search=Phosphorus&searchmode=none

Oenema O, van Liere L, Schoumans O (2005) Effects of lowering nitrogen and phosphorus surpluses in agriculture on the quality of groundwater and surface water in the Netherlands. J Hydrol 304(1–4):289–301. doi:10.1016/j.jhydrol.2004.07.044

Oertli JJ (2008) Fertlizers, inorganic. In: Chesworth W (ed) Encyclopedia of soil science. Springer, Dordrecht, pp 247–248

Ott C, Rechberger H (2012) The European phosphorus balance. Resour Conserv Recycl 60:159–172. doi:10.1016/j.resconrec.2011.12.007

Ott H (2012) Fertilizer markets and their interplay with commodity and food prices. European Commission, Joint Research Centre, Institute for Prospective Technological Studies, Sevilla

Paris Q (1992) The return of von Liebig law of the minimum. Agron J 84(6):1040–1046

Park CM, Sohn HJ (2007) Black phosphorus and its composite for lithium rechargeable batteries. Adv Mater 19(18):2465–2468. doi:10.1002/adma.200602592

Pathak H, Mohanty S, Jain N, Bhatia A (2010) Nitrogen, phosphorus, and potassium budgets in Indian agriculture. Nutr Cycl Agroecosyst 86(3):287–299. doi:10.1007/s10705-009-9292-5

Paustenbach DJ (ed) (2002) Human and ecological risk assessment. Theory and practice. Wiley, New York

Petroianu GA (2010) History of organophosphate synthesis: the very early days. Die Pharm Int J Pharm Sci 65(4):306–311

Pimentel D, Whitecraft M, Scott ZR, Zhao L, Satkiewicz P, Scott TJ, Phillips J, Szimak D, Singh G, Gonzalez DO, Moe TL (2010) Will limited land, water, and energy control human population numbers in the future? Human Ecol 38(5):599–661

Polprasert C (2007) Organic waste recycling—technology and management. IWA Publishing, London

Potter P, Ramankutty N, Bennett EM, Donner SD (2010) Characterizing the spatial patterns of global fertilizer application and manure production. Earth Interact 14(2):1–22. doi:210.1175/2009ei288.1

Prud'homme M (2010) World phosphate rock flows, losses and uses. Paper presented at the British Sulphur Events Phosphates, Brussels, 22–24 March 2010

Prud'homme M (2013) Personal communication. Difference between IFA and IFDC data on losses. Paris, 11 Aug 2013

Quazi A, Islam R (2008) The reuse of human excreta in Bangladesh. In: Beyond construction: use by all—a collection of case studies from sanitation and hygiene promotion practitioners in South Asia. IRC International Water and Sanitation Centre (The Netherlands). London, Delft

Ragnarsdottir KV, Sverdrup HU, Koca D (2011) Challenging the planetary boundaries I: Basic principles of an integrated model for phosphorous supply dynamics and global population size. Appl Geochem 26:S303–S306. doi:10.1016/j.apgeochem.2011.03.088

Regh W (1996) Translator's introduction. In: Habermas J (ed) Contributions to a discourse theory of law and democracy. MIT Press, Cambridge

Reller A (2011) Criticality of metal resources for functional materials used in electronics and microelectronics. Phys Status Solid-Rapid Res Lett 5(9):309–311. doi:10.1002/pssr.201105126

Reller A, Zepf V, Achzet B (2013) The importance of rare metals for emerging technologies. In: Angrick M, Burger A, Lehmann H (eds) Factor X, re-source-designing the recycling society. Springer, Dordrecht, pp 203–220

Richmond L, Stevenson J, Turton A (eds) (2003) The pharmaceutical industry. A guide to historical records. Ashgate, Burlington

Roberts TL (in print) Cadmium and phosphorous fertilizers: the issues and the science. Procedia engineering (2nd international symposium on innovation and technology in the phosphate industry [SYMPHOS 2013])

Rockström J, Steffen W, Noone K, Persson A, Chapin FS, III, Lambin E, Lenton TM, Scheffer M, Folke C, Schellnhuber HJ, Nykvist B, de Wit CA, Hughes T, van der Leeuw S, Rodhe H, Sorlin S, Snyder PK, Costanza R, Svedin U, Falkenmark M, Karlberg L, Corell RW, Fabry VJ, Hansen J, Walker B, Liverman D, Richardson K, Crutzen P, Foley J (2009) Planetary boundaries: exploring the safe operating space for humanity. Ecol Soc 14 (2). http://www.cabdirect.org/abstracts/20103063016.html?freeview=true

Roosevelt FD (1938) Message to congress on phosphates for soil fertility, 20 May 1938. In: G Peters, JT Woolley, The American Presidency Project. http://www.presidency.ucsb.edu/ws/?pid=15643

Rosegrant MW, Ringler C, Msangi S (2008) International model for policy analysis of agricultural commodities and trade (IMPACT): model description. IFPRI, Washington

Roy AH, Hellums D, Scholz RW , Beaver C (2014) Fertilizers change(d) the world. In: Scholz RW, Roy AH, Brand FS, Hellums DT, Ulrich AE (eds) Sustainable phosphorus management: a global transdisciplinary roadmap. Springer, Berlin, pp 114–117

Ruddy BC, Lorenz DL, Mueller DK, Norton GA, Leahy PP (2006) Country level estimates of nutrient inputs to the land surface of the conterminous United States, 1982–2001. U.S. Department of the Interior & U.S. Geological Survey, Reston, VA

Rustad JR (2012) Peak nothing: recent trends in mineral resource production. Environ Sci Technol 46:1903–1906

Ruttenberg KC (2003) The global phosphorus cycle. Biogeochemistry 8:585–643

Sanchez PA, Swaminathan MS (2005) Hunger in Africa: the link between unhealthy people and unhealthy soils. Lancet 365(9457):442–444

Sanders DR, Irwin SH (2010) A speculative bubble in commodity futures prices? Cross-sectional evidence. Agric Econ 41(1):25–32. doi:10.1111/j.1574-0862.2009.00422.x

Sattari SZ (2013) Global P use efficiency, figure produced for this paper. 13 June 2013

Sattari SZ, Bouwman AF, Giller KE, van Ittersum MK (2012) Residual soil phosphorus as the missing piece in the global phosphorus crisis puzzle. Proc Natl Acad Sci USA 109(16):6348–6353. doi:10.1073/pnas.1113675109

Schipanski ME, Bennett EM (2012) The influence of agricultural trade and livestock production on the global phosphorus cycle. Ecosystems 15(2):256–268. doi:10.1007/s10021-011-9507-x

Schipper W (2013) Personal communication. 9 July 2013

Schlesinger P (1991) Biogeochemistry. An analysis of global change. Academic Press, San Diego

Schlezinger DR, Howes BL (2000) Organic phosphorus and elemental ratios as indicators of prehistoric human occupation. J Archaeol Sci 27(6):479–492. doi:10.1006/jasc.1999.0464

Schnee R, Stevens HC, Vermeulen M (2014) Phosphorus in the diet and human health. In: Scholz RW, Roy AH, Brand FS, Hellums DT, Ulrich AE (eds) Sustainable phosphorus management: a global transdisciplinary roadmap. Springer, Berlin, pp 232–236

Schnug E, Haneklaus S, Schnier C, Scholten LC (1996) Issues of natural radioactivity in phosphates. Commun Soil Sci Plant Anal 27(3–4):829–841. doi:10.1080/00103629609369600

Schnug E, Katz S, Stöven K, Godlinski F (2011) Die Nutzung von Schlachtnebenproducten als Dünger. Paper presented at the Die (Wieder-)Nutzung von Schlachtnebenprodukten, Tierärztliche Hochschule Hannover

Scholz RW (1987) Cognitive strategies in stochastic thinking. Reidel, Dordrecht

Scholz RW (2000) Mutual learning as a basic principle of transdisciplinarity. In: Scholz RW, Häberli R, Bill A, Welti M (eds) Transdisciplinarity: Joint problem-solving among science, technology and society. Workbook II: Mutual learning sessions. Proceedings of the international transdisciplinarity 2000 conference. Haffmans Sachbuch, Zürich, pp 13–17

Scholz RW (2011a) Environmental literacy in science and society: from knowledge to decisions. Cambridge University Press, Cambridge

Scholz RW (2011b) The need for global governance of ecosystem services: a human-environment systems perspective on biofuel production. In: Koellner T (ed) Ecosystem services and global trade of natural resources Routledge, Abingdon, pp 57–80

Scholz RW, Binder CR (2004) Principles of human-environment systems (HES) research. In: Pahl-Wostl C, Schmidt S, Rizzoli AE, Jakeman AJ (eds) Complexity and integrated resources management transactions of the 2nd biennial meeting of the international environmental modelling and software society, vol 2. Zentrum für Umweltkommunikation (ZUK), Osnabrück, pp 791–796

Scholz RW, Blumer YB, Brand FS (2012) Risk, vulnerability, robustness, and resilience from a decision-theoretic perspective. J Risk Res 15(3):313–330. doi:10.1080/13669877.2011.634522

Scholz RW, Lang DJ, Wiek A, Walter AI, Stauffacher M (2006) Transdisciplinary case studies as a means of sustainability learning: historical framework and theory. Int J Sustain High Educ 7(3):226–251

Scholz RW, Roy AH, Brand FS, Hellums DT, Ulrich AE (eds) (2014) Sustainable phosphorus management: a global transdisciplinary roadmap. Springer, Berlin

Scholz RW, Schmitt H-J, Vollmer W, Vogel A, Neisel F (1990) Zur Abschätzung des gesundheitlichen Risikos kadmiumbelasteter Hausgärten. [Assessing the health risks from cadmium-contaminated residential gardens]. Das öffentliche Gesundheitswesen 52:161–167

Scholz RW, Stauffacher M (2007) Managing transition in clusters: area development negotiations as a tool for sustaining traditional industries in a Swiss prealpine region. Environ Plann A 39(10):2518–2539

Scholz RW, Tietje O (2002) Embedded case study methods: integrating quantitative and qualitative knowledge. Sage, Thousand Oaks

Scholz RW, Ulrich AE, Eilittä M, Roy AH (2013) Sustainable use of phosphorus: a finite resource. Sci Total Environ 461–462:799–803

Scholz RW, Wellmer F-W (2013) Approaching a dynamic view on the availability of mineral resources: what we may learn from the case of phosphorus? Global Environ Change 23:11–27

Scholz RW, Wiek A (2005) Operational eco-efficiency: comparing firms' environmental investments in different domains of operation. J Ind Ecol 9(4):155–170

Sharpley A, Kleinman P, Weld J (2004) Assessment of best management practices to minimise the runoff of manure-borne phosphorus in the United States. New Zealand J Agric Res 47(4):461–477

Sharpley AN, Weld JL, Beegle DB, Kleinman PJA, Gburek WJ, Moore PA, Mullins G (2003) Development of phosphorus indices for nutrient management planning strategies in the United States. J Soil Water Conserv 58(3):137–152

Sharrock P, Fiallo M, Nzihou A, Chkir M (2009) Hazardous animal waste carcasses transformation into slow release fertilizers. J Hazard Mater 167(1–3):119–123. doi:10.1016/j.jhazmat.2008.12.090

Sheldrick WF, Lingard J (2004) The use of nutrient audits to determine nutrient balances in Africa. Food Policy 29(1):61–98. doi:10.1016/j.foodpol.2004.01.004

Shinh A (2012) The outlook for industrial & food phosphates. Paper presented at the Phosphates 12, El Jadida

Sims JT, Simard RR, Joern BC (1998) Phosphorus loss in agricultural drainage: historical perspective and current research. J Environ Qual 27(2):277–293

Smil V (1999) Crop residues: agriculture's largest harvest—Crop residues incorporate more than half of the world agricultural phytomass. Bioscience 49(4):299–308. doi:10.2307/1313613

Smil V (2000) Phosphorus in the environment: natural flows and human interferences. Annu Rev Energy Env 25:53–88

Smil V (2004) Enriching the earth: Fritz Haber, Carl Bosch, and the transformation of world food production. The MIT Press, Cambridge

Smith P, Martino D, Cai Z, Gwary D, Janzen H, Kumar P, McCarl B, Ogle S, O'Mara F, Rice C, Scholes B, Sirotenko O, Howden M, McAllister T, Pan G, Romanenkov V, Schneider U, Towprayoon S, Wattenbach M, Smith J (2008) Greenhouse gas mitigation in agriculture. Philos Trans R Soc B-Biol Sci 363(1492):789–813. doi:10.1098/rstb.2007.2184

Smith SR (1995) Agricultural recycling of sewage sludge and the environment. Oxford University Press, Oxford

Stumm W, Stumm-Zollinger E (1972) The role of phosphorus in eutrophication. In: Mitchell RC (ed) Water pollution microbiology. Wiley, New York, pp 11–42

Sutton MA, Bleeker A, Howard CM, Bekunda M, Grizzetti B, de Vries W, van Grinsven HJM, Abrol YP, Adhya TK, Billen G, Davidson EA, Datta A, Diaz R, Erisman JW, Liu XJ, Oenema O, Palm C, Raghuram N, Reis S, Scholz RW, Sims T, Westhoek H, Zhang FS (2013) Our Nutrient World. The challenge to produce more food and energy with less pollution. Centre for Ecology and Hydrology (CEH) and the United Nations Environment Program (UNEP), Edinburgh, Nairobi

Sutton MA, Bleeker A, Howard CM, Erisman JW, Abrol YP, Bekunda M, Datta A, Davidson E, de Vries W, Oenema O, Zhang FS, from: ic, Adhya TK, Billen G, Bustamante M, Chen D, Diaz R, Galloway JN, Garnier J, Greenwood S, Grizzetti B, Kilaparti R, Liu XJ, Palm C, Plocq, Fichelet V, Raghuram N, Reis S, Roy A, Sachdev M, Sanders K, Scholz RW, Sims T, Westhoek H, Yan XY, Zhang Y (2012) Our nutrient world The challenge to produce more food & energy with less pollution. In: Centre for ecology and hydrology on behalf of the global partnership on nutrient management (GPNM) and the International Nitrogen Initiative (INI) (ed) Key Messages for Rio+20. Falmouth

Syers JK, Johnston AE, Curtin D (2008) Efficiency of soil and fertilizer phosphorus use: reconciling changing concepts of soil phosphorus behaviour with agronomic FAO, Rome

Tanner EVJ, Kapos V, Franco W (1992) Nitrogen and phosphorus fertilization effects on Venezuelan montane forest trunk growth and litterfall. Ecology 73(1):78–86. doi:10.2307/1938722

The World Bank (2012) Fertilizer consumption (kilograms per hectare of arable land). The World Bank. http://data.worldbank.org/indicator/AG.CON.FERT.ZS/countries?display=default. Accessed 7 Dec 2012

Thompson Klein J, Grossenbacher-Mansuy W, Häberli R, Bill A, Scholz RW, Welti M (eds) (2001) Transdisciplinarity: joint problem solving among science, technology, and society. An effective way for managing complexity. Birkhäuser, Basel

Tillman D, Cassmann KG, Matson P, Naylor R, Polansky S (2002) Agricultural sustainability and intense production practices. Nature 418(6898):671–677

Tilman D, Lehman C (2001) Human-caused environmental change: impacts on plant diversity and evolution. Proc Natl Acad Sci USA 98(10):5433–5440. doi:10.1073/pnas.091093198

Tirado R, Allsopp M (2012) Phosphorus in agriculture. Greenpeace, Amsterdam

Udo de Haes HA, Van der Voet E, Kleijn R (1997) Substance flow analysis (SFA), an analytical tool for integrated chain management. In: Bringezu S, Fischer-Kowalski M, Kleijn R, Viveka P (eds) Regional and national material flow accounting: From paradigm to practice of sustainability. Proceedings of the ConAccount workshop 21–23 January, 1997 in Leiden, The Netherlands. Wuppertal Institute for Climate, Environment and Energy, Wuppertal

Ulex GL (1845) On struvite, a new mineral. Memoirs and proceedings of the chemical society CLXIII:106–110

Ulrich AE (2011) A lake of opportunity. Rethinking phosphorus pollution and resource scarcity. In: Klopfer N, Mauch C (eds) Big country, big issues: Canada's environment, culture and history. LMU Munich, Rachel Carson Center for Environment and Society, Munich, pp 86–100

UN (2006) United Nations conference on trade and development. The emerging biofuel market. Regulatory, trade and development implications. United Nations, New York

UN (2011) Environmental Indicators. United Nations Statistics Division. http://unstats.un.org/unsd/environment/wastewater.htm. Accessed 1 March 2013

UN (2013) World population prospects. The 2012 revision. Key findings and advance tables. United Nations, New York

UNEP and IFA (2001) Environmental aspects of phosphate and potash mining. United Nations Environment 1 Programme (UNEP) and International Fertilizer Industry Association (IFA), Paris

USCB (2009) Data from U.S. Census Bureau, International Database, update 2009, retrieved February 12, 2011

USGS (2000) Mineral commodity summary 2000. Washington

USGS (2010) Mineral commodity summary 2010. Washington

USGS (2012) Mineral commodity summary 2012. Washington

USGS (2013a) Geology research and information. USGS. http://geology.usgs.gov/. Accessed 15 April 2013

USGS (2013b) Mineral commodity summary 2013. Washington

van der Molen DT, Pot R, Evers CHM, van Nieuwerburgh LLJ (eds) (2012) Referenties en Maatlatten voor natuurlijke Watertypen voor de Kaderrichtlijn Water 2015–2021. STOWA Rapport, Amersford

van Kauwenbergh SJ (2010) World phosphate rock reserves and resources. IFDC, Muscle Shoals

van Otterdijk R, Meybeck A (2011) Global food losses and food waste. FAO, Rome

Venterink HO (2011) Does phosphorus limitation promote species-rich plant communities? Plant Soil 345(1–2):1–9. doi:10.1007/s11104-011-0796-9

VFRC (2012) Global research to nourish the world. A blueprint for food security. Virtual Fertilizer Reserach Center, Washington

Villalba G, Liu Y, Schroder H, Ayres RU (2008) Global phosphorus flows in the industrial economy from a production perspective. J Ind Ecol 12(4):557–569

Vogelsang P (1950) New techniques in seed pelleting. In: American society of sugar of beet technologists (ed) Sixth general meeting of the american society of sugar beet technologists, 6–9 Feb 1950, Detroit, MI, 11 Feb 2012 1950. pp 75–78

Vollenweider RA (1970) Scientific fundamentals of the eutrophication of lakes and flowing waters, with particular reference to nitrogen and phosphorus as factors in eutrophication. OECD, Paris

von Liebig J (1840) Die organische Chemie in ihrer Anwendung auf Agricultur und Physiologie [The organic chemistry in its application on agriculture and physiology]. Friedrich von Vieweg, Braunschweig

von Pier JC (2006) History's great untold stories: obscure and fascinating accounts with important lessons for the world. Murdoch, Millers Point

Vu MQ, Le QB, Scholz RW, Vlek PL (2012) Detecting geographic hotspots of human-induced land degradation in Vietnam and characterization of their social-ecological types. In: IEEE (ed) IEEE international geoscience and remote sensing symposium, Munich, 22–27 July 2012

Wagner H (1999) Stoffmengenflüsse und Energiebedarf bei der Gewinnung ausgewählter mineralischer Roshstoffe. teilstudie Phosphat, vol SH 5. Wirtschaftsgeologie, Bereichte zur Roshstofgfwirtschaft. Bundesanstalt für Geowissenschaften und Rohstoffe und Staatliche Geologische Dienste in der Bundesrepublik Deutschland, Hannover

Walther B, Schmid A (2008) Vorsicht vor überhöhtem Phosphorkonsum. Swissmilk. http://www.swissmilk.ch/de/services/ernaehrungs-fachleute/fachbibliothek/-dl-/fileadmin/filemount/studie-vorsich-vor-ueberhoehtem-phosphorkonsum-ernaehrungswissenschaft-de.pdf

Wang F, Sims JT, Ma L, Ma W, Dou Z, Zhang F (2011) The phosphorus footprint of China's food Chain: Implications for food security, natural resource management, and environmental quality. J Environ Qual 40(4):1081–1089. doi:10.2134/jeq2010.0444

Ward J (2008) Peak phosphorus: quoted reserves vs. production history. Energy Bulletin. http://www.resilience.org/stories/2008-08-26/peak-phosphorus-quoted-reserves-vs-production-hist

Weber O, Delince J, Duan Y, Maene L, McDaniels T, Mew M, Schneidewid U, Steiner G (2014) Trade and finance as cross-cutting issues in the global phosphate and fertilizer market. In: Scholz RW, Roy AH, Brand FS, Hellums DT, Ulrich AE (eds) Sustainable phosphorus management: a global transdisciplinary roadmap. Springer, Berlin, pp 275–294

Webster's (1913) basic slag. Webster's Revised Unabridged Dictionary

Webster's (2002a) "loss". Third New International Dictionary, Unabridged. Merriam-Webster

Webster's (2002b) "sink". Third New International Dictionary, Unabridged. Merriam-Webster

Wetzel RG (1983) Limnology, 2nd edn. Saunders College Publishing, Philadelphia

Wilkinson TJ (1982) The definition of ancient manured zones by means of extensive sherd-sampling techniques. J Field Archeology 9(3):1982

Wisniak J (2005) Phosphorus-from discovery to commodity. Indian J Chem Technol 12(1):108–122

Woltering DM (2004) Health risk assessment for metals in inorganic fertilizers: development and use in risk management. In: Hall WJ, Robarge WP (eds) In environmental impact of fertilizer on soil and water. ACS Symposium Series, vol 872. American Chemical Society, Washington, pp 124–147

Wright BD (2011) The economics of grain price volatility. Appl Econ Perspect Policy 33(1): 32–58. doi:10.1093/aepp/ppq033

YARA (2013) Yara International ASA second quarter results 2013. YARA International ASA. http://www.yara.com/doc/86912_20132Q13Webpresentation.pdf . 19 July 2013

Zhang NQ, Wang MH, Wang N (2002) Precision agriculture—a worldwide overview. Comput Electron Agriculture 36(2–3):113–132. doi:10.1016/s0168-1699(02)00096-0

Zhang WF, Ma WQ, Ji YX, Fan MS, Oenema O, Zhang FS (2008) Efficiency, economics, and environmental implications of phosphorus resource use and the fertilizer industry in China. Nutr Cycl Agroecosyst 80(2):131–144. doi:10.1007/s10705-007-9126-2

Zhang YH, Caupert J (2012) Survey of mycotoxins in U.S. distiller's dried grains with solubles from, 2009 to 2011. J Agricultural Food Chem 60(2):539–543. doi:10.1021/jf203429f

Zheng L (2013) Import and export. Phosphorus Industry China Monthly Report 3(1)

Appendix: Spotlight 1

Fertilizers Change(d) the World

Amit H. Roy, Deborah T. Hellums, Roland W. Scholz, and Clyde Beaver

Due to significant advances in agriculture and medicine in the last century, both food production and global population have increased dramatically. The last 3 years have seen particularly significant benchmarks, with Africa reaching one billion people in 2009 and the world population reaching seven billion in 2011. Looking to the future, FAO (High Level Expert Forum, 2009) and other experts have agreed that the population is likely to surpass nine billion by 2050.

The question that remains in the face of that prediction is whether food production can keep pace with population growth to provide food security for all. More effective use of agricultural inputs—improved seeds, crop protection products and chemical and organic fertilizers can tip the scales in that production goal (Mueller et al. 2012).

The argument that chemical fertilizers have dramatically increased cereal production over the last 50 years seems to be irrefutable. Also acknowledged is that these fertilizers help save the lives of over 3.5 billion people who otherwise would starve given lower agricultural production (Smil 1999; Wolfe 2001; Hager 2008). In 1961—effectively the dawn of modern fertilizer use—global cereal production stood at 877 Mt. By 2010, annual cereal production had increased to 2.4 Gt (FAOSTAT/IFDC data 2012).

A. H. Roy · D. T. Hellums · C. Beaver
International Fertilizer Development Center (IFDC) P.O. Box 2040, Muscle Shoals,
AL 35662, USA
email: aroy@ifdc.org

D. T. Hellums
email: dhellums@ifdc.org

C. Beaver
email: cbeaver@ifdc.org

R. W. Scholz
Fraunhofer Project Group Materials Recycling and Resource Strategies IWKS,
Brentanostrasse 2, 63755 Alzenau, Germany
email: roland.scholz@isc.fraunhofer.de

ETH Zürich, Natural and Social Science Interface (NSSI), Universitaetsstrasse 22,
CHN J74.2, 8092 Zürich, Switzerland

This 174 % increase over the past half century was clearly not serendipitous. From 1970 through 2011, global nitrogen, phosphorus and potassium (NPK) mineral fertilizer consumption increased by 154 %, from 69 Mt to 175 Mt—a strikingly clear correlation between increased production and broader use of NPK fertilizers (FAOSTAT/IFDC data 2012). Perhaps the greatest evidence for the effectiveness of fertilizer in intensifying food production can be found in South Asia, where progressive use of fertilizer on roughly the same area of land over the past 50 years has produced a 165 % increase in output (FAOSTAT/IFDC data 2012). While there are numerous examples of excessive and inefficient fertilizer use (typically above recommended rates of application) resulting in negative environmental impacts, the larger issue is that low productivity (in large part due to underuse of fertilizers) is resulting in millions of people suffering with malnutrition. Over the same period, Africa, which is plagued by inherently nutrient-deficient soils and the lack of fertilizer use (averaging only 8 kg input of NPK fertilizer per hectare [Abuja Declaration 2006]), experienced production increases of only 60 %—and not through crop intensification utilizing modern agro-inputs, but by extending the area of land cultivated while almost irreparably mining the soils of their remaining nutrients (FAOSTAT/IFDC data 2012).

Among the primary nutrients, phosphorus deficiency in the world's soils stands out as a major constraint to food crop production in low-input systems such as those in the sub-humid and semi-arid regions of sub-Saharan Africa. Large areas of the developing world's soils are chronically deficient in phosphorus; legumes, a key to low-input agriculture because of their capability to produce plant available nitrogen through biological nitrogen fixation (BFN), are particularly sensitive to phosphorus deficiency (Parish 1993). Unless phosphorus fertilizers are used in these areas, even the best-managed nutrient recycling system will not achieve the minimum soil phosphorus levels required for good yields.

However, the judicious use of our mineral and chemical nutrient resources alone will not allay future agricultural production concerns. In fact, fertilizers alone will not solve the 2050 dilemma. A more balanced approach to agricultural production that focuses on soil nutrient-supplying capacity, while simultaneously maintaining or improving overall soil quality must rise to the top of production agendas. *Integrated soil fertility management* (ISFM), which includes the combined use of organic and inorganic (commercial fertilizers) nutrient inputs and soil amendments, can lead to sustainable nutrient management. This nutrient management approach along with improved germplasm and water management must become the production norm of the future in order to conserve soil and water resources, build soil fertility and improve water quality.

Even with widespread adoption of production techniques utilizing ISFM, demand for fertilizers will remain high in the coming years, but could also remain out of reach for many. In 2010, according to FAO (FAOSTAT 2012), global consumption of the major phosphate fertilizers (P_2O_5) was 45.4 Mt (equivalent to 19.8 Mt on a mineral P fertilizer basis), with the least developed countries, as a group, consuming only 1.5 % of that annual total. Clearly, a focused effort is required by all stakeholders to increase the production, availability and responsible use of phosphorus to advance global food security, particularly in the developing world. According to Sattari et al. (2012), mineral P demand by 2050 may range from 14.6 to 28 Mt annually—a range derived based on the anticipated combination of residual soil P, the supplementary use of manure and P recycling efforts. While this range considers the regional variations in historical P use and current soil P status, the anticipated global consumption of mineral P fertilizers in 2050 is projected to be 20.8 Mt, slightly more than current consumption rates. This estimate was derived based on the assumption that farmers worldwide will be applying best agricultural technologies and management practices.

In the same run-up to 2050, global NPK demand is estimated to be 223.1 Mt in 2030 and 324 Mt in 2050, and thus an increase of fertilizer use by 27 to 85 % (Drescher et al. 2011). This and similar estimates are based on current agricultural practices and may reflect the massive food production requirements at mid-century. However, this projected fertilizer requirement is likely to continue to be revised downward with the advent of more efficient fertilizer technologies and the widespread adoption of nutrient-supplying and resource-conserving approaches such as ISFM.

References

Abuja Declaration: Africa Fertilizer Summit of the African Union Ministers of Agriculture (2006) Africa Fertilizer Summit Proceedings IFDC Special Publication. SP-39, International Fertilizer Development Center

Drescher A, Glaser R, Clemens R, Nippes KR (2011) Demand for key nutrients (NPK) in the year 2050. Report for the European Commission Joint Research Centre, Brussels. 77p. Available at: http://eusoils.jrc.ec.europa.eu/projects/NPK

FAOSTAT/IFDC (2009) Food and agriculture organization of the United Nations (FAO). How to feed the world in 2050, High Level Expert Forum, Rome Available at http://cap2020.ieep.eu/2009/11/3/fao-forum

FAOSTAT/IFDC (2012) Food and agriculture organization of the United Nations (FAO). FAOSTAT database, Rome Available at http://faostat.fao.org

Hager J (2008) The alchemy of air. Harmony Books, New York

Mueller MD, Gerber JS, Johnston M, Ray DK, Ramankutty N, Foley JA (2012) Closing yield gaps through nutrient and water management. Nature 490:254–257

Parish D (1993) Agricultural productivity, sustainability, and fertilizer use. IFDC Paper Series P-18, International Fertilizer Development Center

Sattari S, Bouwman AF, Giller KE, van Ittersum MK (2012) Residual soil phosphorus as the missing piece in the global phosphorus crisis puzzle. PNAS 109(16):6348–6353

Smil V (1999) Long-range perspectives on inorganic fertilizers in global agriculture 1999 Travis P. Hignett Lecture 1 November 1999, International Fertilizer Development Center

Wolfe D (2001) Tales from the underground: a national history of subterranean life. Perseus Publishing, Cambridge

Appendix: Spotlight 2

A Novice's Guide to Transdisciplinarity

Roland W. Scholz and Quang Bao Le

Transdisciplinarity (td) is a key term of the Global TraPs project. All activities of the project on all three levels of the project are transdisciplinary processes: the 'Umbrella project', the Nodes of the P supply chain including the Trade and Finance Node, as well as the case studies which are launched to better define or to close the knowledge gaps on sustainably P management. In this brief, we (1) provide a brief definition of td, (2) outline one of the twenty-five td case studies that have been successfully conducted at ETH NSSI since 1993; and (3) provide a "model" for a brief description of a planned td case study in Vietnam.

What is Transdisciplinarity

Transdisciplinarity is a third mode of doing science complementing **disciplinarity** and **interdisciplinarity**. It was developed during the last two decades in Europe and is now well accepted in the European academic community.

Whereas **interdisciplinarity** means the integration of concepts and methods from different **disciplines, td** integrates additionally different epistemics (i. e., ways of knowing) from science/theory and practice/stakeholders. Td starts from the assumption that scientists and practitioners are experts of different kinds of knowledge where both sides may benefit from a mutual learning process. Thus, co-leadership among science and practice based on equal footing on all levels of the project (i. e., the umbrella project, the nodes and the case studies) are needed to assure that the interests and capacities of theory and practice are equally acknowledged.

R. W. Scholz
Fraunhofer Project Group Materials Recycling and Resource Strategies IWKS,
Brentanostrasse 2, 63755 Alzenau, Germany
email: roland.scholz@isc.fraunhofer.de

Q. B. Le · R. W. Scholz
ETH Zürich, Natural and Social Science Interface (NSSI), Universitaetsstrasse 22,
CHN J74.2, 8092 Zürich, Switzerland
email: Quang.le@env.ethz.ch

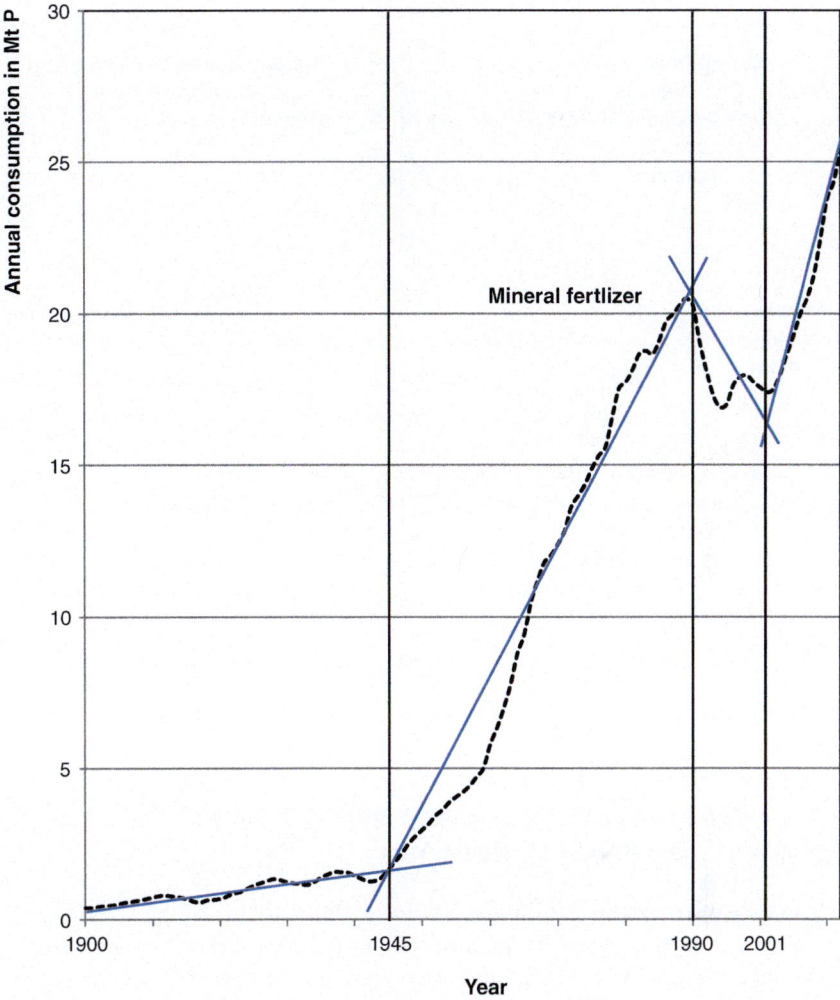

Fig. 28 Different trends of phosphorus use in different periods (*x*-axis: Mt P, data from USGS Mineral Commodity Summaries; the graph is generated by unweighted moving average statistics to smooth annual fluctuations using a five-year time window)

Transdisciplinary processes (td-processes) target the generation of knowledge for a sustainable transition of complex, societally relevant real-world problems.

Td-processes include joint (1) problem definition, (2) problem representation and (3) preparation for sustainable transitions (see Scholz 2000). In general, td-processes provide an improved problem understanding and robust orientations on policy options or business decisions for the practitioners. Experts from science and practice benefit by getting in-depth insight into the dynamics of complex systems and mechanisms of sustainable transitions.

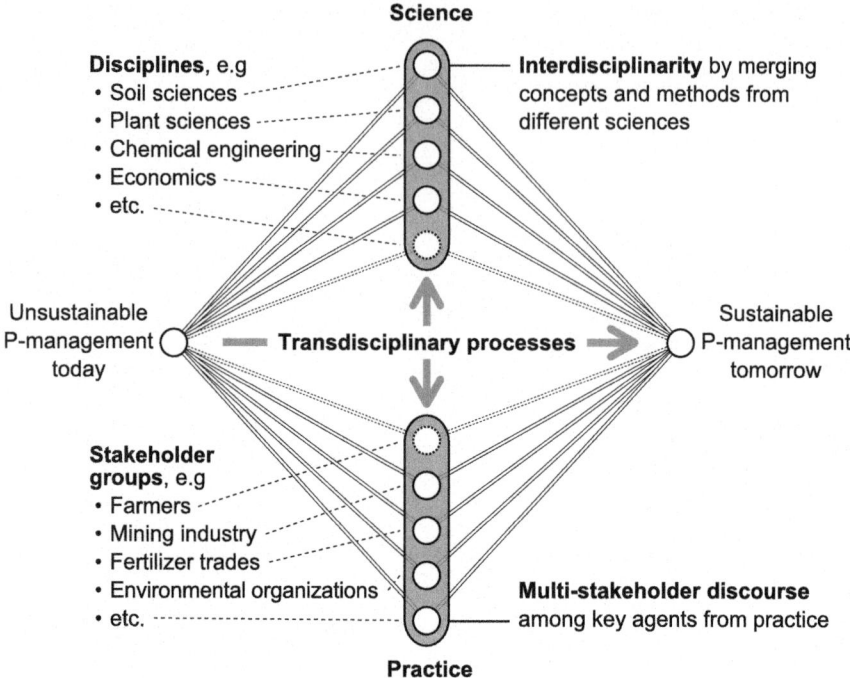

Fig. 29 Disciplines, interdisciplinarity, multi-stakeholder discourses, and transdisciplinarity

---

A Successful Example: "Sustainable Future of Traditional Industries" in a Rural Pre-Alpine Area

Building partnership[5]: Both, the president Hans Altherr of the small pre-alpine Swiss state Appenzell Ausserrhoden (AR), and ETH professor Roland W. Scholz, were interested in understanding mechanisms of sustaining traditional industries in rural regions. Jointly, they decided to run a transdisciplinary case study and to take *co-leadership* on equal footing for a td-process.

1. **Joint problem definition**: Key representatives (e. g., presidents of industry associations and unions as well as representatives of the communities) formed the steering board. A challenge was to negotiate and define the **guiding question**. It reads: *What are the prerequisites for a sustainable regional economy meeting environmental and*

---

[5] Please note that this step also should include a thorough actor analysis identifying "legitimized decision makers" who may become co-leaders of the case study and of the stakeholders who should be involved in the case study.

socioeconomic needs? Further, three industries, i. e., textile, dairy and sawmilling industry, were selected for in-depth understanding of key mechanisms of sustainable transitions. In addition a Knowledge Integration Group was built to identify communalities and specificities of the three industries.

2. **Preparing for sustainable transitions**: By means of a scientific method (i. e., formative scenario analysis), for each industry a *set of different business strategies* (including state, community, and multi-stakeholder activities) were constructed. These strategies were evaluated by the different stakeholder groups to gain insights into dissent and consent within and between them. Scientists analyzed these evaluations, compared them to a "data-based multi-criteria sustainability assessment," and discussed the results with key stakeholders and further interested people. For each industry meaningful business options as well as related latent conflicts (between companies, economic and environmental impacts) were identified. Based on this, a process of mutual understanding was moderated so that consensus could be formed on many issues. The Knowledge Integration Group integrated these results and—together with the head officials of AR—identified potential policy options for the state. The results were published in a book targeting practitioners at regional and national level (Scholz et al. 2003).

3. **Outcomes and follow-ups**: The knowledge generated in the process was used by the practitioners involved in their daily business and policy decisions. Based on the AR-study, the Swiss textile industry launched a study to utilize the favorite strategy from the td-process for new business models (Scholz and Kaufmann 2003). Various concrete projects such as a new wastewater treatment plant for the textile industry, new cooperatives for the dairy industry and (cantonal) forest management followed the study. The study allowed for robust scientific publications on sustainable regional wood flows (Binder et al. 2004), business strategies of traditional industries (Scholz and Stauffacher 2007) or the methodology of transdisciplinary case studies (Scholz et al. 2006)

References

Binder CR et al. (2004) Transition towards improved regional wood flows by integrating material flux analysis and agent analysis: the case of Appenzell Ausserrhoden, Switzerland. Ecol Econ 49(1):1–17

Scholz RW, Kaufmann D (2003) Zukunft der Schweizer Textilindustrie. Erkenntnisse einer gesamtschweizerischen Analyse aufbauend auf den Ergebnissen der ETH-UNS Fallstudie 2002, Appenzell Ausserrhoden—Umwelt Wirtschaft

Scholz RW et al. (eds) (2003) Appenzell Ausserrhoden Umwelt Wirtschaft Region. ETH-UNS Fallstudie 2002, Zürich: Rüegger und Pabst

Scholz RW et al. (2006) Transdisciplinary case studies as a means of sustainability learning: historical framework and theory. Int J Sustain Higher Educ 7(3):226–251

Scholz RW, Stauffacher M (2007) Managing transition in clusters: area development negotiations as a tool for sustaining traditional industries in a Swiss prealpine region. Environ Planning A 39(10):2518–2539

Scholz RW (2000) Mutual learning as a basic principle for transdisciplinarity. In: Scholz RW et al (eds) Transdisciplinarity: joint problem-solving among science, technology and society, Workbook II: Mutual learning sessions. Haffmans Sachbuch, Zürich, pp 13–17

Appendix: Spotlight 3

The Yen Chau—Hiep Hoa Case Study: Avoiding P Fertilizer Overuse and Underuse in Vietnamese Smallholder Systems

An Example of How a Transdisciplinary Case Study in the Use Node May be Developed

Quang B. Le and Roland W. Scholz

The problem

Globally, unsustainable P fertilizer management challenges for farmers fall primarily into two P use regimes. The first regime is representative of farmers engaged in intensified production to meet the global demand for food. These farmers often apply P fertilizer at higher than recommended rates in order to reduce risks that could limit production. However, if they fail to utilize best soil management practices significant P losses can result from surface water run-off and soil erosion. Included in this group are smallholder farmers engaged in intensified agricultural production of cereals, fruits and vegetables, who often produce two to three crops per year on the same land area. This overuse scenario often occurs in peri- urban agriculture where the smallholder farmers have good access to local traders and markets.

The second regime is characterized by subsistence smallholder farmers who may or may not have access to fertilizers, but cannot afford the inputs.

Q. B. Le · R. W. Scholz
ETH Zürich, Natural and Social Science Interface (NSSI), Universitaetsstrasse 22, CHN J74.2, 8092 Zürich, Switzerland
email: Quang.le@env.ethz.ch

R. W. Scholz
Fraunhofer Project Group Materials Recycling and Resource Strategies IWKS, Brentanostrasse 2, 63755 Alzenau, Germany
email: roland.scholz@isc.fraunhofer.de

Here P fertilizer is underused, leading to nutrient mining and soil degradation which exacerbates poverty. In both cases, viable options for economically and environmentally efficient P resource use and recycling in smallholder agro-ecosystems require special attention. Vietnam's smallholder systems in the Red River Delta (fertilizer-overuse, market-oriented) and in the Northwest Mountain Region (fertilizer-underuse, subsistence) will be used as example cases for contrasting two P use regimes.

1. *Building partnership and Td organization*

Science-practice co-leaders from the Province's People Committees in Son La and Bac Giang provinces, and researchers from science (ETH, University of Zürich etc.) are interested in understanding mechanisms of sustaining traditional industries in rural regions. Jointly, they will decide to run a Td case study and to take *co-leadership* on equal footing for a td-process. The co-leaders preside over the steering group.

Project groups—Reference groups: The scientific work will take place in project groups spanning the case facets (see session Case Faceting below). The project groups are counterbalanced on the case side with the "so-called" reference groups, which are the committees of stakeholders relevant for the respective case facet (Stauffacher et al. 2008). The reference group regularly meets their corresponding project group to discuss the results and subsequent steps of the work.

Steering group: The group consists of representatives of the scientific disciplines involved in the study topic (e. g., soil and crop scientists, environmental chemists, human-environment system scientists), as well as the representatives of the two provinces. During the problem definition process, representatives of a few (2–3) selected districts/communes will join the steering group. As the study progresses, the steering group will identify additional participants who should be involved in each phase of the project based on the nature of the work (Stauffacher et al. 2008). For this, it seems meaningful/necessary that both locations, i. e., Yen Chau and Hiep Hoa about 12–16 farmers make commitment to be involved in the study and the mutual learning process. These farmers will be key members of the reference groups.

2. *Joint problem definition*

The steering group members (which include the main stakeholder groups) will negotiate and define guiding question, goal and the case areas (system boundaries).

Guiding questions: As a result of science-practice discussion, examples of possible guiding questions could be:

Project year 2013:

What are science-based and society-relevant strategies for P resource use that help improve soil fertility, food productivity and profitability for Vietnamese smallholders of two contrasting P use regimes? What options/

pathways/means are available for the transition of current smallholders' P use to a sustainable use of P?

Goal: Based on these questions, the goal of the case study can be defined so as to provide (strategic) orientations for future development of smallholders regarding P use.

Case definition: The case study should allow to better understand "overuse" and "underuse" of P under certain constraints. The case's characteristics and contextual factors should allow some generalization for other cases (we are investigating cases for something of general interest). Based on reviewing the existing classification of world farming systems (Dixon et al. 2001), the global pattern of agronomic P balance (MacDonald et al. 2011), and national patterns of climate, soil, demography and land uses, the steering group—presumably interacting with regional case stakeholders—identifies case areas in the Hiep Hoa and Yen Chau districts. Characteristics of these areas are in Table 8. Based on extensive farm survey across the selected areas, a limited number of farms (about 6–8 farms/site) representing major farm types will be selected for further considerations.

Case faceting: The goal of faceting is the formulation of a research concept, which is written by the scientists in collaboration with practitioners. Together with the stakeholders, the involved scientists create a general model of smallholder farming system in the two districts with a focus on P use, which allow the application of relevant disciplinary fields and their theories. In order to reduce the complexity and to better analyze the farming practices a 'faceting' of the case should be done. Facets (which have to be discussed) could be: 'Crop-Livestock Production including P fertilizer use and flows', 'Household Decision', and 'Policy, Finance and Market'. Consequently, three corresponding project groups should be formed. For each case facet, P-use related scientific tasks (subprojects) will be identified. In the presented case, there may be an additional project group which focuses on integrating/synthesizing the results from the subprojects, i. e. the so-called "Integrated Assessment" group. It is expected that the case faceting will jointly identify a couple (common) disciplinary sub-tasks with particularly disciplinary foci, such as:

Crop-livestock production, P fertilizer use and flows

- Current state of P use and cycle in the study of smallholder systems,
- problems in P fertilizer use, P-cycle management with respect to sustaining soil fertility and crop/livestock production,
- potential alternatives for P use technology/practice and (on-farm) recycling
- household decisions,
- social-policy, economic, ecological factors that affect farmers' decision about nutrient use and management,

Table 8 Regional settings of the two study areas

Aspect	Hiep Hoa district (P fertilizer overuse)	Yen Chau district (P fertilizer underuse)
World regional farming system/climate zone (Dixon et al. 2001)	Lowland rice-based farming system in Eastern Asian tropical monsoon climate	Highland extensive mixed farming system in Eastern Asian tropical monsoon climate
	About 71 million ha, mainly located in flood-plains of South and Central East China, Korean peninsula and Southeast Asia	About 8 million ha, mainly located in mountains of Southeast Asia
Cultivation area	28.5 thousand ha	201 thousand ha
Agricultural population	216 thousand people (95 % of total population) (2008)	65 thousand people (95 % of total population) (2009)
Main soil (FAO-UNESCO) and land form	Plinthic Acrisols	Ferrasols, Acrisols
	River floodplain	Complex mountain
Key components of smallholder farming system	Crop: Paddy rice (80 % of total crop area), maize, beans, vegetables. Livestock: pig (high density), poultry, cattle. Aquaculture: fish ponds	Crop: Maize (70–60 %), paddy and upland rice, cassava, beans, vegetable, fruit trees. Livestock: pig, cattle (open raising, extensive care). Aquaculture: fish ponds
General livelihood strategy	Market-oriented. Both crop and livestock productions are important sources of cash income. Vegetable is increasingly grown to meet the increasing market demand	Subsistence. Maize, rice and cassava are important food crop. No/weak market links for some marketable crop (maize and fruits)
Existing fertilizer use and nutrient management	Intensive uses of inorganic fertilizers, combined with some manures. Nutrient loop between crop-livestock-fish in some households. About 80 % of animal manure is discharged to the environment	Compound NPK, urea and K fertilizers are used only for a very small share of cropland. Almost no P fertilizer for hillside crops. Manure is seldom used. Nutrient recycling or soil conservation practice is hardly observed
Prevalence of food insecurity and poverty	Medium	Very high (poverty hotspot in Vietnam)
Key problems	(1) lost yield if no or less use of fertilizer, (2) low fertilizer use efficiency, (3) high livelihood vulnerability to increase in fertilizer cost, (4) water pollution	(1) degraded soil and declining crop yield, (2) very low household income, (3) knowledge, cultural and labor constraints for nutrient recycling practices, (4) lack of access to fertilizer and food market, financial services

- interferences between farmer's decision-making and other important human agents at higher levels (e. g., provincial department of agriculture and rural development, rural credit agencies, traders).

Policy, finance and market

- Constraints in policy (e. g., subsidy), finance institution (e. g., rural loans/credit institution) and market (e. g., prices of farming inputs and outputs) with respect to smallholder's P uses,
- potential alternatives for improving these factors.

Integrated Assessment

- Integrated Assessment including conceptual and parameterized system model that integrates the above-mentioned facets,
- scenarios of soil fertility, food productivity & profitability versus P use strategies, evaluation of trade-offs.

3. *Joint problem representation*

System analysis: A special challenge of the td-process is to collaborate with the decision makers and the stakeholders in a way that the system model and what is focused can be understood by all key (practice) case agents. The system model should represent the smallholder agro-ecosystem in a way that all can understand the dynamics of soil fertility and food production in response to changes in P use and other related drivers (e. g., fertilizer subsidies, market prices). The system model will serve as a basis for the construction of case scenarios in the next steps. Moreover, the joint system model construction will result in a shared representation of the constructed case study. This shared constructive aspect should greatly enhance the mutual learning process.

Scenario construction: For each case area, through Td workshops, stakeholders will jointly identify a set of different alternative P use strategies that they would like to evaluate. By means of either the computerized system model or a scientific participatory method so-called "formative scenarios analysis" (Scholz and Tietje 2002), future scenarios of identified outcome variables corresponding to the alternative strategies will be constructed. The scenarios will also be presented in a verbal or visual form that can be understood by the stakeholders and all case agents involved.

4. *Assessment and preparing for sustainable transitions*

The above-mentioned strategies will be evaluated by referring to scientific data to gain insights into trade-offs, e. g., between costs and benefits or environmental impacts and economical cost driven by the alternative strategies. Further and complementary to that, participatory multi-criteria assessment (Scholz and Tietje 2002) will be used. Different stakeholder groups will evaluate the different scenarios. This will serve to identify tradeoffs between the stakeholder groups and between different aspects of p-use that may be improved. For each meaningful scenario, tradeoffs (between social, economic and environmental impacts; between different preference systems of stakeholder groups) will be identified. Based on this, a process of mutual understanding will be moderated and consensus can potentially be formed on many issues.

5. *Outcomes and follow-ups*

In the final Td workshops, stakeholders will discuss how the knowledge generated in the process should be used for different societal processes, such as farming practices, policy decisions, sustainability learning in higher education systems, framing of follow-up research activities. As one important follow-up, written products will be prepared both for practice partners (e. g. practice manuals, policy briefs) and scientists (articles in academic journals).

References

Dixon J, Gulliver A, Gibbon D (2001) Farming systems and poverty—improving farmers' livelihoods in a changing world. FAO, Rome and Washington D.C.

MacDonald GK, Bennett EM, Potter PA, Ramankutty N (2011) Agronomic phosphorus imbalances across the world's croplands. PNAS 108(7):3086–3091

Scholz RW, Tietje O (2002) Embedded case study methods: Integrating quantitative and qualitative knowledge. Sage Publications, Thousand Oaks

Stauffacher M, Flüeler T, Krütli P, Scholz RW (2008) Analytic and dynamic approach to collaborative landscape planning: a transdisciplinary case study in a Swiss pre-alpine region. Syst Prac Action Res 21(6):409–422

Chapter 2
Exploration: What Reserves and Resources?

David A. Vaccari, Michael Mew, Roland W. Scholz and Friedrich-Wilhelm Wellmer

Abstract The Exploration Node focuses on the search for assessment and quantification of phosphate reserves and resources in relation to the geopotential (i.e., the undiscovered reserves and resources). The Exploration Node encompasses all aspects of the predevelopment stages of phosphate deposits from initial discovery of deposits to the involved feasibility studies required to obtain funding for the development of a mine. The feasibility of producing phosphate rock (PR) can be broadly defined in terms of technical feasibility and economic feasibility. In order for potential ores to be classified as *reserves*, consideration must be given to issues of grade, quality, operating, and investment costs which include studies of the accessibility and availability of financing. Details about these considerations are often proprietary, making it difficult to publically assess the resource picture. Phosphorus is the eleventh most abundant element. P is essential for life and cannot be substituted by other elements in food production. The given knowledge

---

D. A. Vaccari (✉)
Stevens Institute of Technology, Castle Point on Hudson, Hoboken, NJ 07030-5991, USA
e-mail: dvaccari@stevens.edu

M. Mew
CRU International-Consultant, Polignac, Route de Fources, 32100 Condom, France
e-mail: michaelmew@wanadoo.fr

R. W. Scholz
Fraunhofer Project Group Materials Recycling and Resource Strategies IWKS,
Brentanostrasse 2D, 63755 Alzenau, Germany
e-mail: roland.scholz@isc.fraunhofer.de

ETH Zürich, Natural and Social Science Interface (NSSI), Universitaetsstrasse 22,
CHN J74.2, 8092 Zürich, Switzerland

F.-W. Wellmer
Formerly Bundesanstalt für Geowissenschaften und Rohstoffe (BGR),
Geozentrum Hannover, Stilleweg 2, 30655 Hannover, Germany
e-mail: fwellmer@t-online.de

F.-W. Wellmer
Neue Sachlichkeit, 30655 Hannover, Germany

about *reserves* and *resources*, the *accessibility*, and *scarcity* of phosphorus and phosphate rock may depend on available technologies and is finally an economic question. We discuss a number of parameters which may indicate whether scarcity of a resource may be an increasing concern. These include the resource/consumption ratio, Hubbert-curve-based peak predictions, trends in ore grade, new resource discovery rates, and resource pricing as they are important for understanding exploration efforts. We further discuss estimates of reserves and the trends in actual estimates of phosphate reserves. P reserve estimates are dynamic and will increase for some time. Nevertheless, at some time in the long-term future, there will be a peak such as there will be a point in time that mined P becomes less economical than conservation and recovery.

Keywords Phosphorus reserves · Indicators for resources scarcity · Uneven distribution of reserves · Supply–demand dynamics

Contents

1 Introduction	130
2 Indicators and Causes of Resource Scarcity	132
2.1 R/C Ratio	133
2.2 Hubbert Curve Modeling	135
2.3 Price	136
2.4 Grade	137
2.5 Purity	139
2.6 Discovery and Growth of Reserves	140
2.7 Distribution	141
3 Resources in Light of Global and National Supply–Demand Dynamics	142
3.1 Sovereignty Over Natural Resources	142
3.2 Demand Dynamics and Exploration Efforts	144
4 Work in Global TraPs	146
4.1 Knowledge and Critical Questions	146
4.2 Role, Function, and Kind of Transdisciplinary Case Process	147
4.3 Suggested Case Studies	148
References	148

1 Introduction

The Exploration Node focuses on the search for assessment and quantification of phosphate reserves and resources. The key question to be assessed is: What knowledge is available about the nature, location, quality, quantity, and accessibility of P deposits in various world regions? As we will show, the answer to this question is often depending on economic issues.

2 Exploration: What Reserves and Resources?

At the present time, for the purpose of this study, *phosphorus reserves* are considered phosphate rock (PR) that can be economically produced at the time of the determination using existing technology, reported as (metric) tons of recoverable P_2O_5 concentrate (which includes 43.6 % pure P). Resources are considered PR of any grade, including reserves that may be produced at some time in the future, reported as tonnage and grade (in terms of the mean percentage of P_2O_5 contained in that rock which is assigned to phosphate processing) in situ. The Exploration Node covers sedimentary and igneous deposits and includes the assessment of all forms of mining as exploitation methods, in particular surface, underground, and offshore mining for establishing the viability of a deposit from a geophysical accessibility and projecting perspective. The exploitation itself is part of the "Mining Node." These deposits may be "conventional" deposits containing the mineral apatite, or nonconventional containing P in some other mineral forms.

The Exploration Node encompasses all aspects of the predevelopmental stages of phosphate deposits from initial evaluations to the involved feasibility studies required to obtain funding for the development of a mine. This generally includes a detailed geologic, mining engineering, and economic studies to quantify reserves, production and investment costs, and consequently optimal capacity. This would also include preliminary viability and prefeasibility studies to the final feasibility study including a market assessment. If positive, an investment decision normally follows. The Mining Node picks up from that point on. However, as P is an essential element, from a sustainable development perspective, also the (real) long-term availability of P is of interest. As we will show, this asks for comprehensive considerations on supply–demand dynamics.

The feasibility of producing PR and other mineral deposits can be broadly defined in terms of *technical feasibility* and *economic feasibility*. Determination of technical feasibility includes geologic, chemical, mineralogical, textural, and other studies of the potential ore, beneficiation studies, processing studies of raw ore or concentrates, and mining studies. Determination of economic feasibility is performed at increasing levels of complexity and cost data from opportunity studies to very detailed bankable feasibility studies. Engineering cost estimations are performed at varying levels of expense and accuracy, and these analyses rely heavily on supporting technical studies. Associated technical studies often have to be performed at higher levels of complexity and cost. Economic and financial analyses further depend heavily on the comparative cost of raw materials or fertilizer products on the world market delivered to a production facility or distribution point, the grade, quality, characteristics of potential products, and the cost of borrowing money.

To enable a financial analysis, not only does the cost side have to be investigated but also the potential financing and revenue aspects must be assessed. Therefore, assumptions about realistic future price levels and price developments have to be made. All of these factors, and others, must be addressed before potential ores can be called reserves, i.e., they should address that part of the total resources that can be economically extracted with available technology under environmentally and socioeconomically acceptable conditions (the modifying factors according to JORC or CIMM codes, which today are the worldwide standards; JORC 2004). The Canadian National Instrument 43-101 standard is even more restrictive, requiring technical disclosure.

The phosphate Exploration Node is a supply chain node, wherein the body of knowledge is largely contained in mining companies (often integrated with fertilizer companies), consulting companies, and independent consultants specialized in such studies. Other organizations that may have an in-depth understanding of the issues associated with geological exploration and mining are geological surveys and mining bureaus (often at the national level; e.g., the US Geological Survey (USGS), the German Geological Survey (BGR), the French Geological Survey (BRGM)) and international organizations (IFA, IFDC). Universities with mining schools (University of Nevada's MacKay School of Mines, McGill University, etc.) may have general knowledge, but there are only few special exploration programs on a national scale, which focused exclusively on phosphorus (Tilton 1977). It should be noted that large international mining companies that are primarily interested in large, long-lived deposits with prospectively low costs—so-called tier-1 projects—show growing interest in the fertilizer minerals potash and phosphate (Crowson 2012).

2 Indicators and Causes of Resource Scarcity

Scholz and Wellmer (2013) distinguish between *physical scarcity* and *economic scarcity*. Phosphorus is common throughout the Earth's crust. Obtaining the mineral resources that can meet the demand into the future may require a significantly increasing share of our economic resources or new technology as usually the highest grade most accessible material near the surface on the continents is consumed first and the average grade of the PR ore that is being mined is going to slowly decline (see Watson et al. 2013, this volume). This requires, in turn, that there be adequate levels of reserves that are sufficiently accessible, located in surface rock, high enough in grade, relatively free of impurities which can be removed economically and/or that recovery technology improves to compensate for changing ore characteristics.

Scarcity can also be the result of factors that are not physical or economic. Competing land use options wherein the land is more valuable for other uses may preclude development as phosphate mines. While environmental awareness and activism can have very positive results on avoiding mining in environmentally sensitive areas, missing trust in mining companies and state agencies, sometimes linked with political opposition, can result in putting meaningful mining areas off limits to exploration and development. Large prospective areas have been excluded as areas within preexisting mining areas. Examples of such areas can be found within the USA such as the Georgia coastal area or the North Slope area of Alaska. Environmental laws concerning reclamation or mitigation may make production economically infeasible, therefore excluding deposits or portions of deposits from consideration as reserves.

Phosphate producers may have to accept, or may perceive they have to accept, conditions imposed on them by importing countries imposing their environmental and socioeconomic values and standards through environmentally based trade-related laws. This would include laws concerning impurities in products and the way the materials are processed and by-products are disposed of. Naturally, sometimes these laws are motivated by other interests; they may actually be disguised trade barriers which may promote local or regional production. An example of trade restrictions based on environmental standards was the European water solubility requirements for triple superphosphate (TSP), which were launched in 1975 (EU 1976). Reduced solubility is largely caused by impurities in the precursor PR and processing conditions. The EU restrictions excluded North American TSP product from the European market for some time (this law got changed in 2004, EU 2004). Another case would be the EU cadmium laws that have been drafted (Chemicals Unit of DG Enterprise 2004). Such laws can impose thresholds for cadmium in fertilizers imported into the EU and thus affect what PR will be mined. Of course, such European laws would give companies producing phosphorus fertilizers from PR with a low cadmium contents advantage in the European market. Feasibility studies of P mines have to take these aspects into account.

Is exploration keeping up with the demand for phosphorus? Teitenberg (2003) lists several potential indicators that could signal increasing scarcity, including the resource/consumption ratio (R/C ratio), trends in resource pricing, trends in ore grade and purity, and trends in new resource discovery rates. When discussing these indicators, we will reflect both on the general trends of these indicators for minerals and metals (which may emerge after wars or economic crises) and on specifics. To this, we could also add indications of uneven distribution, such as the Herfindahl–Hirschman-Index (HHI), which is defined as the sum of the squares of the shares of the P production or P reserves.

2.1 R/C Ratio

The R/C ratio is sometimes called the (static) resource lifetime, as it has units of years and would represent the lifetime of the resource if no new deposits were put into production and consumption rates did not change. Given those limitations, Scholz and Wellmer (2013) point out that the appropriate use of this ratio is not to predict ultimate depletion of a resource, but as an "early warning indicator" calling for action, for instance increased exploration efforts. The R/C ratio for PR is commonly based on data maintained by the USGS.

Scholz and Wellmer compared the R/C ratio of 34 minerals and metals in 2002/2003 with that of 2010. In 2003/2003, the R/C ratio for PR was 130. In the ranking of this ratio, phosphorus was tenth of the 32 elements (which could be compared in this year). However, in 2010, the R/C ratio turned to 370, which brought it to rank 4 out of 34. On a first view, this increase has been due to an increase in the documented reserves in Morocco from 5,700 Mt PR to 50,000 Mt, which has been provided in an International Fertilizer Development Center (IFDC) report reassessing the Moroccan reserves (van Kauwenbergh 2010b). However, even if the Morocco reserves are not considered, the USGS estimates of world reserves increased from 9,300 Mt P in 2009 to 15,000 Mt P in 2011 (Jasinski 2009, 2010, 2011; USGS 2004), mainly because of increased documented reserves and reassessments of phosphate reserves from countries in the North African phosphate belt. This indicates exploration efforts in many countries though Morocco holds around three quarter of all reserves.

If the *R/C* ratio is best used as an "early warning indicator" calling for action, as described above, then the data indicate that society has ample time to make such efforts. This is not to argue that action should be delayed. On the contrary, it could be interpreted as a call to phase in action before criticality is approached. However, one may ask on what basis data on reserves are altered and substantiated and how valid are the USGS data. The USGS is one of the public institutions, which provides free world resource and reserve data for many commodities since more than 130 years. The 2010 change in USGS estimates was primarily based on a comprehensive secondary analysis by IFDC (van Kauwenbergh 2010a) using a 1998 report on Morocco reserves (Gharbi 1998) as well as numerous other reports and references. But it often is unclear (also for other countries in the case of phosphorus and for other minerals as well) on which geological and drilling data the conclusions are made.

Well-documented studies of commodities which have a political dimension for comparison might shed light on the factors which influence the sum of country and world reserves. It is therefore subject of a case study of the Global TraPs project described below in Sect. 2.4. Three levels of detail for estimating reserves can be distinguished: companies, regional planning authorities, and the highest political level. For *companies*, reserves are their future working inventory. They, therefore, normally only gather data and estimate reserves for as many years of production as the cost associated with obtaining the data and their preference for business planning justify. These reserves normally have to comply with the JORC code as mentioned above. The *regional planning authorities* (e.g., municipal or state authorities) have to look farther into the future. So what are reserves for them are resources that can be technically exploited, but their environmentally and socioeconomically acceptable conditions (the modifying factors) have not yet to be defined. *The highest political level* is normally involved when for the national budget, significant tax contributions are expected or—for social reasons or for reasons of security of raw material supply—significant subsidies are required. A good example of the interaction between these three levels is the German coal production history. Coal occurs in seams and from its geometry is comparable to phosphate.

The rapid increase in PR resources after 2007 may be due to various reasons, especially the price signals in metals and fertilizer. The increase in prices allows a lowering of the cutoff boundary between reserves and resources. A linear decrease in grade normally has the consequence of a nonlinear (in excess of being proportional) increase in reserves or resources. A reduction in grade by 10 % relatively would increase the reserves by more than 10 % as there normally is more rock with lower content of phosphorus than with more phosphorus.

2.2 Hubbert Curve Modeling

The Hubbert curve is a modeling method originally proposed to analyze US oil production. It assumes that the rate of production is proportional to the product of the amount of the resource that has been consumed times the amount remaining. The model that predicts the shape of the production versus time curve will be "bell-shaped," with two parameters: the peak year of production and the amount of ultimately recoverable resource (URR). Hubbert (1962) successfully predicted the peak of the national US oil production (Brandt 2007). However, subsequent use of Hubbert's approach was not always successful and has engendered considerable controversy (Sorrell and Speirs 2010).

Déry and Anderson (2007) tested the Hubbert model, among others, on global phosphorus production data. They just adjusted a symmetric curve to the historic production curve and predicted that the global peak occurred in 1989. More significantly, the URR predicted by the method was only 35 % even of the USGS reserve numbers of the time and only 10 % of the global reserves documented in 2010. A similar attempt has been reported by Ward (2008). So this approach clearly seems inappropriate for predicting URR of a global resource such as phosphorus. Other applications of the Hubbert model were presented by Cordell et al. (2009, 2011, April 4). They adjusted a Gaussian symmetric curve but used the USGS estimates of reserves to fix the value of the URR (and hence the area under the Hubbert curve and the so-called peak of 1989). Using the 2008 USGS reserve estimates of 16 Gt PR, the date for peak global PR production was estimated to be in 2035. But the amount of reserves is dynamic. When they incorporated the increase in reserves of 60 Gt PR (Smit et al. 2009), a new peak year was found to around 2,071, plus or minus about 20 years.

An analysis similar to that of Cordell et al. (2009) shows a peak year of production in 2087 at 442 Mt PR per year, to be followed by a decline. This peak value is about 2.5 times the 2010 value. When compared to the median UN population projection for that year, this value results in a per capita PR production about 87 % higher than the 2010 figure of 25.8 kg/capita/yr. This corresponds to an annual rate of 0.82 % per year. From 2001 to 2012 the per capita production increased at a rate of 3.4 % per year following a decline of world production after the decline of the Soviet Union (see Chap. 1 Figs. 7 and 28). Extrapolating the current trend suggests that the amount of phosphorus consumption at the peak is not implausible.

Vaccari and Strigul (2011) have shown that the peak timing of Hubbert curve-fits is very sensitive to the stage of the production data sequence. In an example using the historical record of US PR production, the method was shown to have significant difficulty identifying the peak unless the peak was already past. Furthermore, they suggest that Hubbert curve extrapolation may not be appropriate for global resource modeling. It could be appropriate for local resource fields because consumers could switch to other sources. Strictly speaking, there are alternatives to global reserves: Previously, uneconomic resources can be converted to reserves, and conservation and recovery methods may become competitive with extraction.

The Hubbert curve also predicts that production will be symmetrical in time, declining once half of the initial reserves have been consumed. Again, there may be reason to question the validity of this conclusion for global production of a resource with "no alternative." In such a case, it may be expected that demand is "inelastic," meaning that consumption would not be greatly sensitive to price. In such a case, high production levels would be maintained until very high costs force the adoption of much more difficult technical and societal options such as intensive recycling or by attempting to reduce per capita consumption via changes in diet. This shows the importance of exploration and recycling but also for a comprehensive monitoring of phosphorus reserves for avoiding that the reserves becomes critically small.

As with the R/C ratio, the Hubbert modeling approach suffers from the limitation that it does not account for the expansion of the URR that could be brought about by changes in economics and/or technology. The predictions of the Hubbert modeling approach might best be examined as a conceptual model to describe one scenario for how future production might play out. It could also be useful as a stage in the development of such models, as criticism may lead to the incorporation of more factors, so one may judge what the potential impact of them may be. However, one should reflect whether the assumption of a symmetric rise and fall of production is adequate (as it may be the case for single deposits such as Nauru on a supply market) or whether other types of production curves (such as logistic functions with saturated supply on a demand market) are adequate. As may be learned from shale gas, also new groundbreaking technologies of mining may change the reserves.

2.3 Price

As that of all metals and minerals, the price of PR has been showing increasing volatility and growth in the last 5 years. Some analysis of what caused the 2008 peak and what level the phosphate price will take is provided by Weber et al. (2013). They elaborate that the price of one commodity such as P should not be considered in isolation. The price of most commodities has also increased substantially over this period (see Weber et al. 2013: Figs. 2 and 4 in Chap. 7, this book). There are many factors causing the price peak and the potential transition to a new plateau including speculation and too low prices (lack of investments in mines and exploration) over a long period. A number of fertilizer plants have closed in North America, while new production facilities have been added in Morocco and other countries. Naturally we have to reflect and to investigate whether the price volatility of phosphorus including the 2008 price is an indicator scarcity of production means, accessibility of reserves or changed constraints of production (e.g. lower phosphate rock ore grades). This should become a matter of research (Chap. 7 is starting a discussion here). However, in general, the current short-term price dynamics are affected by many factors and are related to economic scarcity, i.e., short-term overshoot of demand, and therefore are not an indicator of physical scarcity. Furthermore, there is evidence that high market

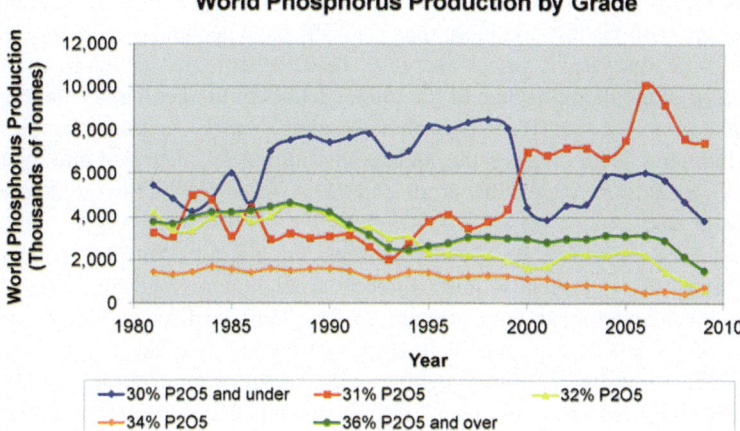

Fig. 1 Trends of grades of beneficiated P [in general, which departs P mines after beneficiation; *source* IFDC (2010)]

prices, together with concerns about physical short- and mid-term scarcity, may motivate policy makers and (financial) business agents to induce exploration efforts and the building of new mines. This could lower PR prices in the midterm.

2.4 Grade

There are no reliable data on how the average grade of mined PR was altered in the past. However, grade is only one out of many aspects which affect the economic value of PR concentrate, and there are many other aspects such as location and ore characteristics. We should also note that igneous PR is often very low in ore grade (even sometimes less than 5 %). But by beneficiation, it is upgraded. Figure 1 presents the trends of grades of beneficiated PR concentrate which shows rather a decline in low-grade (beneficiated) PR (below 30 % P_2O_5) and high-grade beneficiated P (above 34 % P_2O_5) is increasing and makes about 31 % in 2010 (though statistics on phosphorus consumption usually refer to 30 %). In general, we were facing a decline of the PR concentrate in the last 30 years (see Chap. 3, Fig. 6) which may ask for an improvement of beneficiation technologies. Of special interests are the trend in the ore grades. According to our knowledge, no data have been published on this issue so far.

In 2011, the USGS added five new countries (Algeria, India, Iraq, Mexico, and Peru) to its list of major reserve holders. Their figures indicate that Iraq (5.8 Gt PR) holds the second largest deposits of PR reserves in the world, after Morocco [50 Gt PR (Jasinski 2010, 2011)]. However, the reported grades of Iraq's reserves are below 22 % P_2O_5 (Al-Bassam et al. 2012). And most of these reserves were taken out in the 2013 accounting as the classification has been wrongly classified according to Russian instead of the US criteria of reserves. Thus, Iraq reserves are 0.46 Gt PR, and more than 5 Gt PR have become resources again and may become reserves if prices increase, new technologies develop, or market change. To judge the relevance of the grade, details of the geology of the ore have to be taken into account. If the phosphate content is easily separated (i.e., it lies in distinct mineral assemblages rather than being intimately integrated into the rest of the rock), a medium-grade concentrate can be produced from a low-grade ore at an acceptable cost.

Igneous PR of 4–5 % P_2O_5 can sometimes be upgraded to 39 % concentrate. Sedimentary PR is typically quite different from igneous phosphate rocks. The Central Florida sedimentary phosphate ore is composed of a varying proportion mixture of "pebble" at around 29 % P_2O_5 and "feed" at under 9 % P_2O_5 (whole ore average around 11 % P_2O_5). Pebble is recovered or removed by washing and screening, while sand-sized "feed" is put into the flotation circuits to raise its grade to around the same as the pebble. The pebble content of the ore was one of the major variables in production cost in Florida and the pebble content decreased as the mining district moved south, ending at around 4–5 % in the new prospects in the southern portion of the field.

PR beneficiation has been the subject of considerable research. The question in what way the decline in the ore grades may be compensated by extraction or beneficiation technologies may become a matter of research. Beneficiation usually starts with dry and/or wet sizing. Flotation may be utilized if water is available and economics justify its use. As mentioned, all igneous PR is subject to floatation because of the low ore grade, often around 5 %, and the question is what floatation technology may contribute to transfer of resources to reserves. The Crago process, the basis for many PR floatation processes, was implemented in Florida in the 1920s and 1930s. For an overview of PR beneficiation, see IFDS/UNIDO (1998). Overall recovery has improved over the years (Al Rawashdeh and Maxwell 2011), for example, report that phosphate flotation recovery in some plants has increased from 60 to 90 %, thereby compensating for lower-grade ores. In general, multiple efforts are made by PR companies to improve recovery, reduce waste, and conserve resources. With increased prices for PR concentrates, it is possible to apply more resources to beneficiation infrastructure and processing to improve recovery of already mined materials or the utilization of beds that may not have been previously utilized due to cost or technical considerations. Van Kauwenbergh (2010a) has pointed out that flotation was not used in Jordan or Morocco until the late 1990s due to cost considerations. The cost-effective implementation of this beneficiation technique has put more resources in the reserve category in these countries. The use and development of floatation is one technological means to reduce the comprehensive losses of phosphorus before processing (VFRC 2012).

2.5 Purity

In order for a PR to be suitable for a wide range of processing options, the R_2O_3/P_2O_5 ratio (i.e., $(Fe_2O_3 + Al_2O_3)/P_2O_5$ ratio) must generally be below approximately 0.10. The $(R_2O_3 + MgO)/P_2O_5$ ratio must generally be below 0.12 (Jasinski 2010). When phosphate rocks have impurity contents at or above these levels, problems can occur in processing. For example, products may not dry properly or the products may contain phosphates that are not water soluble. For a review of the effect of impurities on fertilizer processing, see van Kauwenbergh (2006). According to Prud'homme (2010), around 90 % of current PR production is used for phosphoric acid production mainly by the wet process, requiring a feedstock with fairly low impurities. PR used in some fertilizer processes may contain more impurities than the material used for wet-process phosphoric acid.

Common potentially hazardous elements associated with PR include arsenic, cadmium, chromium, lead, selenium, mercury, uranium, and vanadium. Cadmium and uranium show the greatest enrichment in sedimentary PR, relative to crustal rocks, by a factor of 60–70 in concentration for Cd and about 30 for U (van Kauwenbergh 1997, 2012). Igneous PR, which is about 13 % of world production, is mostly enriched in arsenic (by a factor of about 67) and selenium (by a factor of 76), but also in U (22 times) and Cd (7.5 times).

Uranium for use as nuclear fuel can be extracted from phosphoric acid produced during acid-based fertilizer manufacturing. The market for uranium is expected to grow. When making diammonium phosphate (DAP) and monoammonium phosphate (MAP), most of the uranium from PR tends to be found in the fertilizer product, with a small fraction in the gypsum by-product. Among the decay products of U is radium, and radon gas is a radium daughter. Most of the radium in PR winds up in the gypsum by-product (van Kauwenbergh 2012). Making fertilizer via the dihydrate route in effect cleans the fertilizer product. Due to regulations imposed by the USEPA concerning radioactivity of the phosphogypsum product, the use and the disposal of the gypsum by-product from Florida PR are prohibited and handling and storage are very costly. Due to regulations in the USA, Florida-produced phosphogypsum is stacked in large piles requiring monitoring and maintenance in perpetuity. There has been considerable debate concerning the EPA-imposed radioactivity limits for phosphogypsum. Florida-produced phosphogypsum is below IAEA-suggested limits for naturally occurring radioactive matter (NORM; van Kauwenbergh 2010a).

Cadmium, a natural element in the Earth's crust, is among the most toxic heavy metals. It accumulates in the soil from several natural and anthropogenic sources including zinc smelting, burning coal, and weathering of coal and base metal sulfides, and potentially by extensive application of fertilizers. It may critically accumulate in humans from the food supply, breathing fumes and dust, and smoking. Interestingly, it also interferes with the use of uranium as a fuel, since it is a neutron absorber. A number of European Union countries have placed limits on Cd in fertilizer (ranging from 21.5 to 90 mg/kg P_2O_5), or on input to agricultural soil (0.15–150 g/ha/yr), or on Cd content in agricultural soils (0.5–3.0 mg/kg dry soil, van Kauwenbergh 2010a). When PR is used to manufacture single superphosphate, essentially all the Cd from the PR is found in the fertilizer. With other processes, the fraction captured in the fertilizer may be less, but can still range from 55 to 90 %. Technologies for removal of cadmium (decadmiation) are available. However, costs for removal and disposal or processing unwanted by-products or wastes have not been established.

Regulatory limitations on impurities are usually expressed relative to P_2O_5 content of the fertilizer. Soil accumulation can be controlled by regulating agricultural practices. Pending European legislation is predicated on European studies, indicating that fertilizers with 20 mg Cd/kg P_2O_5 are not expected to produce long-term accumulation in most soils (Hutton and Meeûs 2001). But those with more than 60 mg Cd/kg P_2O_5 are expected to do so (Nziguheba and Smolders 2008). Regulations in the USA are based on scientific risk-based assessments, and the allowable Cd, and other metals, levels are much higher than European standards (van Kauwenbergh 2012). The risk-based limits recommended in the USA fall below WHO-recommended limits. Van Kauwenbergh (2010a, b) has pointed out that while technologically advanced and affluent countries may be able to set high standards with respect to cadmium levels in fertilizer products, less affluent countries may be prone to take a more pragmatic approach.

2.6 Discovery and Growth of Reserves

Although major discoveries are few and infrequent, Sheldon (1987) presented an analysis showing that discovery of new PR resources in the 20th century outstripped consumption over the same period not taking reserve growth into account. And there have been new discoveries after 1987. A large share of about 74 % of this increase is from a few discoveries, primarily from Morocco. This highlights the regional inhomogeneity of phosphate reserves in the world. Concerning sedimentary deposits which occur in seams such as phosphate, salt, or coal, the major replacement of mined inventory (reserves) occurs by so-called reserve growth. As explained above, reserves are the working inventory for companies. They, therefore, normally only gather data and estimate reserves for as many years of production as the cost associated with obtaining the data and their preference for business planning justify. So by detailed exploration, resources of the same system of seams or layers in the forefront of the mining area are converted into reserves. For example, for Israel, the reserves were 180 Mt PR in 2005 and the same number as in 2010. The figures for Jordan were 900 Mt PR in 2005 and 1.500 in 2010 (Prud'homme 2011; Jasinski 2006). Even for commodities which do not occur in seams, reserve growth, once a deposit is exploited and offers the possibility of better understanding the deposit, is a common feature. Wagner (1999) showed that, for example, for lenticular Pb/Zn deposits in limestones—so-called Mississippi Valley type deposits—often a reserve growth factor of two could be achieved.

An essential aspect of future availability of P is that resources may become reserves. Today, USGS (2012) records 71 Gt PR reserves and 200 Gt PR resources. The latter do not include the western phosphate fields in the USA, which include besides 8 GT PR reserves at an average grade of 24 % P_2O_5 507 GT of "subresource-grade phosphatic material … at a depth greater than 305 m" (Moyle and Piper 2004). Though it may become more expensive to mine these phosphates, underground mining is technically and economically feasible (note that today an average world citizen consumes about 30 kg PR per person and year which costs about 6 USD, see Scholz and Wellmer 2013). It seems economically feasible, from the perspective of the cost for an average world citizen, to double the price and correspondingly the reserves. Naturally, these are heuristic like inferences that may rely on experience with coal mining. Given the essentiality of phosphorus, this issue may ask for closer investigation.

Nevertheless, as with most minerals and metals, there is a large regional heterogeneity of phosphate reserves in the world. There are more than 1,600 world phosphate deposits compiled (Orris and Chernoff 2004). Libya, which is also located in the North African phosphate belt, is not yet documented in the USGS Mineral Commodity Summaries. If we acknowledge that various offshore mining projects got explored (Midgley 2012), the new discovery has continued to outstrip consumption.

In some areas of the world, reserves may not be established for tax purposes. When the material in the ground is quantified and resources are changed to reserves, which have a monetary value, the land can be reclassified. Unimproved unoccupied land or agricultural lands are taxed at different rates than mine lands with valuable contained reserves. This sometimes prevents P seams to be classified as reserves.

2.7 Distribution

Scholz and Wellmer (2013) proposed using the HHI as an indicator of uneven distribution of *reserves* and of *production*. The German Raw Materials Agency (DERA 2011) divided the possible range of the HHI from 0 to the maximum of 10.000 into three ranges: 0–1.000 low, 1.000–2.000 middle, and 2.000 to the maximum of 10.000 as highly critical from a supply security perspective. They assessed the HHI for production in the year 2010 at nearly 2.000, which is in the middle range in comparison with other metals and minerals. However, the HHI for reserves was a high 5.000 in 2010. They note that if the HHI for *reserves* is much higher than for *production*, it may be due to the incomplete recording of the reserves, lack of exploration activities in certain countries, or political reserve corrections. Whether a high HHI index is indicating a critical distribution, in which sufficient supply is dependent on few countries, very much depends on the total volume of the reserves compared to the demand. Here, as phosphorus has a high static lifetime globally compared to other minerals and metals and some overcapacities in production (USGS 2004), disturbances with some major suppliers may be compensated by suppliers with smaller (production) and reserves. Thus, the relatively high HHI indices are less critical.

Since local markets for commodities even for low-value commodities are only existing in exceptional cases, and we, therefore, have to consider the world market as one commodity market, there is a regional imbalance for most of the commodities. Commodities such as phosphate or metals are mined from deposits which were formed as a result of enrichment processes which are tied to paleoecologic conditions and geologic processes. Since paleoecologic and geologic conditions varied around the world, most industrial mineral and metal deposits are unevenly distributed geographically. Even in cases where deposits of certain minerals or energy-bearing materials (like in the case of uranium, see Wedepohl 1995) formed in a wide variety of geological environments, the geologic conditions where the best deposits occur are few. Morocco has a special position. A large share of the world sedimentary PR reserves is found in this area, and Morocco area embodies a large share of resources though no complete picture about the reserves is available yet. Naturally, large mining companies concentrate their efforts on the best deposits due to the investment costs for new mines. Companies only invest if the projected operating costs of a future project are better than average, preferably in the lower third quartile (Wellmer et al. 2008). A consequence is a growing concentration of mining in countries with a very favorable geology for a specific commodity (Stephenson 2001). This may lead to high HHI indices signaling a dependency of supply on a few countries. This is a consequence of globalization. This appears to be an unwanted consequence, depending on the nations involved and their intentions.

The consequence is that international politics has to develop strategies such as to secure for certain diversity and or/security of supply to support sufficient and free flow of raw materials. Another aspect concerning high HHI is to understand the relationship between producer and customer. Certainly, a producer can try to "squeeze" the customer in a seller's market. However, historically, sellers' markets never lasted for long. This in particular holds true if the supply is shaped in a way that "blocked supply" by main providers may be compensated by other providers for a long time. This is, in principle, possible for PR. For PR, in time, the market would swing back to a buyer's market.

3 Resources in Light of Global and National Supply–Demand Dynamics

3.1 Sovereignty Over Natural Resources

Within the "New International Economic Order," the sovereignty of a nation over its natural resources is a key element. It became international law with the UN resolution 1803 of December 14, 1962, "Permanent Sovereignty over Natural

Resources" (UN 1962). The nation state has become the dominant political organization after World War II (Parsons 1951). As the case of the European Union shows, new organizational schemes such as supranational societies who take joint action with respect to supply security may emerge. And there are also critical voices which consider the nation state principle as inefficient with respect to challenging global questions such as climate change, biodiversity, resources, and environmental pollution management (Beck 2000; Scholz 2011). Nevertheless, it is self-understood that natural resources have to contribute to a nation's prosperity within the intrageneration fairness concept of sustainable development as defined in the Agenda 21 of the Rio Declaration at the UN Conference on Environment and Development in Rio de Janeiro in 1992 (1) to enable all people to achieve economic prosperity, (2) to strive toward social justice and (3) to conserve the basic needs of life. This concept is developed further in the Intergovernmental Forum on Mining, Minerals, Metals, and Sustainable Development, which was one of a number of Partnership Initiates launched at the World Summit on Sustainable Development in Johannesburg/South Africa in 2002 (UNCTAD 2012).

Since resources in the ground have only a hypothetical value, it is necessary to turn them into real value for the benefit of a nation by mining, beneficiating the resource into a marketable product, and bringing the product to markets. Therefore, capital and know-how for the exploitation of the natural resources are required. The sovereign can either decide to exploit the resources himself or through national mining companies alone and develop the necessary know-how on its own or decide to open the resources to foreign mining companies which often are large multinational companies.

Exploration and mining is a high-risk business, more risky than normal industrial business (Wellmer 1986). Even when a phosphate seam is known in its general geology, the chances to develop a profitable mine are not better than 1:2 or even 1:3 judging from worldwide exploration success statistics of similar deposit types (Sames and Wellmer 1982). So there is a potential conflict of interest between the sovereign who wants to maximize the revenues of mining operations for the benefit of his nation and the mining companies. Besides the normal profit possibilities for a business, independent mining companies require a reward for their risk taking, besides the assurance for stability, i.e., that the legal and financial framework is not changed during the lifetime of the mining operation to their disadvantage.

Since the 1960s, a wide spectrum of arrangements, often specific for a commodity, have been developed worldwide to come to a fair distribution of resource revenues for the nation and the mining risk taker. Options range from the USA where in the eastern states, the surface property owners also own the subsurface mineral rights, but the miners have a special depletion allowance for this tax calculation as examples in the North African countries in which mining is done by the sovereign. In between, there exists a wide arrangement of tax and royalty agreements and carried interests, i.e., a free share in the mining operation for the state without having to carry exploration and investment costs. Vielleville and Vasani

(2008) state: "Those rights (over natural resources in their territories) ... have been given shape over time. They have been limited by bilateral and multilateral investment treaties, free trade agreements and customary international law."

3.2 Demand Dynamics and Exploration Efforts

Exploration efforts depend on the known reserves and resources and available geopotential as well as on the future demand. The world demand of phosphorus has continuously increased over the last decades after an abrupt decrease caused by the decline in the Soviet Union and reduction in fertilizer use in Japan, North America, and Europe in the early 1990s by improving the technology and efficiency of fertilization (Smil 2000). There are two main factors causing the recent trend. One is the ongoing population growth. Between 1992 and 2011, the world population grew by 27.3 %, whereas the global P production increased by 30.8 % from 146 Mt PR in 1992 (Llewellyn 1993) to 191 Mt PR in 2011 (Prud'homme 2011). This suggests a second factor driving recent trends, which is the increase in per capita consumption. This is likely due to changing standard of living, especially in China. Here, it is worth noticing that according to the US Mineral Commodity Summaries, China's production more than tripled in the last two decades from 23 to 72 Mt PR and in 2011 was by far the world's largest producer at 38 % of the global total.

It took until 2009 for the global total to fully recover from the drop in the late 1980s and early 1990s due to the collapse of the Soviet Union. Figure 2 shows the per capita production, based on USGS data for PR production (Jasinski 2012) and global population from the US Census (2012). We can see that the per capita production has not yet recovered, although it has been increasing steadily for the past decade.

Special attention should be paid to the rapid increase in the last 5 years. The one-year increase in per capita PR production from 2010 to 2011 was 7.1 %. For the five-year period ending in 2011, it was 4.9 % per year, and for the ten-year period ending in 2011, it was 3.0 % per year. This could indicate an accelerating trend in per capita demand, possibly associated with changes in diet or agricultural practices in some developing countries.

From an overall system theory perspective, besides population growth, consumption may be expected to respond to the changing diets (primarily in countries which turn from vegetable-based to high meat consumption), P dynamics in soil (there are large P deposits in the soils of many areas), and increasing efficiency in agrarian application and in mining and processing, recycling, etc. (van Vuuren et al. 2010; US Geological Survey 2002).

The future demand of P is difficult to predict as there are many competing factors. Mid- and large-scale recycling is starting, and agriculture may become more efficient, whereas high demands may emerge in some parts of the world with highly weathered soils. Increasing reserves and increasing efficiency may, at least

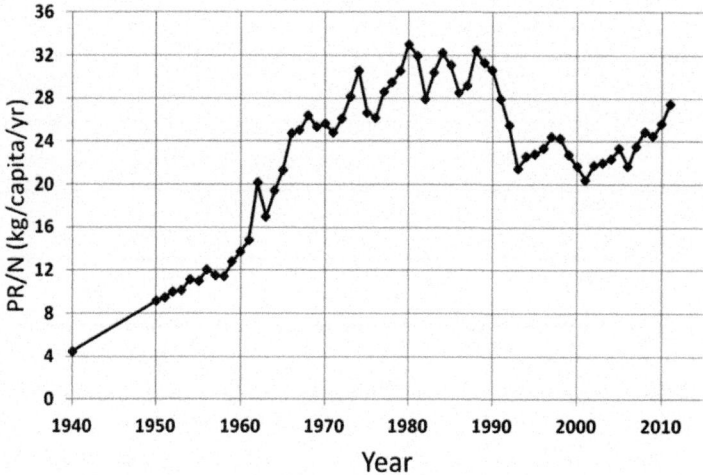

Fig. 2 Per capita global phosphate rock production (PR rock/capita)

in principle, balance the further demand. Global learning curves on efficient P use may reduce the P use per capita.

A challenging question for prospecting and exploration is in what way the feedback control cycle of mineral supply and demand functions. This could include mechanisms such as that increasing prices increase resources, recycling, innovation, and exploration efforts. (Wellmer and Becker-Platen 2002) and the fact that prices that are too high for society may develop for phosphorus, putting downward pressure on demand. Though phosphorus is a low-price commodity (see above), current and prospective price increases may be critical for certain parts of society.

A refinement of the use of the *R/C* ratio would be to make forecasts assuming current per capita rates, coupled with UN population forecasts. Figure 3 shows these forecasts under three UN population scenarios: high, medium, and low growth. Under the peak modeling assumption, when half of the reserves have been consumed, then production will decline and shortages will become critical. However, at least at the present time, production of PR is governed more by demand than by availability of reserves. In such a case, production would not decline unless population and/or per capita demand do.

The prediction of the constant per capita demand model can be compared to Hubbert curve predictions by showing the points at which half of the current reserves would have been used and the point at which current reserves will have been completely used. The latter point is shown in Fig. 3 as a solid diamond. For the medium scenario, this will occur in the year 2350, sooner than predicted using the *R/C* ratio. The point of 50 % depletion would occur for the medium population scenario in 2,145. But under this assumption, production does not begin to decrease significantly. Of course, various factors could shift this scenario either way. If we would face an increase in per capita consumption (see Fig. 2), we can expect on the

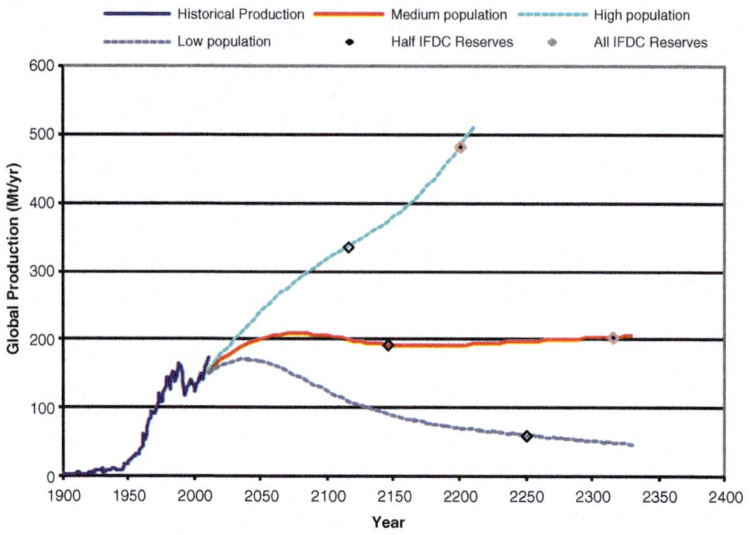

Fig. 3 Projected PR production based on fixed per capita rate coupled with UN projections of population growth

one hand that this point could be reached much sooner than that. On the other hand, reserve growth, i.e., reserves grow in parallel with production and consumption as discussed above under "Discovery," will continue for the foreseeable future and would push this point further into the future. As with the *R/C* ratio, this analysis is not intended to predict actual depletion, but rather as a warning indicator.

Figure 4 shows a scenario according to Scholz and Wellmer (2013) that the production curve will go through a series of plateaus before 1 day a peak is reached. The reduction in consumption equivalent to a plateau between 1990 and 2009 was due to learning effects in the industrialized countries (using less fertilizer per area without reducing the crop output) and reduction in consumption in the Eastern Bloc. The next plateau will probably be reached when the Chinese learn to use fertilizers as efficiently as the developed nations. China's croplands show the same P imbalance as those in North America or Europe (MacDonald et al. 2011).

4 Work in Global TraPs

4.1 Knowledge and Critical Questions

The most critical question for the Exploration Node is a better *understanding of the reserve and resources figures*. A case study was decided in the Global TraPs project. The USGS and the German Geological Survey, which are both in leadership positions in the Global TraPs project (Steering Board, Practice Leader of the Exploration Node), will support these studies.

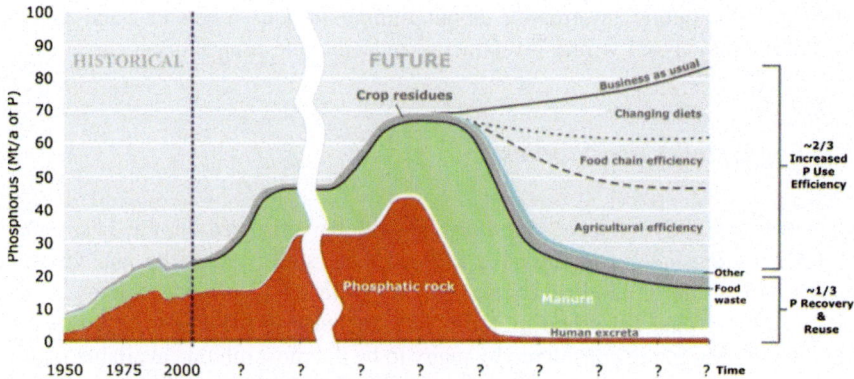

Fig. 4 Scenario for PR production and consumption (Scholz and Wellmer 2013, historical data, and exit strategy after Cordell et al. 2009)

The study will examine (1) what standard methods may be used for assessing availability and the recording of reserves in the USGS Mineral Commodity Summaries, a question which has already been dealt with before (Orris and Chernoff 2004). Further, the questions (2) of how would reserve estimates respond to changes in the price of PR (which may depend on many aspects which are difficult to predict such as innovation of exploration, mining, and beneficiation technologies) and (3) how does the elasticity of demand look like, i.e., how sensitive is consumption to price, are subject of investigation. With the latter question, particularly, whether prices stress the system is of overarching interest. Finally, (4) the risk due to the uneven global distribution and midterm vulnerabilities of phosphorus supply and how it may be compensated is an issue, which is also of interest for exploration efforts (e.g., from the perspective of increasing the potential diversity of reserves and supply).

4.2 Role, Function, and Kind of Transdisciplinary Case Process

The questions dealt with in the Exploration Node are of genuine interest for *private and state-run mining companies* and *geological surveys* whose task is—among others—to provide timely and relevant information on the natural resources we rely on. From the science perspective, these questions are of interest for geological economics and sustainability and all sciences which deal with supply security. Thus, also institutions such as UNEP International Resources Panel and similar institutes are interested in these questions.

Td process is precompetitive and may include competing companies for the purpose of developing a joint view on scarcity and mining options. In principle, this is delicate as owning and not revealing information is a business advantage, whereas valid public information about mining options, demands, and market constraints may avoid wrong investments and motivate proper ones.

4.3 Suggested Case Studies

Many of the above critical questions should be answered in the case study *Phosphorus Resource and Reserves Figures*. The study should be launched via master or PhD thesis in close cooperation with the US Geological Survey.

The reserve data published by the USGS annually in the Mineral Commodity Summaries (MCS) are considered by many to be the most reliable available figures in the world. Many users of reserve data use the MCS without acknowledging the dynamic character of these figures and their limitations. The USGS does not actively determine the reserves and resources of the world as a world authority but collects information from a variety of more or less publicly available sources which are examined and screened. The sources of information for reserve figures are company data or data from local or regional government bodies up to national governments, but also international or multinational institutions such as the IFDC can be a source of valuable information. The different interests of these organizations in data may be a source of biasing data. Data may be working and planning inventory for *companies* and serve as basis for receiving credits from banks, and they are the foundation of policy means (e.g., taxes, resources conservation, recycling, subsidies) for governmental bodies. As also nations are in economic competition, special attention should be paid that the Global TraPs project provides a forum that allows all participants to equally benefit.

Acknowledgments We want to thank Debbie T. Hellums, Steve van Kauwenbergh, and Michel Prud'homme for their detailed feedbacks to a previous version of this chapter.

References

Al Rawashdeh R, Maxwell P (2011) The evolution and prospects of the phosphate industry. Min Econ 24(1):15–27. doi:10.1007/s13563-011-0003-8

Al-Bassam K, Fernette G, Jasinski SM (2012) Phosphate deposits of Iraq. In: Phosphates 2012, El-Jadida, Morocco, 20 Mar 2012

Beck U (2000) The cosmopolitan perspective: sociology of the second age of modernity. Br J Sociol 51(1):79–105

Brandt AR (2007) Testing Hubbert. Energy Policy 35(5):3074–3088. doi:10.1016/j.enpol.2006.11.004

Chemicals Unit of DG Enterprise (2004) Draft proposal relating to cadmium in fertilizers. European Commission, Brussels

Cordell D, Drangert JO, White S (2009) The story of phosphorus: global food security and food for thought. Glob Environ Change-Hum Policy Dimensions 19(2):292–305. doi:10.1016/j.gloenvcha.2008.10.009

Cordell D, White S, Lindström T (2011) Peak phosphorus: the crunch time for humanity? Sustain Rev 2(2):3

Crowson P (2012) Solving the minerals equation? Demand, prices and supply. In: Life and innovation cycles in the field of raw materials supply and demand—a transdisciplinary approach, Orléans, France, 19–20 Apr 2012

DERA (2011) Deutschland—Rohstoffsituation 2011. DERA Rohstoffinformation 13. Deutsche Rohstoffagentur, Hannover

Déry P, Anderson B (2007) Peak phosphorus. Energy Bull (Retrieved September 22, 2011)

EU (1976) Council directive of 18 December 1975 on the approximation of the laws of the member states relating to fertilizer. Off J Eur Comm

EU (2004) Commission regulation (EC) No 2076/2004 of 3 December 2004 adapting for the first time annex I of regulation (EC) No 2003/2003 of the European parliament and of the council relating to fertilisers (EDDHSA and triple superphosphate). The commission on the treaty establishing the European community, Brussels

Gerst MD (2008) Revisiting the cumulative grade-tonnage relationship for major copper ore types. Econ Geol 103:615–628

Gharbi A (1998) Les Phosphates Marocains. Chronique de la Recherche Minière 531–532:127–138

Hubbert MK (1962) Energy resources a report to the committee on natural resources. National Academy of Sciences-National Research Council, Washington, DC

Hutton M, Meeûs D (2001) Analysis and conclusions from member states' assessment of the risk to health and the environment from cadmium in fertilisers. Contract no. ETD/00/503201. European commission—enterprise DG. Environmental Resources Management, London

IFDC (2010) World phosphate rock reserves and resources. IFDC, Muscle Shoals

IFDC/UNIDO (ed) (1998) Fertlizer Manual. Kluwer, Dordrecht

Jasinski SM (2006) Phosphate rock. In: US Geological Survey (ed) Mineral commodity summaries. USGS, St. Louis, pp 122–123

Jasinski SM (2009) Phosphate rock. In: US Geological Survey (ed) Mineral commodity summaries. USGS, St. Louis, pp 120–121

Jasinski SM (2010) Phosphate rock. In: US Geological Survey (ed) Mineral commodity summaries. USGS, St. Louis, pp 118–119

Jasinski SM (2011) Phosphate rock. In: US Geological Survey (ed) Mineral commodity summaries. USGS, Mineral commodity summaries, pp 120–121

Jasinski SM (2012) Phosphate rock. In: US Geological Survey (ed) Mineral commodity summaries. USGS, Mineral commodity summaries, pp 118–119

JORC (2004) Australasian code for reporting of exploration results, mineral resources and ore reserves. The JORC code 2004. http://www.jorc.org/main.php

Llewellyn TO (1993) Phosphate rock. Minerals yearbook. US Department of the Interior, Bureau of Mines, Washington, DC

MacDonald GK, Bennett EM, Potter PA, Ramankutty N (2011) Agronomic phosphorous imbalances across the world's cropland. PNAS 108(7):3086–3091

Midgley JF (2012) Sandpiper project. Proposed recovery of phosphate enriched sediments from the marine mining licence area no. 170 off. Appendix 5. Namibian Marine Phosphate (PTY) LTD. J Midgley & Associates, Highett

Moyle PR, Piper DZ (2004) Western phosphate field—depositional and economic deposit models. In: Hein JR (ed) Handbook of exploration and environmental geochemistry. Vol. 8. Life cycle of the phosphoria formation—from deposition to the post-mining environment. Elsevier, Amsterdam, pp 45–71

Nziguheba G, Smolders E (2008) Inputs of trace elements in agricultural soils via phosphate fertilizers in European countries. Sci Total Environ 390(1):53–57. doi:10.1016/j.scitotenv.2007.09.031

Orris GJ, Chernoff CB (2004) Review of world sedimentary phosphate deposits and occurrences. In: Hein JR (ed) Handbook of exploration and environmental geochemistry. Vol. 8. Life cycle of the phosphoria formation—from deposition to the post-mining environment. Elsevier, Amsterdam, pp 559–573

Parsons T (1951) The social system. The Free Press, New York

Prud'homme M (2010) World phosphate rock flows, losses and uses. Paper presented at the British sulphur events phosphates, Brussels, 22–24 Mar

Prud'homme M (2011) Global phosphate rock production trends from 1961 to 2010. Reasons for the temporary set-back in 1988–1994. IFA, Paris

Sames CW, Wellmer F-W (1982) Exploration, part I: nothing ventures, nothing gained—risks, strategies, costs, achievements. Glückauf 117(10):267–272

Scholz RW (2011) Environmental literacy in science and society: from knowledge to decisions. Cambridge University Press, Cambridge

Scholz RW, Wellmer F-W (2013) Approaching a dynamic view on the availability of mineral resources: what we may learn from the case of phosphorus? Glob Environ Change 23:11–27

Sheldon RP (1987) Industrial minerals, with emphasis on phosphate rock. In: McLaren DJ, Skinner BJ (eds) Resources and world development. Wiley, New York, pp 347–361

Smil V (2000) Phosphorus in the environment: natural flows and human interferences. Annu Rev Energy Env 25:53–88

Smit AL, Bindraban PS, Schröder JJ, Conijn JG, van der Meer HG (2009) Phosphorus in agriculture: global resources, trends and developments, vol 282. Plant Research International B. V, Wageningen

Sorrell S, Speirs J (2010) Hubbert's legacy: a review of curve-fitting methods to estimate ultimately recoverable resources. Nat Resour Res 19(3):209–229

Stephenson PR (2001) The JORC code. Trans Inst Min Metall Sect B-Appl Earth Sci 110:B121–B125

Teitenberg T (2003) Environmental and natural resource economics. Pearson Education Inc, New Jersey

Tilton JE (1977) The future of nonfuel minerals. The Brookings Institution, Washington, DC

UN (1962) Permanent sovereignty over natural resources. UN Codification Division Office of Legal Affairs, New York

US Census Bureau (2012) Las Vegas (city), Nevada. http://quickfacts.census.gov/qfd/states/32/3240000.html

UNCTAD (2012) Intergovernmental forum on mining, minerals, metals and sustainable development. UN. http://unctad.org/en/Pages/MeetingDetails.aspx?meetingid=151

US Geological Survey (2002) Rare earth elements—critical resources for high technology, Fact sheet 087–02

USGS (2004) Mineral commodity summaries 2004. US Geological Survey, Washington, DC

USGS (2012) Mineral commodity summary 2012. US Geological Survey, Washington, DC

Vaccari DA, Strigul N (2011) Extrapolating phosphorus production to estimate resource reserves. Chemosphere 84(6):792–797. doi:10.1016/j.chemosphere.2011.01.052

van Kauwenbergh SJ (1997) Cadmium and other minor elements in world resource of phosphate rock, No 400. The Fertilizer Society, London

van Kauwenbergh SJ (2006) Fertilizer raw material resources of Africa. IFDC, Muscle Shoals

van Kauwenbergh SJ (2010a) World phosphate rock reserves and resources. IFDC, Muscle Shoals

van Kauwenbergh SJ (2010b) World phosphate rock reserves and resources. Paper Presentation at fertilizer outlook and technology conference hosted by the Fertilizer Institute and the Fertilizer Industry Roundtable, Savannah, 16–18 Nov 2010

van Kauwenbergh SJ (2012) Heavy metals and radioactive elements in phosphate rock processing. In: 4th global traps meeting, El-Jadida, Morocco, 17–18 Mar 2012

van Vuuren DP, Bouwman AF, Beusen AHW (2010) Phosphorus demand for the 1970–2100 period: a scenario analysis of resource depletion. Glob Environ Change-Hum Policy Dimensions 20(3):428–439. doi:10.1016/j.gloenvcha.2010.04.004

VFRC (2012) Global research to nourish the world. A blueprint for food security. Virtual Fertilizer Research Center, Washington, DC

Vielleville DE, Vasani BS (2008) Sovereignty over natural resources versus rights under investment contracts: which one prevails? OGEL 5(2):1–22

Wagner M (1999) Ökonomische Bewertung von Explorationserfolgen über Erfahrungskurven. Geologisches Jahrbuch, vol SH 12

Ward J (2008) Peak phosphorus: quoted reserves vs. production history. Energy Bull (Retrieved, September 21, 2011). http://energybulletin.net/node/33164. http://www.resilience.org/stories/2008-08-26/peak-phosphorus-quoted-reserves-vs-production-history

Watson I, Stauffacher M, van Straaten P, Katz T, Botha L (2014) Mining and concentration: what mining to what costs and benefits? In: Scholz RW, Roy AH, Brand FS, Hellums DT, Ulrich AE (eds) Sustainable phosphorus management: a global transdisciplinary roadmap. Springer, Berlin, pp 153–182

Weber O, Delince J, Duan Y, Maene L, McDaniels T, Mew M, Schneidewid U, Steiner G (2014) Trade and finance as cross-cutting issues in the global phosphate and fertilizer market. In: Scholz RW, Roy AH, Brand FS, Hellums DT, Ulrich AE (eds) Sustainable phosphorus management: a global transdisciplinary roadmap. Springer, Berlin, pp 275–299

Wedepohl KH (1995) The composition of the continental-crust. Geochim Cosmochim Acta 59(7):1217–1232

Wellmer F-W (1986) Risk elements characteristic of mining investments. In: 13th CMMI congress, 11–16 May 1986, Singapore. pp 17–24

Wellmer F-W, Becker-Platen JD (2002) Sustainable development and the exploitation of mineral and energy resources: a review. Int J Earth Sci 91(5):723–745

Wellmer F-W, Dahlheimer M, Wagner M (2008) Economic evaluations in exploration. Springer, Heidelberg

Chapter 3
Mining and Concentration: What Mining to What Costs and Benefits?

Ingrid Watson, Peter van Straaten, Tobias Katz, and Louw Botha

Abstract This chapter presents the activities in the Mining Node of Global TraPs, a multi-stakeholder project on the sustainable management of the global P cycle. The scope of the Mining Node is the extraction, primary processing of ore to produce phosphate rock concentrates and transportation to a port or processing plant. Phosphate ore deposits are primarily sedimentary in origin, although igneous, and, to a much lesser extent, guano deposits are also mined. A range of mining methods is used, and a significant amount of mining is by opencast or strip mining methods. Following the extraction of the ore, it undergoes a process of concentration before being transported to a plant for further processing or a port for export. World phosphate rock concentrates are produced in 37 countries, with the top ten accounting for 90 % of world production. Global phosphate production

---

Contributions from members of the Mining Node and participants at the Global TraPs March 2012 workshop are gratefully acknowledged.

I. Watson (✉)
Centre for Sustainability in Mining and Industry, University of the Witwatersrand,
Private Bag 3, WITS, Johannesburg 2050, South Africa
e-mail: ingrid.watson@wits.ac.za

P. van Straaten
School of Environmental Sciences, University of Guelph, Guelph, ON, Canada
e-mail: pvanstra@uoguelph.ca

T. Katz
GHH Fahrzeuge GmbH, Gelsenkirchen, Germany
e-mail: tobias.katz@ghh-fahrzeuge.de

L. Botha
Foskor, Moshate House, 27 Selati Road, P.O. Box 1, Phalaborwa 1390, South Africa
e-mail: louwb@foskor.co.za

has increased to a 2011 level of 195 million metric tons and is set to increase still further to meet, primarily, the growing demand for fertilizers for food, feed, and biofuel production. The overarching objective of the Mining Node, aligned to that of the Global TraPs project, is to address the question: *How can we contribute to sustainability in the phosphorus mining sector through promoting resource efficiency and innovation to avoid and mitigate negative environmental and social impacts, and contribute to food security?* Based on this question, areas of focus for further research and case studies include (1) the optimization of mining and beneficiation recovery rates; (2) addressing environmental and social impacts; (3) small-scale phosphate mining; and (4) mining costs. Transdisciplinary projects teams, involving both science and practice, will work on these focus areas.

Keywords Phosphate mining · Phosphate beneficiation · Environmental impacts of phosphate mining · Phosphorus losses in mining and beneficiation

Contents

1	Mining Activities on the Global Scale	154
	1.1 Phosphate Mining and Beneficiation	155
	1.2 Production Trends	157
	1.3 Potential Medium-Term Phosphate Rock Developments	159
2	Critical Issues	159
	2.1 Mining Extraction and Beneficiation Recovery Rates	160
	2.2 Environmental and Social Impacts	163
	2.3 Small-Scale Phosphate Mining and Agriculture	166
	2.4 Mining Cost Structure	168
3	Work in Global TraPs	168
	3.1 Knowledge Gaps and Critical Questions	168
	3.2 Role, Function, and Kind of Transdisciplinary Process	169
	3.3 Suggested Case Studies	170
References		171

1 Mining Activities on the Global Scale

The Mining Node's focus in the phosphate life cycle includes mine planning and development, extraction, primary processing of ore to produce phosphate rock (PR) concentrates and transportation to a port or processing plant (Scholz et al. 2011).

Mine planning and development is initiated following the discovery and evaluation of a resource during the exploration process and is, in turn, followed by downstream processing for fertilizer and non-fertilizer production.

1.1 Phosphate Mining and Beneficiation

Phosphate ore deposits are primarily igneous or sedimentary in nature, with the majority of phosphate ore (85 %) currently mined from sedimentary deposits, which account for about 90 % of world known phosphate reserves (IFA 2012). Sedimentary phosphate deposits are exploited in China, the United States, Morocco, Tunisia, Jordan, Syria, Saudi Arabia, Peru, India, Vietnam, Iraq, and Australia. Igneous phosphate deposits are currently being exploited in Brazil, China, Finland, Russia, South Africa, and Zimbabwe (IFA 2012). Guano-derived phosphate deposits account for only a small fraction of mined phosphates. Deposits on islands related to guano were particularly important in the mid- to late 1800s.

Phosphate ore is mined using both surface (opencast or strip mining) and underground mining methods (van Kauwenbergh 2010), with marine dredging having been investigated, for example off the coast of Mexico, the Eastern coast of the United States and currently off the Namibian coast. The mining method employed depends largely on the type, size, and depth of the deposit and economics. Phosphate mines vary in size and degree of mechanization with mining ranging from labor-intensive methods to highly mechanized operations (van Kauwenbergh 2010).

The proposed development of the Namibian Sandpiper marine phosphate project off the west coast of southern Africa provides an interesting example of planned marine dredging. The Sandpiper deposit occurs as unconsolidated sea floor sediments, located approximately 60 km off the coast of Namibia, in water depths of 180–300 m.

Phosphate sediments will be recovered using a trailing suction hopper dredge (Fig. 1), which includes a cutter head linked to a suction pipe. This is trailed along the seabed removing the target sediments and pumping it into the cargo hold (hopper). The dredger transports the sediments to an offshore discharge buoy pipeline, transferring the material to holding ponds on shore from where it is beneficiated (see Fig. 2; Midgley 2012).

Beneficiation plants are commonly associated with a mine. Once the phosphate ore is extracted, it typically undergoes a process of concentration (beneficiation), which may include primary screening, wet or dry screening, washing, flotation, magnetic separation and drying to produce what is commercially referred to as phosphate rock concentrate. Phosphate rock concentrates typically range between 28 and 40 % P_2O_5 (vs. phosphate ore which may have a grade anywhere between 5 and 39 % P_2O_5; IFA 2012).

Phosphate rock may be transported to a port for export or a domestic plant for direct use or further downstream processing. A limited amount of phosphate rock

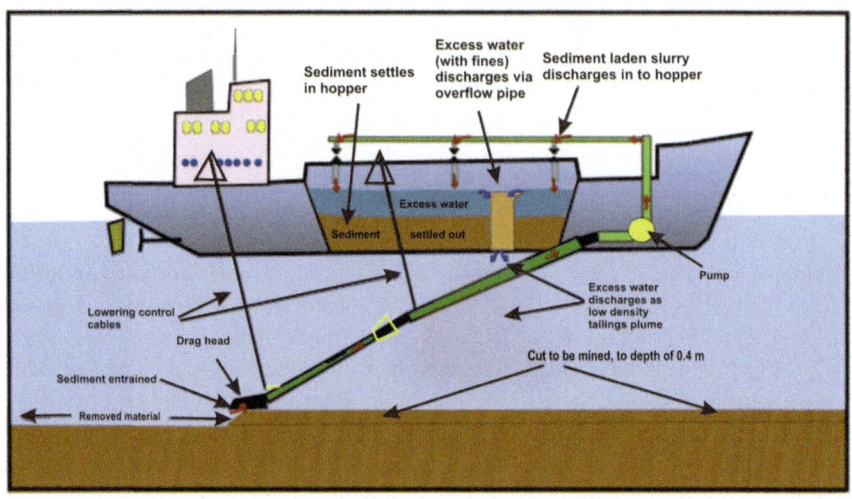

Fig. 1 A schematic of a trailing suction hopper dredge (TSHD) determined to be the optimal method by which the Sandpiper deposit can be developed (Midgley 2012)

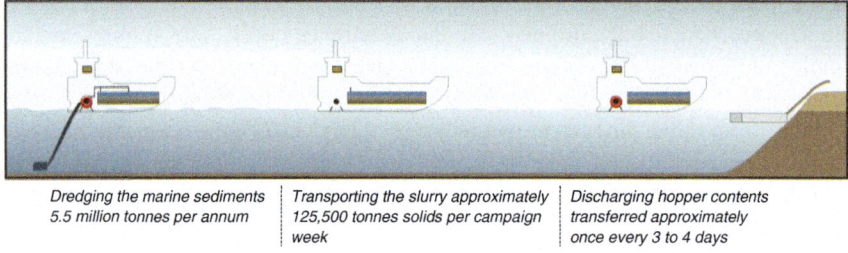

Fig. 2 The proposed dredging cycle (Midgley 2012)

may be sold or used without primary processing (Prud'homme 2012). Transportation is usually accomplished by trucks, railways, conveyor belts, or pipelines (as slurry). OCP in Morocco is currently investing in 235 km of slurry pipelines to transport all phosphate ore mined (in the form of pulp) from Khouribga, down gravity, to the chemical plants and port at Jorf Lasfar. This pipeline will replace transportation by train, saving water and energy and reducing the cost per tonne of phosphate delivered to Jorf Lasfar from the current 8 dollars per tonne to less than 1 dollar (OCP 2011).

According to the International Fertilizer Industry Association (IFA), phosphate rock concentrates are produced in 37 countries. The top ten producing countries account for 90 % of world production, with the top four contributing 72 %. Global reserves are similarly concentrated with Morocco holding around 70 % (USGS 2012). In terms of geographical distribution, close to 16 countries are producing phosphate rock in Africa and West Asia; 10 countries in Asia and Oceania; 7 countries in the Americas; and 4 countries in Europe and Central Asia.

1.2 Production Trends

Production of modern fertilizer started in the 1840s with James Murray becoming the first commercial vendor to treat phosphate-containing material with dilute sulfuric acid. The expansion of the industry stimulated the search for phosphate deposits. Extraction of high-quality apatite started in 1851 in Norway, followed by phosphate mining in the United States in the late 1860s (Smil 2000).

Current (2011) global phosphate rock production is estimated at 195 Mt (million metric tonnes), a significant increase since the 1961 level of 42.4 Mt (IFA 2011). As 80–85 % of phosphate rock is used in the fertilizer sector (IFA 2011), increases are attributable to a growing population and the need for phosphate fertilizer to grow crops for food, feed, and biofuels (Jasinski 2011).

Between 1992 and 2011, the global production of phosphate rock rose by an overall 40 %. This equates to an average annual growth rate of 2.1 %. Much of the net increase in production was driven by increased demand from the domestic market (home deliveries) (Fig. 3). Home deliveries of phosphate rock include material used in-country to make fertilizers, or other products. While the global export trade of phosphate rock remained relatively stable at around 30 Mt during the period from 1992 to 2011, home deliveries increased by 52 Mt, to reach 170 Mt in 2011. The share of home deliveries over total sales grew from 77 % in 1992 to 85 % in 2011 (IFA 2012).

Although demonstrating a general increase in output, world phosphate rock data indicate a temporary, conjectural decrease in production between 1988 and 1994, prompting some to propose a theory of "peak phosphate." According to the IFA (2011), this decrease was in fact related to a dramatic drop in use, which is directly related to the drop in fertilizer consumption following the collapse of the former Soviet Union. Phosphate fertilizer consumption in Eastern Europe and Central Asia (EECA) decreased at a rate of 35 % per annum between 1988 and 1994, which was paralleled by the drop in phosphate rock production of a rate of 21 % per annum for this period. The EECA contributed three-quarters of the 40 Mt decline in world phosphate rock output between 1988 and 1994. Since 1994, phosphate fertilizer consumption in the EECA has gradually recovered, with phosphate rock production remaining fairly stable (IFA 2011).

The world's top four producing countries, China, the United States, Morocco, and Russia, accounted for a stable share of 72–73 % of global production between 1992 and 2011 (IFA 2012). The following production trends are evident for these top four producers and are illustrated in Fig. 4 (IFA 2012):

- **China**, the world's largest producer at close to 77 Mt in 2011, has seen massive growth in production driven by a national investment policy to encourage domestic phosphate fertilizer production and reduce its prevalent heavy import reliance. China's phosphate rock exports have decreased to less than 0.8 Mt in 2011, from 5 Mt in 2001. This is due to the implementation of export restrictions on raw materials in order to increase the life span of this resource.

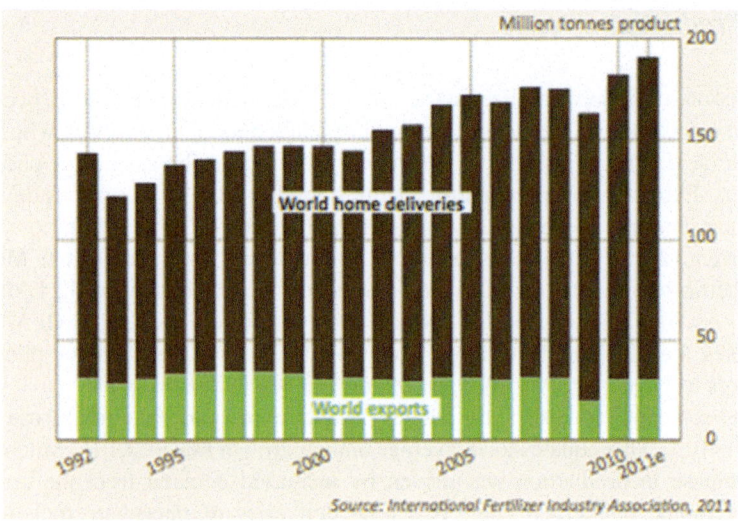

Fig. 3 Global deliveries of phosphate rock (IFA 2012)

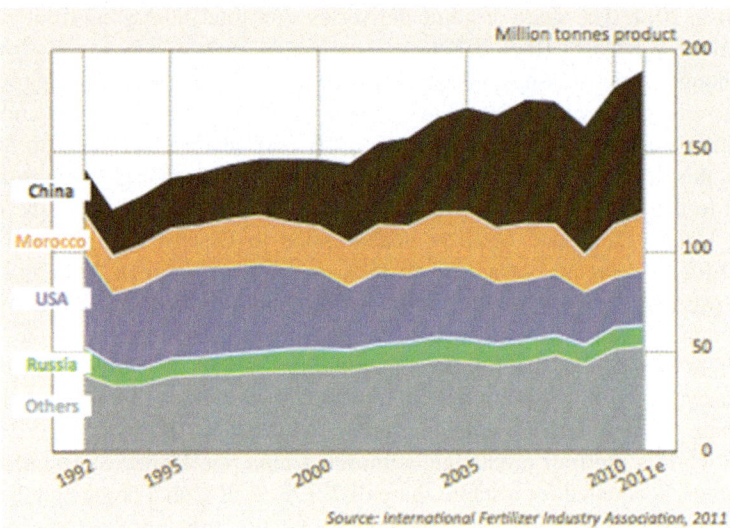

Fig. 4 Production of phosphate rock in the top four producing countries (IFA 2012)

- The **United States** has seen a gradual decline in production from 47 Mt in 1992 to 27 Mt in 2011. Factors contributing to this are a reduction in the exports of processed phosphates and phosphate rock due to exhaustion of higher-grade and quality reserves in the Central Florida Phosphate District, market factors due to rising domestic supply in large importing countries such as China, a decline in

the production of other P-based products in the USA and increased costs and unavailability of phosphate bearing lands due to environmental restrictions and legal challenges based on environmental regulations (van Kauwenbergh 2012).
- **Moroccan** phosphate production has increased to a current level of 29 Mt (for 2011), for the downstream production of phosphate fertilizers earmarked for the export markets. Morocco remains the world's largest exporter of phosphate rock with a market share ranging between 35 and 45 % of global phosphate rock trade.
- As already discussed, production in **Russia** fell between 1988 and 1994. Exports of phosphate rock have gradually declined, from close to 5 Mt in 1998 to less than 1.5 Mt. During the past decade, rock production remained stable at around 10–11 Mt, due to a gradual recovery in home deliveries of phosphate fertilizers.

Emerging production in Egypt, Algeria, Australia, Syria, and Peru has offset the gradual decline in formerly large phosphate rock producing countries such as Kazakhstan, Togo, Senegal, and Nauru (IFA 2012).

1.3 Potential Medium-Term Phosphate Rock Developments

Based on the 2013 IFA survey of future phosphate rock supply, world phosphate rock capacity would increase by an overall 11 %, from 261 Mt in 2012 to 290 Mt in 2017, assuming the realization of the ongoing projects. With the exception of North America potential supply is projected to increase in all areas with Africa, China, and West Asia accounting for most of the growth. These projections clearly demonstrate that there are sufficient reserves under development in order to meet the growth of P demand in the near term. The large number of prospective projects shows the abundance of accessible reserves for several decades (Prud'homme 2013).

2 Critical Issues

Based on this overall situation review, critical issues for the Mining Node were identified by a team representing industry, industry bodies, and academic institutions. These issues focus on different aspects that, we believe, contribute to sustainable phosphate management, as follows:

- Assessing mining and beneficiation recovery and identifying possible areas for improved and efficient extraction of phosphate rock resources.
- The environmental and social impacts and costs associated with phosphate rock mining.

- Artisanal and small-scale phosphate mining with the focus on local provision of "agrominerals" to enhance crop production.
- Understanding current cost structures in mining and likely scenarios for future mining costs when recognizing and incorporating environmental and social costs.

Each of these issues is introduced in the discussion below, to provide background to the critical questions identified.

Mining is just one aspect of the phosphate life cycle, and hence, the issues identified may be crosscutting, incorporating other nodes, or the whole life cycle. In reality, the boundaries between exploration, mining, and, in some cases, processing are blurred and are often undertaken by a single organization. In addition to this, issues such as environmental and social concerns may be relevant throughout the life cycle. Mining should therefore not be seen in isolation and the linkages between the critical issues identified for mining and those for the other nodes must be identified and understood.

2.1 Mining Extraction and Beneficiation Recovery Rates

Mining recovery is best described as a ratio between the amount of valuable commodity (ore) extracted and the total amount of valuable commodity determined by ore resource evaluation. Typically, only a portion of any resource is recovered by the mining process. Total mining recovery of the reserve would involve recovering 100 % of the reserve, normally not an economical or practically viable option. Mining recovery will vary and depend largely on geological factors of the deposit (shape, thickness, weathering zone, and overburden), mining method, skill of operators and economics. These factors are described as modifying factors and can vary tremendously during the life of the mine.

According to a survey conducted by the IFA in 2010, mining extraction efficiencies would average 82 % for the 93 % of the world phosphate producers surveyed (Prud'homme 2010). Two-thirds of producers operate above this weighted average ratio (Fig. 5).

There are numerous examples of phosphate rock mining in the world where substantial amounts of resources were spoiled or passed over during the initial mining phase, due to the fact that at the time it was not economic or sustainable to mine, process, and recover these resources (van Kauwenbergh 2012). Under different economic conditions and/or with improved technology (modifying factors), mining these resources would become viable. The Namibian Sandpiper marine phosphate project is such an example, where although the deposit was delineated during the 1970s, it has only now become economic to mine (Drummond 2012).

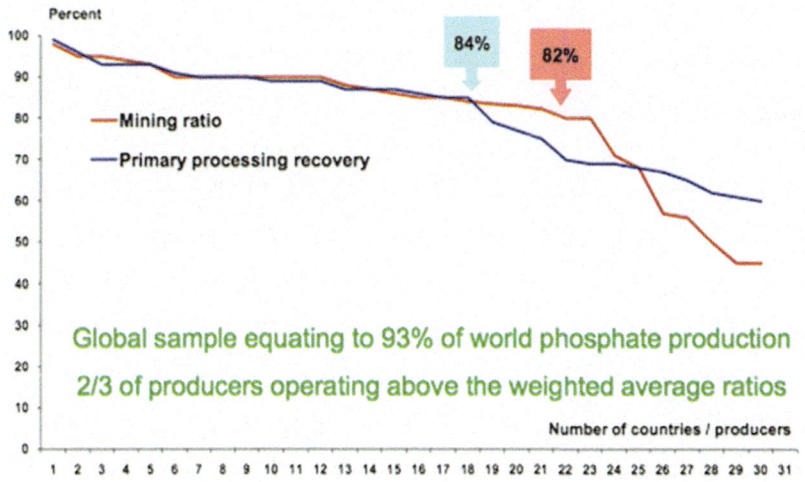

Fig. 5 Phosphate mining and primary processing recovery (Prud'homme 2010)

Beneficiation recovery refers to the quantity of P that is recovered during beneficiation. This would refer to the removal to saleable P expressed as a ratio to the ore mined portion only. Normally waste or mining discard does contain valuable commodity, but below a predetermined economically viable level. Again, the optimum extraction level of recovery would depend on the processes and technology used and economics. Depending on how the waste is disposed, and the remaining concentration of P, this may become a future source of P, given the right economic conditions and technological advances.

Historical trends have signaled a slow declining average grade (P content) for mined phosphate ore as well as derived concentrate (Fig. 6; see Prud'homme 2010; Smil 2000). Van Kauwenbergh (2010) has indicated that some of this effect is due to producers using lower-grade ores to meet growing demand and to recover more of the P. Recovery of P is inversely proportional to product grade. If a lower-grade concentrate (necessitating more P recovery) can be used to produce an intermediate such as phosphoric acid, and eventually high-grade fertilizers, lower-grade concentrate can be used, if economically profitable. The net effect is a better utilization of in situ ore resources toward beneficiable ore, thus prolonging the life of the mine. This strategy is typically implemented when phosphate rock production is integrated with localized fertilizer production. When phosphate rock is transported over long distances for fertilizer or other types of processing, the higher the grade, the less cost per tonne of P. This also leads to lower volumes that have to be processed to produce the same amount of acid or fertilizer.

Some of world supplies are more easily accessible, less costly to produce, and higher-grade reserves have been exploited. Some producers have been moving on

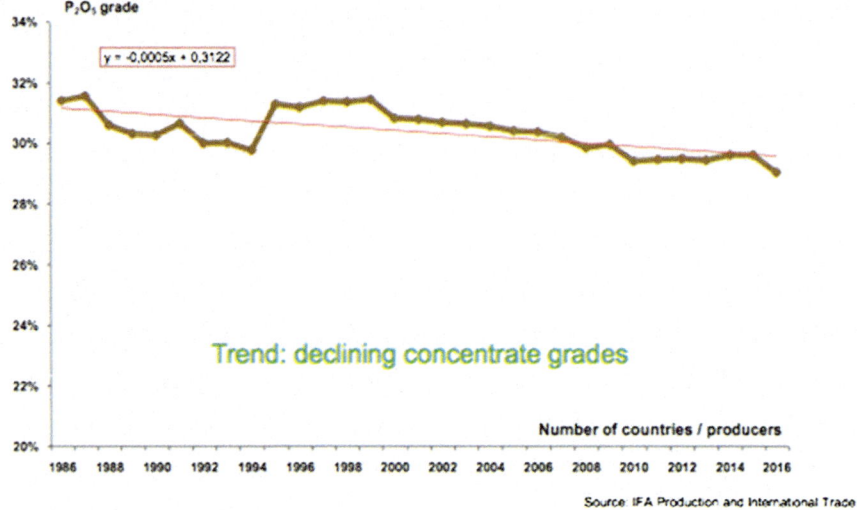

Fig. 6 Phosphate rock concentrate—trend of grade (Prud'homme 2010)

to lower-grade ore or reserves that are more expensive to exploit (van Kauwenbergh 2012). The current high market price of P makes such moves economically viable. Phosphate rock deposits under thicker overburden will be more costly to mine from the surface or by underground methods, phosphate rock deposits offshore may be more costly to mine, and lower-grade ores may be more expensive to beneficiate to make water-soluble products. Mining phosphate rock offshore or in environmentally sensitive areas may dictate higher costs for mitigation of potential environmental impacts.

Understanding mining recovery and the most efficient use of available resources in the phosphate industry is an area that requires further research. Ideally improving recovery should not come at the cost of increasing resource input (water, energy, and land) consumption, additional detrimental environmental and social impacts and limiting possible contributions to sustainability. Optimal recovery may be a preferred option.

Initiatives to address recovery should not only consider improving existing methods ("system optimization") but also focus on innovation and developing new methods (radical "system innovation") of mining and concentration (Rotmans et al. 2001). The factors driving efficiency, both endogenous (e.g., business necessity or corporate responsibility of individual companies) and exogenous (e.g., global commodity market forces, actions by political pressure groups, norms and standards of industry associations, national or international laws, requirements by purchasers down the supply chain), should be considered.

2.2 Environmental and Social Impacts

Phosphate mining, as with all mining, impacts the environment and society. The extent of impacts depends on the resource mined, the sensitivity of the receiving environment, social situation, formal and informal governance, and the practices and performance of the mining operation concerned. Impacts may be as a result of the construction, operation, and/or closure of a mine. Activities associated with the mine, such as transportation and settlements, also have a direct, indirect, or cumulative impact on the environment and broader society.

2.2.1 Environmental Impacts

Environmental impacts resulting from land-based mining operations can be generally grouped as environmental pollution, landscape and environmental change, and resource consumption. Resources consumed would include water, energy, and land. The impacts for marine dredging would be similar; however, landscape and environmental change would include disturbance of the sea floor, impact on fish habitat and marine organisms; and environmental pollution would largely be limited to water pollution (from sediments, spills, and waste). Marine dredging would however normally also have a land-based footprint (plant, waste facilities, storage, etc.).

Landscape and Environmental Change

Surface mining (open-pit or strip mining) is the most utilized method by far for mining phosphate deposits (van Kauwenbergh 2010; UNEP and IFA 2001). Surface mining generally has a larger physical footprint than underground mining. In addition to the physical mining area, the mine footprint may include processing facilities, overburden storage sites, waste impoundments, concentrate stockpiles, and other infrastructure, further increasing the surface area disturbed, and altering the topography. Within the working mine footprint, vegetation and topsoil are removed (UNEP and IFA 2001), which may impact on habitat, biodiversity, ecosystem functioning, and land productivity. Surface and underground mining may require the dewatering of large areas, leading to altered unconfined or confined aquifer characteristics.

The impacts of marine dredging are similar, although in a different receiving environment. Removing sediments from the seabed will disturb and change the habitat available, impacting on benthic flora and fauna, the water column and the broader functioning of the ecosystem (Midgley 2012). As on land, there are various mining systems that can be used in the marine environment, with differing impacts on the environment. Marine dredging will have an associated land-based footprint, including a storage facility, beneficiation plant, and waste impoundments.

Environmental Pollution

The major waste streams from phosphate mining and the adjacent mineral processing comprise waste rock, phosphatic tailings, and process water (UNEP and IFA 2001; Lottermoser 2007).

The main wastes in common phosphate mining operations include overburden, rock materials below cutoff grade, and zero-grade rock waste. Mine wastes from igneous phosphate deposits are clearly different from sedimentary phosphate deposits. Igneous phosphate deposits are in general more massive and homogeneous than layered sedimentary phosphate deposits. Mining of sedimentary phosphates requires not only the removal of overburden but also requires the physical removal of intercalated strata, such as clay-stones, carbonates, and other sedimentary rocks. Zero-grade material is "waste rock" material that has to be removed during mining operations, benching and accessing the deposit. Like in all mining operations, the overburden/soil-to-rock ratio (stripping ratios) varies considerably depending on local conditions. In PR mining operations, the stripping ratio varies from 0.5 to ≥ 3.

The assessment of what is "waste" and what is "phosphate ore" is subject to continuous revision in response to economic parameters, such as mining costs and market conditions (refer to Sect. 2.1). Some "wastes" generated from igneous phosphate mines might be re-assessed on their economic merits as they may contain valuable coproducts such as vermiculite, fluorite, Ba and Ti minerals as well as rare earth element (REE) minerals; and various clay and carbonate minerals in sedimentary phosphate deposits. Some of the coproducts of phosphate mining, e.g., vermiculite and carbonates, might become interesting targets for low-cost use in nearby agricultural communities, as well as for use in road construction.

Possible negative impacts arising from waste rock disposal include elevated radioactivity and radon levels as well as dump stability issues. Impacts are highly site specific; in most of the currently active mining operations, waste rock does not pose environmental threats (Lottermoser 2007).

Disposal of tailings, which may still contain some phosphate, has included discharge to rivers or other water bodies, and disposal to engineered storage impoundments or mined-out areas (UNEP and IFA 2001) as well as reclamation of sand particles as backfill material in mine workings (Lottermoser 2007). The disposal by discharge may pollute surface and ground water, soils and impact on ecosystem functioning. Kuo and Muñoz-Carpena (2009) cite an example from central Florida where mining activities degraded water quality in the upper Peace River basin, where the average dissolved phosphorus concentration of runoff water from waste impoundments exceeded the maximum allowable total phosphorus concentration discharging into a river established by the US Environmental Protection Agency, reinforcing the need for effective rehabilitation and ongoing management of mine wastes.

Currently, recycling and reuse efforts for waste from phosphate mining operations focuses on the utilization of waste rock for landscaping purposes

(Lottermoser 2011). In many operations, process water is recovered and reused (UNEP and IFA 2001).

2.2.2 Social Impacts

Social impacts may relate to competition for resources, health and safety concerns, in some cases resettlement, conflict resulting from unequal distribution of benefits and concerns around identity, community and way of life. At the general level, the research of social change in natural resource-based rural communities offers a broad account of the various opportunities and threats before, during the most intense activities (construction and operation), and after these (see, e.g., Krannich 2012; Freudenburg and Gramling 1992; Gramling and Freudenburg 1992). At the more specific level of P mining, the research base is rather restricted.

Phosphate rock naturally contains uranium and radionuclides of uranium; similarly, associated heavy metals such as cadmium can also be present (van Kauwenbergh 1997, 2001, 2002, 2009; Cordell et al. 2009). The impact of these elements on health and the environment at individual mining and processing sites will vary.[1] According to Othman and Al-Masri (2007) based on a case study in Syria, the phosphate industry is considered to be the main source of enhancement of naturally occurring radionuclides in the Syrian environment and that these elevated levels were found to be generally located around the workplaces in the mine, fertilizer factory and export platforms, increasing the exposure of workers.

2.2.3 Mitigation and Management

The legacy of some mining projects and processing operations on human health and the environment has tainted the public perception of the industry. However, the trend among mining companies globally is toward better management of impacts and improved environmental and social performance. This is also in large part due to having environmental professionals working on site, the use of improved scientific knowledge, an improved and formalized governance framework, greater public involvement and community expectations, and voluntary initiatives adopted by industry (Lottermoser 2011). Environmental management systems are now commonly implemented at mines, with many committing to industry codes of practice and regular reporting on their performance. Mine planning, closure, and rehabilitation have improved, in many cases allowing for the sustainable use of post-mining landscapes.

---

[1] Numerous environmental health and safety studies have been conducted in Florida by the Florida Institute of Phosphate Research (FIPR), now known as the Florida Industrial and Phosphate Research Institute (FIPR Institute) www.fipr.state.fl.us.

Environmental and social performances are just two aspects of what is commonly referred to as Social Responsibility (also termed corporate social responsibility (CSR) or corporate citizenship). The ISO26000: 2010 Guidance on social responsibility defines social responsibility as "the responsibility of an organization for the impacts of its decision and activities on society and the environment, through transparent and ethical behavior that; contributes to sustainable development, including health and the welfare of society; takes into account the expectations of stakeholders; is in compliance with applicable law and consistent with international norms of behavior; and is integrated throughout the organization and practiced in its relationships." Issues addressed under this banner include human rights, labor practices, the environment, fair operating practices, consumer issues, and community involvement and development.

The social responsibility of phosphate mining companies would vary between different companies. Annual sustainability reports, where these are produced, would provide an indication of what a particular company is doing in this regard. Within the mining sector, the International Council on Mining and Metals (ICMM), through their Sustainable Development Framework, provides guidance on what would be considered good practice. The 10 principles of the United National Global Compact, although not specific to the sector, are also indicative of what is considered responsible behavior.

The concept of decoupling may be useful in the context of both mining and the broader phosphate life cycle. Decoupling means using fewer resources and less of a resource per unit of economic output and reducing the environmental impact of any resources that are used or economic activities that are undertaken (UNEP 2011a). From a mining perspective, impact decoupling (increasing phosphate extraction while reducing negative environmental impacts) is possible and relates back to the issue of recovery and reducing resource intensity, discussed earlier. As regards the phosphate life cycle, opportunities to reduce phosphate demand (resource decoupling) through more efficient use, recycling and the use of substitutes are being addressed in other nodes. Reducing demand could however have a significant impact on the phosphate mining sector.

2.3 Small-Scale Phosphate Mining and Agriculture

In addition to the large phosphate deposits that are exploited commercially, there are many small- and medium-scale deposits (van Kauwenbergh 1987; van Straaten 2002). These deposits may present an opportunity to artisanal and small-scale miners and an alternative source of phosphate supply for local farmers. This is especially relevant to developing countries and in particular sub-Saharan Africa where soil nutrients have been depleted and imported commercial fertilizers are generally not available to under-resourced farmers.

In most of sub-Saharan Africa, more than 50 % of the population relies on agriculture for their livelihood. Agriculture is the major source of income, employment, food security, and survival (van Straaten 2002). However, crop yields are a quarter of the global average, due to the depletion of soil nutrients, with more nutrients being removed each year than are added in the form of fertilizer, crop residues, and manure (UNEP 2011b). A high proportion of African farmers are resource poor in terms of capital, land, labor, and livestock (van Straaten 2002), and this together with high costs and low accessibility prevents many African farmers from acquiring fertilizers (UNEP 2011b). Poor transport, low trade volumes, and lack of local production or distribution capacity result in farm-gate imported fertilizer prices two to six times higher than the world average (UNEP 2011b).

Additional phosphate, as well as other nutrients, is needed to achieve adequate sustainable crop yields. "Agrominerals" have the potential to, at least partially, address this need. Agrominerals, such as phosphate rock, liming materials, and gypsum, are naturally occurring geological materials in both unprocessed and processed forms that can be used in crop production systems to enhance soil productivity (van Straaten 2002, 2011). Unlike conventional, chemically processed "industrial" fertilizers, which are derived from chemically processed rocks, agrominerals are commonly only physically modified, by crushing and grinding (van Straaten 2002, 2011), and can be applied directly. Where they exist, local small to medium phosphate deposits, mined by artisanal and small-scale miners, using labor-intensive methods and appropriate technologies may have the potential to contribute to agricultural production.

It is however recognized that the effectiveness of agromineral use and economics of small-scale mining pose special challenges. The phosphate rock that is locally available may not be sufficiently reactive, or the local soils may not be suitable for use of direct application phosphate rock. The use of phosphate rock for direct application is dictated by apatite mineralogy, soil, crop and agroclimatic conditions. There are innovative ways to make PR resources more reactive using local modification techniques (van Straaten 2002, 2011).

Small-scale mines are often not economically viable. Small deposits may be located far inland or far from agricultural areas; transportation can be cost prohibitive. Due to the nature of the activity, small-scale mines generally have a low recovery rate (usually below 30 % in the case of China), selecting only the high-grade P rock. This results in a significant loss of P resources and has contributed to governmental restrictions on small-scale P mining (Ma et al. 2012). Artisanal and small-scale mining is also renowned for its unsafe work practices and often results in significant pollution and land degradation. A further concern is possible health impacts associated with the use of unprocessed rock, mainly sedimentary phosphate rock, exposing users to heavy metals and radioactive nuclides.

As part of the Global TraPs project, it would be important to develop an understanding of the extent and local importance of such mining and based on this, opportunities to enhance viable small-scale phosphate mining and processing in sub-Saharan Africa, for example in Tanzania and Burkina Faso, and in Bolivia and Indonesia.

2.4 Mining Cost Structure

Information on existing mining costs is difficult to obtain in the public domain; however, the cost structure information for new projects is generally in the public domain. An analysis of these would give an indication of mining costs and how these differ based on the geology, mining method employed and mine location and how mining costs may be changing to address the increasing costs of resources (water and energy in particular) and the "cost of compliance." It is proposed that such an assessment be included as a case study in the Global TraPs project.

3 Work in Global TraPs

Not all the identified critical issues will be dealt with within the Global TraPs project, mainly for the following three reasons. Firstly, all research activities need focus to allow for the necessary depth of analysis; secondly, the duration of Global TraPs with an expected final global conference in 2015 necessitates that those questions that can be investigated in this time period will be prioritized, the remainder will be recommendations for research plans; and thirdly, the transdisciplinary setting of Global TraPs requires potential benefit of the followed research both for practice and science, which is not given with all the issues.

The focus of this node is summed up in the guiding question defined jointly by practice and science—*How can we contribute to sustainability in the phosphorus mining sector through promoting resource efficiency and innovation to avoid and mitigate negative environmental and social impacts, and contribute to food security?*

3.1 Knowledge Gaps and Critical Questions

Based on this overall guiding question for the Mining Node, a number of critical questions have been identified by academia, industry, and further stakeholders in Global TraPs, as follows:

- Mining extraction and beneficiation recovery
  - What is the current recovery rate in mining operations and what are the main factors impacting on this?
  - What changes or innovations may have the greatest potential impact on optimizing recovery rates?
- Environmental and social impacts
  - What are the most promising avenues for decreasing or rationalizing the environmental and social impacts and costs in future?

– What is the typical water and energy use at different types of phosphate mining operations and what is the potential to reduce it (if warranted)?
– How do societal factors (perceptions of mining in society) impact eventual resource availability?

- Small-scale phosphate mining

    – How can the viability of small-scale phosphate rock mining and appropriate primary processing in sub-Saharan Africa be enhanced?

- Mining cost structure

    – What is the general current cost structure in mining, and likely scenarios for future mining costs, including energy and water, and environmental mitigation costs?
    – What would the impact of internalization of environmental costs be?
    – How does mining react to price changes over time?

An essential overarching process-related question of Global TraPs, namely "how to organize transdisciplinary processes at the global level?", will be addressed in parallel to these substantive questions in the Mining Node.

3.2 Role, Function, and Kind of Transdisciplinary Process

Transdisciplinarity understood here as mutual learning process between science and practice (Scholz 2000) necessitates balancing between various perspectives, interests and expertise of the involved actors. To this end, the Mining Node of Global TraPs tries to achieve inclusiveness at three levels.

Within Global TraPs, the Mining Node regularly presents its plans and results to the plenary for critical review by various stakeholders involved, for example representatives from various industry organizations, international organizations like UNEP, smallholder farmers, non-governmental organizations like Greenpeace and academic institutions from around the world. This further allows to coordinate with the other nodes and the cross-cutting issue of Trade and Finance.

The Mining Node itself is again composed of actors from practice (primarily industry and industrial associations) and science with various disciplinary backgrounds (e.g., environmental and social sciences, industrial ecology, geology). Here, the current situation of P mining was reviewed, critical issues derived and the research scope of the Mining Node defined. In a joint process between science and practice, pertinent case studies were selected (see below for some first sketches) and respective results will be exchanged and critically reviewed.

Transdisciplinary case studies will be developed locally. At local case study level, it is envisaged to build transdisciplinary project teams comprising not only experts from academia and industry but also representatives from additional actor groups, such as local communities, environmental NGOs. These project teams will first try to build a common problem understanding and formulate a jointly agreed upon concrete and case-study-specific guiding question. The project team will also discuss whom to involve in each project phase based on the nature of the work (Stauffacher et al. 2008). Subsequent project steps will be implemented in close collaboration within the project team. The project team will aim at producing tangible results in the form of orientations for future action of the various actors involved, entailing different possible future pathways of the analyzed concrete case and their respective potential positive and negative outcomes. In addition, the whole process of implementing the case studies should also lead to a mutual learning process of all the participating actors and likewise help building capacity locally and trust among the different parties.

3.3 Suggested Case Studies

Based on discussions between science and practice in Global TraPs, case studies to address the critical questions have been suggested, at the various levels, as follows:

Global TraPs Level

- **Extraction Rates and Beneficiation Recovery Rates**. Assess current mining extraction rates and beneficiation recovery rates in mining operations to determine what the main factors impacting these are. It is proposed that a survey of mining companies, based on the correct questions with the right terminology, will provide this information. There is a possibility for expanding this survey to simultaneously address critical questions from the Exploration and Processing Nodes. Core stakeholders that could be involved would comprise industry associations and scientists with a background in mining and primary processing.
- **Environmental and Social Impacts**. Determine what the most promising avenue for decreasing or rationalizing environmental and social impacts and costs in the future. The objective of this case study is to raise awareness of what is currently being done by mining companies by documenting their experiences. Again there may be an opportunity to expand the scope of this case study to include environmental and social impacts during the exploration process. Again, core stakeholders will be representatives from industry and scientists from various disciplines like for instance environmental and social sciences. Further, interactions with respective NGOs should be envisaged.

Mining Node Level

- **Innovation**. Document current innovation in mining and beneficiation. There are examples of innovation on improving mining and beneficiation recovery, reducing waste and increasing resource efficiency. The objective of this case study would be to collect information on these and write them up, with the possibility of further investigation and sharing within the sector. Industry and academia should be involved in this case study.

- **Mining Costs**. Documenting and analyzing mining costs. Information on existing mining costs is difficult to obtain; however, the cost structure information for new projects is generally in the public domain. An analysis of these costs could give an indication of mining costs and how these differ between projects and locations, what the basis for these differences are, and whether mining costs are changing to incorporate the increasing costs associated with more stringent legislation, as an example. Again, mainly industry and academia are concerned by this project.
- **Mining Impacts**. Assess and compare mining impacts between different types of deposits (sedimentary and igneous, land and marine) as part of an exercise to benchmark for sustainable mining. Industry and academia are core stakeholders concerned by this project.

Case Study Level

- **Small-Scale Mining**. Determine how the viability of small-scale phosphate rock mining and appropriate processing in developing countries can be enhanced. A comparative analysis between small-scale phosphate mining in a sample of developing countries is proposed, in particular focusing on opportunities for innovation and possibilities for "up-scaling" these. A comprehensive variety of different stakeholders has to be involved in the different countries, namely small-scale miners, farmers, community leaders, regional and national administration, NGOs, and academia.
- **Mine Waste**. Building on work already done, investigate the use of mine waste for other purposes, such as use in agriculture, as building material or for the extraction of other minerals. This will require an analysis of the mine waste streams and identification of opportunities for reuse. Besides industry and academia, potential users of mine waste (farmers, building sector, and processing industry) certainly need to be involved. In addition, local and regional policy makers and administration are to be integrated.

Acknowledgments The authors thank Michel Prud'homme, Executive Secretary Production and International Trade Committee, International Fertilizer Industry Association, and Stephen M. Jasinski, W. David Menzie, and Joyce A. Ober of the USGS for their critical feedback on previous versions of this text.

References

Cordell D, Drangert J, White S (2009) The story of phosphorus: global food security and food for thought. Global Environ Change 19:292–305

Drummond A (2012) The Namibian Sandpiper marine phosphate project: almost ready to go. Paper presented at the phosphates 2012 conference and exhibition, Morocco, March 2012

Freudenburg WR, Gramling R (1992) Community impacts of technological change: toward a longitudinal perspective. Soc Forces 70(4):937–955

Gramling R, Freudenburg WR (1992) Opportunity-threat, development, and adaptation: toward a comprehensive framework for social impact assessment. Rural Sociol 57:216–234

International Fertilizer Industry Association (IFA) (2011) Global phosphate rock production trends from 1961 to 2010. Reasons for the temporary set-back in 1988–1994. Feeding the earth series October 2011

International Fertilizer Industry Association (IFA) (2012) Debunking ten myths about phosphate rock production. Trends from 1992 to 2011. Feeding the earth series February 2012

International Organization for Standardization (ISO) (2010) ISO 26000:2010—Guidance on social responsibility

Jasinski SM (2011) Phosphate rock. 2010 minerals yearbook. Geological Survey, USA

Krannich R (2012) Social change in natural resource-based rural communities: the evolution of sociological research and knowledge as influenced by the contributions of William R Freudenburg. J Environ Stud Sci 2:18–27

Kuo Y-M, Muñoz-Carpena R (2009) Simplified modelling of phosphorus removal by vegetative filter strips to control runoff pollution from phosphate mining areas. J Hydrol 378:343–354

Lottermoser BG (2007) Mine wastes—Characterization, treatment, environmental impacts. Springer, Berlin

Lottermoser BG (2011) Recycling, reuse and rehabilitation of mine wastes. Elements 7:405–410

Ma D, Hu S, Chen D, Yourun L (2012) Substance flow analysis as a tool for the elucidation of anthropogenic phosphorus metabolism in China. J Cleaner Prod 29(2012):188–198

Midgley J (2012) Sandpiper project. Proposed recovery of phosphate enriched sediments from the marine mining licence area no. 170 off Walvis Bay Namibia. Environmental impact assessment report for the marine component. Final report. http://www.envirod.com/draft_environmental_impact_report2.html. Accessed 27 Feb 2012

OCP (2011) Annual report 2010. www.ocpgroup.ma

Othman I, Al-Masri MS (2007) Impact of phosphate industry on the environment: a case study. Appl Radiat Isot 65:131–141

Prud'homme M (2010) World phosphate rock flows, losses and uses. Paper presented at the phosphates 2010 conference and exhibition, Brussels, Mar 2010

Prud'homme M (2012) Personal communication

Prud'homme M (2013) Fertilizers and raw materials global supply, 2013–2017. Confidential report presented at the 81st IFA annual conference, Chicago, USA, May 2013

Rotmans J, Kemp R, van Asselt M (2001) More evolution than revolution: transition management in public policy. Foresight 3:15–31

Scholz RW (2000) Mutual learning as a basic principle for transdisciplinarity, in Transdisciplinarity. Joint problem-solving among science, technology and society. In: Proceedings of the international transdisciplinarity 2000 conference, Zürich. In: Scholz RW et al. (eds) Workbook II: Mutual Learning Sessions, Haffmans Sachbuch, Zürich, pp 13–17

Scholz RW, Roy AH, Ulrich AE, Eilittä M, Hellums DT (2011) Global TraPs' workshop III setting the stage for P research: identifying guiding questions, critical issues and case studies, Zürich 29–30 Aug 2011

Smil SV (2000) Phosphorus in the environment: natural flows and human interferences. Annu Rev Energy Env 25:53–88

Stauffacher M, Flüeler T, Krütli P, Scholz RW (2008) Analytic and dynamic approach to collaborative landscape planning: a transdisciplinary case study in a Swiss pre-alpine region. Syst Pract Action Res 21(6):409–422

United Nations Environmental Programme (UNEP) and International Fertilizer Industry Association (IFA) (2001) Environmental aspects of phosphate and potash mining. United Nations

United Nations Environmental Programme (UNEP) (2011a) Phosphorus and food production. UNEP Year Book 2011

United Nations Environmental Programme (UNEP) (2011b) Decoupling natural resource use and environmental impacts from economic growth, A report of the working group on decoupling to the international resource panel. Fischer-Kowalski M, Swilling M, von Weizsäcker EU, Ren Y, Moriguchi Y, Crane W, Krausmann F, Eisenmenger N, Giljum S, Hennicke P, Romero Lankao P, Siriban Manalang A, Sewerin S

US Geological Survey (2012) Mineral commodity summaries 2012: U.S. Geological Survey, p 198
van Kauwenbergh SJ (1987) Overview of the development of phosphate deposits in East and Southeast Africa. In: Proceedings of the East and Southeast African fertilizer management and evaluation network workshop. In: Ssali H, Williams LB (eds) Nairobi, Kenya, 27–29 May 1987, pp 1–6
van Kauwenbergh SJ (1997) Cadmium and other minor elements in world resources of Phosphate Rock. In: Proceedings no. 400, The Fertilizer Society, York, UK
van Kauwenbergh SJ (2001) Cadmium and other potential hazards. Fertilizer Int 380:51–65
van Kauwenbergh SJ (2002) Cadmium content of phosphate rocks and fertilizers. In: Proceedings of the IFA technical conference, Chennai, India, 24–27 Sep 2002, pp 1–39
van Kauwenbergh SJ (2009) Heavy metals and radioactive elements in phosphate rock and in fertilizer processing. Paper presented at the IFDC phosphate fertilizer production technology workshop, Oct 19–23, 2009, Marrakech, Morocco, pp 57
van Kauwenbergh SJ (2010) World phosphate rock reserves and resources. IFDC, Muscle Shoals, AL
van Kauwenbergh SJ (2012) Personal communication
van Straaten P (2002) Rocks for crops: agrominerals of Sub-Saharan Africa. ICRAF, Nairobi, Kenya, p 338
van Straaten P (2011) The geological basis of farming in Africa. In: Bationo et al. (eds) Innovations as key to the green revolution in Africa, Springer, Berlin, pp 31–47. doi:10.1007/978-90-481-2543-2_3

Appendix: Spotlight 4

Phosphorus Losses in Production Processes Before the "Crude Ore" and "Marketable Production" Entries in Reported Statistics

Roland W. Scholz, Friedrich-Wilhelm Wellmer, and John H. DeYoung, Jr.

Since the dawn of the industrial revolution in the nineteenth century, both food production and the world's population have experienced dramatic increases. Recent years have seen particularly significant benchmarks, with Africa reaching one billion people in 2009 and the world population reaching seven billion in 2011. Looking to the future, the United Nations' Food and Agricultural Organization (FAO High-Level Expert Forum 2009) and other experts have agreed that the population is likely to surpass nine billion by 2050. Increasing efficiency by avoiding losses, in particular if the latter are irreversible, is a basic concept of sustainable resources management (Pearce and Turner 1990). In the case of phosphorus, losses have traditionally only been looked at in the use phase. Two types of flows that have not been thoroughly investigated are unintended flows in the processing of other mineral commodities and pre-production losses. Unintended flows of phosphorus in metal processing take place owing to the trait that phosphates bind with metals. These flows have also been called virtual or hidden

R. W. Scholz
Fraunhofer Project Group Materials Recycling and Resource Strategies IWKS, Brentanostrasse 2, 63755 Alzenau, Germany
e-mail: roland.scholz@isc.fraunhofer.de

ETH Zürich, Natural and Social Science Interface (NSSI), Universitaetsstrasse 22, CHN J74.1, 8092 Zürich, Switzerland

F.-W. Wellmer
Formerly Bundesanstalt für Geowissenschaften und Rohstoffe (BGR)—Geozentrum Hannover, Stilleweg 2, 30655 Hannover, Germany
e-mail: fwellmer@t-online.de

John H. DeYoung, Jr.
National Minerals Information Center, USGS, 988 National Center, Reston, VA 20192, USA
e-mail: jdeyoung@usgs.gov

flows (Matsubae et al. 2011). The second flows are the losses in processes that take place before the extraction of crude ore and subsequent marketable production of phosphate rock.

Although it is impossible to provide a precise number, evidence exists that human activities may triple the global phosphorus (P) flows (Carpenter and Bennett 2011; Paytan and McLaughlin 2007; Ruttenberg 2003). Anthropogenic phosphorus flows start with the mining of phosphate rock and continue to the final use of all phosphate products and are influenced by the changes in land use where fertilizer products are applied. The flow of agricultural phosphorus during production and use is a result of the need to nourish an increasing human population. Because the main phosphorus mineral, apatite, is found in many countries, there is a very large geopotential to provide a source of the necessary phosphorus for animal and plant nourishment from nature. Open-field agriculture may accelerate the dissipation of phosphorus to marine systems by runoff and erosion of topsoil. Phosphorus is present in many mineral or metal ores. Thus, significant phosphorus flows are linked to many industrial processes, such as steel production, which have been analyzed in detail in Japan (Matsubae et al. 2011; Matsubae-Yokoyama et al. 2009). These industrial flows have been denoted as *hidden flows*, unintentional flows of phosphorus in the flows of other commodities, such as metals or goods. Some insights into intentional and unintentional losses are provided in Chaps. 1 and 6 of this book.

Phosphorus in by-product flows (the example of phosphorus in converter slags of steel production). Formerly, a large proportion of phosphorus for agricultural use came from phosphorus-containing iron ores, which were processed with the Thomas steel process. This process resulted in a slag rich in phosphorus that could be used as fertilizer and was thus a sought-after by-product. More than 100 years ago, it was stated that this slag is (or may be) "rich in lime and contains 14–20 % of phosphoric acid" (Porter 1913). The trend today is to smelt only iron ore that is as pure and as high grade as possible; thus, Thomas steel plants have gone virtually out of existence. The only places where phosphorus-containing Thomas slag is still sold as fertilizer is in some countries such as France or Luxembourg (Jasinski 2013b) where this material comes from stockpiles from former operations. The average phosphorus content of world iron ore production today is less than 0.1 % P. The trend is to use only those iron ores that contain as little phosphorus as possible. The minimal amount of phosphorus which still must be accepted in iron ore (and in coal, coke, and, in minute amounts, flux) is removed in the converters, which produce a slag with some phosphorus. In Japan, as in most other industrial countries, slag is no longer used as fertilizer. However, research is underway to use higher-grade converter slags as fertilizer again (Ohtake 2013). In Germany in 2011, 7.7 % of slags from the iron and steel industry (LD converter slag) still contained nearly 2 % P_2O_5 and were also rich in calcium oxide (CaO). These slags could be applied as fertilizer and soil amendment, similar to the highest grade slags in Japan (Drissen 2004; Stahlinstitut VDEh, Wirtschaftsvereinigung Stahl 2013; Nippon Slag Association, undated). Most iron and steel industry slag, however, is going directly (or indirectly, via cement production) into the construction industry as construction aggregate. Such phosphorus flows are lost and cannot be recovered with current technology.

Losses of phosphorus in "hidden flows" have been addressed by students of the exhaustibility of resources; for example, in the Hotelling rule (Hotelling 1931). Instead of locking up phosphorus forever, should the material be stockpiled to wait for times when recycling phosphorus from low-grade slags becomes economic? This approach could minimize losses for the benefit of future generations. The starting point for calculating losses from the resource to a marketable product is reserves determined during exploration work by mining companies and, if reported, compiled, and published by geological surveys or other government agencies, trade associations, or research institutions; for example, the US Geological Survey (USGS) Mineral Commodity Summaries (USGS 2013). There are also potential losses of phosphorus that are not included in the determination of reserves (see the Total Resources Box figure that illustrates how the concepts of reserves, resources, and geopotential are used in this text).

As pointed out by Roy et al. (2013), mineral fertilizer increased cereal crop production of the last 60 years. Estimates on lower agricultural production indicate that about 3.5 billion people would have starved if the increase in mineral fertilizer production and use had not taken place (Smil 1999; Hager 2008); over the last 50 years, global cereal crop production has almost tripled to 2.4 Gt (FAOSTAT/IFDC data 2012) (Fig. 7).

Not using *by-product* phosphorus may or may not be viewed as a loss. Potential by-product phosphorus could be a resource for future use if known or, if still unknown, geopotential. By definition, resources are known (to various levels of certainty), but may not be economically recoverable at present. Looking toward the future, resources or geopotential could be of more interest. The geopotential is not yet known but, by geologic reasoning, it can be expected to contain deposits that will be discovered by modern exploration technologies. Reserves are essential for the supply of phosphorus from primary sources today and in the future. Reserves are defined as the category of total resources that can be economically extracted with proven technology and current economic conditions (including having available energy and using environmentally and socioeconomically acceptable practices).

Phosphorus from primary deposits. Because phosphorus from by-product sources is generally no longer used, primary phosphorus must come from phosphate rock deposits. There are mainly two types of these deposits. The dominant type is of sedimentary origin; the other type is of magmatic origin. Both types are formed by geologic processes that result in deposits

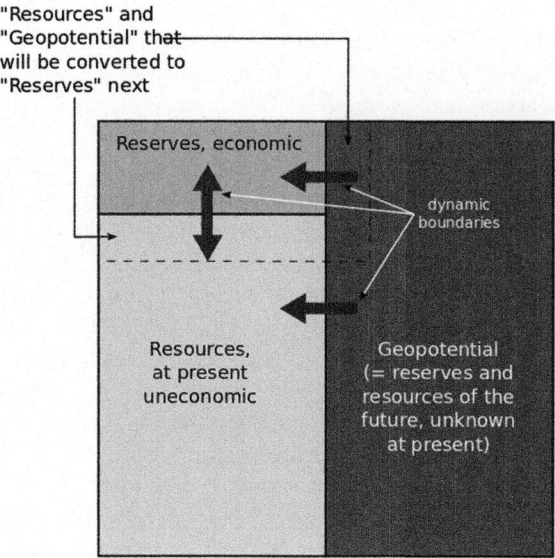

Fig. 7 The Total Resource Box illustrates the interrelationship and dynamics of reserves, resources, and geopotential; the area included by the dashed line outside the reserves box marks the resources and geopotential that will be converted to reserves next. This diagram (modified from that in Wellmer 2008) follows the convention of earlier depictions (Thom 1929, 1940, 1964; McKelvey 1972; Zwartendyk 1972) of using geologic/physical knowledge and economic viability as the horizontal and vertical dimensions of the defined "boxes"

enriched in phosphorus. The sedimentary deposits form on continental shelves or slopes with high biological activity, often stimulated by nutrient-rich upwelling currents. The magmatic deposits owe their existence to sudden and rare igneous events in which phosphate-rich magmas formed. For such mineral commodities, which are extracted from deposits where enrichment processes have taken place, Skinner (1976) postulated a bimodal distribution—the main peak being the normal distribution of an element in the earth's crust with the mean value (the background, or the clarke) and the second or minor peak being the mean of all enriched deposits, the "deposit peak." Here, the term *deposit* denotes all such parts of the total resources with unusually high concentration, regardless of whether the material may currently be economically extracted at a profit (Skinner 1976, 1979).

What part of the area under the "deposit peak" can be considered reserves (that is, can be economically extracted)? This depends on the cost structure (operating and investment costs, tax and royalty regime, etc.) to produce a saleable product. The boundary between ore (reserves) and sub-ore-grade material is called the cutoff boundary. Numerous papers address selection of the cutoff grade to optimize the economic return (e.g., Lane

1988; Bascetin and Nieto 2007). The minimum boundary is an operating cost cutoff, meaning that the grade of this cutoff just covers the operating costs of extraction and processing. This is a cutoff applied frequently in practice (Wellmer et al. 2008); it maximizes the reserves by minimizing losses.

Losses before selection of a mining site. Another type of potential *loss* may occur *before mining*. As the result of insufficient exploration, economic deposits and their reserves may not have been identified. Here, general economic arguments are important because, given the large phosphate reserves, exploration only pays if it provides added value from the company's or the nation's perspective. During times when it was perceived that there was a large amount of high-quality reserves of phosphate rock (equal to more than 300 years of current production), there was no urgent pressure to invest into exploration from a business perspective, but a company can always benefit from better understanding of deposits gained from continuously exploring the deposit and adjacent favorable areas in detail to find the optimal strategy of extraction (Wagner 1999). In recent years, there has been very active exploration for phosphate rock deposits with prices in the $150 to $200 per metric ton range. This may be because exploration has become more attractive for companies owing to increasing demand for phosphorus fertilizers and anticipated higher prices.

The starting point for calculating losses after the delineation of reserves. The actual or potential losses described above remain resources or geopotential because they have not been identified. As stated above, the starting point for defining *losses* is the identification of phosphate rock reserves and the subsequent annual production (crude ore and marketable production), such as those published in the annual USGS Minerals Yearbook and Mineral Commodity Summaries. For instance, 198 Mt of phosphate rock were mined worldwide in 2011 (Jasinski 2013a, b). Note that this "marketable production" tonnage does not refer to the phosphate rock which has been extracted or which was the subject of mining activities, but to a saleable product which was about 30 % P_2O_5. The USGS also publishes data about the P_2O_5 content of marketable production, which was 60.9 Mt in 2011, meaning a worldwide average grade of 30.8 % P_2O_5 in saleable products (Jasinski 2013b, Table 10). P_2O_5 content is a standard way of accounting for the phosphate that leaves the mine or a related production facility after primary beneficiation. The grades of reserves are generally reported as percent P_2O_5, providing a starting point for calculating losses after the delineation stage of reserves. The USGS publishes data on the mine production and P_2O_5 content of crude ore, but only for the United States. In 2011, US production of crude phosphate rock ore was 129 Mt containing 13.3 Mt P_2O_5 (Jasinski 2013b, Table 3). The reduction from crude ore to marketable production (129 to 28.1 Mt) resulted in a loss of 5.14 Mt of P_2O_5

(from 13.3 Mt of P_2O_5 in crude ore to 8.16 Mt of P_2O_5 in marketable production). This amount of P_2O_5 that is not produced amounts to 39 % of P_2O_5 in mined crude ore (or 63 % of P_2O_5 in US marketable production); if similar amounts remain in waste elsewhere in the world, that amounts to over 36 Mt of P_2O_5 each year.

Critical losses during the mining and beneficiation stages. Evidence indicates that a portion of these losses may be preventable. Some losses of phosphorus from runoff, erosion, food waste, etc., might be controlled better, but are these losses *critical* from an economic or environmental perspective during mining and beneficiation on the way from reserves to saleable product? Producers have strong economic incentives to avoid such losses as long as the increase in value of the material recovered exceeds the marginal cost of recovery.

Mining can be done either by open-pit or by underground methods. For open-pit planning, a marginal stripping ratio (i.e., the ratio between waste to be removed and ore) must be defined. The highest marginal stripping ratio comparable to the minimum cutoff boundary defined above is the stripping ratio for which the last ton of ore just covers the operating cost. As noted earlier, resources that cannot bear the higher mining costs must be left in the ground. Because sedimentary phosphate rock deposits are layered deposits, normally with a sequence of phosphate seams and interlayered waste, waste-to-ore ratios are the result of incremental decisions. After mining each seam of phosphate rock, a new decision must be made about whether the next phosphate seam below can carry the removal cost of the interlaying waste layer.

The cut of the open pit "moves" so that the waste removed is stacked in the mined-out areas. Because of this, after a decision has been made not to mine deeper, the deeper resources are, in general, practically lost. In only a few cases, lower layers may be left accessible for some years. This situation is not comparable to resources left in the ground owing to being below the cutoff grade, described above.

For reserves outlined for an open pit, the mining recovery rate is normally about 95 %; for underground mines, recoveries vary from 80 % to only around 50 % because pillars must be left in place to maintain rock stability. According to a recent survey (Prud'homme 2010), overall recovery rate was 82 %. These are final losses, which, under normal circumstances, may never be recovered.

After the ore is extracted from the orebody, *beneficiation* begins. Here, the extracted material is subjected to various types of *mineral processing*, including a process which is generally called *mineral dressing*. This process separates the *gangue*, i.e., the commercially worthless material which is mixed with the phosphate rock, from the economically processable material. In sedimentary deposits, "ore" material with low grades may be stockpiled, and the higher-ore-grade material may be subjected to *comminution* by *crushing* and *grinding*, *desliming*, or *flotation* (Zhang et al. 2006). Normally, the recovery rates in simple washing processes are lower than for flotation, but flotation is a more expensive process. Which ore-grade material is subjected to the various types of processing depends on the chemistry and mineralogy of the ore, technology, market prices, and the specific constraints imposed by the mining plan. In all cases, mineral processing produces tailings, which are the waste materials from the processes that are used to separate the gangue from the ore mineral(s). These *tailings* include some phosphorus; in the United States, this waste material is used to reclaim the mine and clay slimes in tailings ponds are a potential source of future P_2O_5 production. How much and whether this material may be subjected to reprocessing depends on many factors, including the technology of the mining facility. For other mineral commodities, especially base and precious metals, old tailings have been reprocessed owing to price increases and technological developments. Losses during processing can be either intended or unintended. Intended losses are accepted in order to achieve economic optimization. Unintended losses are the result of economic suboptimal operation. Some losses may be economically acceptable and others not; the latter may vary significantly depending upon the decision criteria and timeframe being used.

In 1988, the US Bureau of Mines examined phosphate availability and supply worldwide and concluded that the worldwide losses in the beneficiation stages varied between 21 and 60 % (Fantel et al. 1988). Using 1994 data, the German Geological Survey and the German Federal Environmental Office undertook a comprehensive material flow analysis for eight mineral commodities, including phosphate rock. For 61 % of world phosphate production, the study concluded that losses in the mining and processing stage amounted to 36 % (Kippenberger 2001). Industry sources report that, in 2010, total losses before processing are about 35 % (see Scholz et al. 2013).

References

Bascetin A, Nieto A (2007) Determination of optimal cut-off grade policy to optimize NPV using a new approach with optimization factor. J S Afr Inst Min Metall 107:87–94

Carpenter SR, Bennett EM (2011) Reconsideration of the planetary boundary for phosphorus. Environ Res Lett 6(1):014009

Drissen P (2004) Eisenhüttenschlacken—industrielle Gesteine. Report of FEhS-Institute (Institute for construction material research) Duisburg/Germany 1/2004

Fantel RJ, Hurdelbrink RJ, Shields DJ, Johnson RL (1988) Phosphate availability and supply, a minerals availability appraisal. US Bureau of Mines Information Circular 1987, Washington DC

FAO High-Level Expert Forum (2009) How to feed the world—Global agriculture toward 2050. FAO, Rome

FAOSTAT/IFDC (2012) Food and agriculture organization of the United Nations (FAO). FAOSTAT database, Rome. Available at http://faostat.fao.org

Hager J (2008) The alchemy of air. Harmony Books, New York

Hotelling H (1931) The economics of exhaustible resources. J Polit Econ 39(2):137–175

Jasinski SM (2013a) Phosphate rock. In: US Geological Survey mineral commodity summaries 2013. Reston VA

Jasinski SM (2013b) Phosphate rock. In: US Geological Survey minerals yearbook 2011, v. I. Reston VA

Kippenberger C (2001) Materials flow and energy required for the production of selected mineral commodities—Summary and conclusions. Geol Jahrbuch H, Heft SH 13, Hannover

Lane KF (1988) The economic definition of ore: cut-off grades in theory and practice. Mining Journal Books, London

Matsubae K, Kajiyama J, Hiraki T, Nagasaka T (2011) Virtual phosphorus ore requirement of Japanese economy. Chemosphere 84(6):767–772

Matsubae-Yokoyama K, Kubo H, Nakajima K, Nagasaka T (2009) A material flow analysis of phosphorus in Japan—The iron and steel industry as a major phosphorus source. J Ind Ecol 13(5):687–705

McKelvey VE (1972) Mineral resource estimates and public policy. Am Sci 60(1):32–40

Nippon Slag Association (undated) Chemical characteristics of iron and steel slag. http://www.slg.jp/e/slag/character.html. Accessed 5.6.2013

Ohtake H (2013) The promotion Council of Phosphorus Recycling of Japan—Motivation and prospects. Paper presented at the Global Conference on Nutrient Management, Beijing, 18–20 June 2013

Paytan A, McLaughlin K (2007) The oceanic phosphorus cycle. Chem Rev 107(2):563–576

Pearce DW, Turner KR (1990) Economics of natural resources and the environment. Harvester Wheatsheaf, Hertfordshire

Porter N (ed) (1913) Merriam-Webster's third new international dictionary of the English language, unabridged. G. and C. Merriam Co., Springfield MA

Prud'homme M (2010) World phosphate rock flows, losses and uses. Paper presented at the British Sulphur Events Phosphates 2010, Brussels

Roy AH, Hellums DT, Scholz RW, Beaver C (2014) Fertilizers change(d) the world. In Scholz RW, Roy AH, Brand FS, Hellums DH, Ulrich AE (eds) Sustainable phosphorus management—A sustainable roadmap. Springer, Berlin, pp 114–117

Ruttenberg KC (2003) The global phosphorus cycle. Treatise on Geochemistry. Elsevier, New York, pp 585–643

Scholz RW, Roy AH, Hellums DT (eds) (2014) Sustainable phosphorus management—A global transdisciplinary roadmap. In: Scholz RW, Roy AH, Brand FS, Hellums DT, Ulrich AE (eds) Sustainable phosphorus management: A global transdisciplinary roadmap. Springer, Berlin, pp 1–128

Skinner BJ (1976) A second iron age ahead? Am Sci 64:258–269

Skinner BJ (1979) Earth resources. In: Proceedings National Academy of Sciences, USA, 76(9):4212–4217

Smil V (1999) Long-range perspectives on inorganic fertilizers in global agriculture. Paper presented at the Hignett Lectures. International Fertilizer Development Center, Muscle Shoals, AL

Stahlinstitut VDEh, Wirtschaftsvereinigung Stahl: Jahrbuch Stahl 2013, Düsseldorf

Thom WT, Jr. (1929) Petroleum and coal, the keys to the future. Princeton University Press, Princeton, NJ

Thom WT, Jr. (1940) The goal of democracy or the road to prosperity and peace. The Summer School of Geology and Natural Resources, Princeton, NJ

Thom WT, Jr, (1964) The discovery, development and constructive use of world resources. In: Mudd S (ed) The population crisis and the use of world resources. Dr. W. Junk Publishers, The Hague, pp 496–535

USGS (2013) Mineral commodity summaries. Washington, CC

Wagner MKF (1999) Ökonomische Bewertung von Explorationserfolgen über Erfahrungskurven. Geol. Jahrbuch H, Heft SH 12, Hannover

Wellmer F-W (2008) Reserves and resources of the geosphere, terms so often misunderstood. Is the life index of reserves of natural resources a guide to the future? Zeitschrift Der Deutschen Gesellschaft für Geowissenschaften 159 (4):575–590

Wellmer F-W, Dalheimer M, Wagner M (2008) Economic evaluations in exploration. Springer, Heidelberg

Zhang P, Wiegel R, El-Shall H (2006) Phosphate rock. In Kogel JE, Trivedi NC, Barker JM (eds) Industrial minerals and rocks: commodities, markets, and uses. Society for Mining, Metallurgy, and Exploration, Littleton, CO, pp 703–722

Zwartendyk J (1972) What is "mineral endowment" and how should we measure it? Mineral Bulletin MR 126, Department of Energy, Mines and Resources (Mineral Resources Branch), Ottawa, Canada

Chapter 4
Processing: What Improvements for What Products?

Ludwig Hermann, Willem Schipper, Kees Langeveld and Armin Reller

Abstract This chapter describes the current activities of a multi-stakeholder project known as the "Processing Node of Global TraPs" which focuses on the sustainable management of the global phosphorus cycle. The node team will outline the current state on phosphorus processing (rock phosphate concentrate and phosphorus-rich secondary materials to fertilizers, feed phosphates, and non-agricultural products), identify knowledge gaps as well as critical questions and sketch areas for potential transdisciplinary case studies. The node's critical questions refer to efficiencies, losses, and the environmental footprint of the various manufacturing processes as well as the effects of applying products in terms of fertilizing value, spreading/accumulation of pollutants, and eutrophication as a result of excessive application. Further issues involve the future of local, not fully integrated processing and identification of potential knowledge gaps. The guiding question is, *How to improve the energy, water and material flow balances during the production of fertilizers and other P-based products?* Currently, phosphate processing primarily concerns chemical processing (91 % of concentrates) with acids. Only 5 % of rocks

---

L. Hermann (✉)
Outotec GmbH, Ludwig-Erhard-Strasse 21, 61440 Oberursel, Germany
e-mail: ludwig.hermann@outotec.com

W. Schipper
Willem Schipper Consulting, Middelburg, Zeeland, The Netherlands
e-mail: willemschipper@wsconsulting.nl

K. Langeveld
ICL Fertilizers Europe C.V., Eurocil Holding B.V., P.O. Box 313, 1000 AH, Fosfaatweg 48, 1013 BM Amsterdam, The Netherlands
e-mail: langeveld@iclfertilizers.eu

A. Reller
Resource Strategy, University of Augsburg, Universitätsstr. 1a, 86159 Augsburg, Germany
e-mail: armin.reller@wzu.uni-augsburg.de

are thermally processed to elemental phosphorus. If the latest technologies are employed, P losses during chemical processing generally do not exceed 5 %. The widely used phosphoric acid route (72–78 % of concentrates) transfers impurities to the product or to phosphogypsum, a massive by-product/waste flow amounting to five tonnes per tonne of P_2O_5 in phosphoric acid. About 82 % of rock phosphates are processed to fertilizers, 6–8 % to feed phosphates and the rest to non-agricultural products for a wide variety of applications. Rock processing is usually located near a phosphate mine in highly integrated manufacturing plants designed to process low-impurity rocks to water-soluble phosphate fertilizers with high nutrient concentrations. However, changing natural, societal, and environmental framework conditions challenge the prevailing paradigms. Benefits and drawbacks of high nutrient concentrations and water solubility will be investigated in transdisciplinary case studies, preferably in cooperation with an integrated global phosphate industry. Even though 82 % of rock phosphates are eventually used as fertilizers, they represent only 36 % of P inputs to European soils, by far outnumbered by the P inputs from secondary resources, such as manure, which account for 63 %. Excessive P application in regions with high livestock density and nutrient mining in regions with neither relevant animal husbandry nor access to mineral fertilizers represent a global environmental and food security problem.

Keywords Innovation in phosphate processing · Wet chemical processing of phosphate · Thermal processing of phosphate

Contents

1	Current Status of Knowledge on P Processing	184
	1.1 Wet Chemical Processing of Phosphate to Fertilizer	186
	1.2 Thermal Processing of Phosphate Rock	188
	1.3 Phosphate Processing to Feed Supplements and Detergents	189
	1.4 Current Status Summary	190
2	Work in Global TraPs	191
	2.1 Knowledge Gaps and Critical Questions	191
	2.2 Role, Function and Type of Transdisciplinary Process	193
	2.3 Suggested Case Studies	194
	2.4 Expected Outcome	201
References		202

1 Current Status of Knowledge on P Processing

The processing node of the Global TraPs (global transdisciplinary processes for sustainable phosphorus management) project covers the industrial conversion of phosphate rock and phosphorus-rich secondary materials by wet or thermochemical processes to a variety of liquid and solid concentrates that are used as fertilizers or as intermediates for P-based industrial products. Wet chemical processes encompass (a) reaction of rock phosphate with sulfuric acid to partly acidulated rock phosphate or single superphosphate as solid fertilizers and to phosphoric acid as an intermediate for high-analysis phosphate fertilizers and industrial products; (b) reaction of rock phosphate with nitric acid to produce nitrophosphate fertilizers. Thermo-chemical processes are used to produce (a) elementary P as an intermediate for food, detergent, clean room as well as other high-purity products and (b) different solid P concentrates, largely by melting the feed material.

The processing node is a supply chain node with the *body of knowledge largely found in fertilizer companies, consulting companies* and *independent consultants* specializing in intelligence and studies for the phosphate industry. Other organizations which may have an in-depth understanding of the issues associated with processing include a limited number of international organizations such as International Fertilizer Development Center (IFDC), International Fertilizer Industry Association (IFA), Center for Excellence in Fertilizers, Brazil (CFErt), Fertilizers Europe (the former European Fertilizer Manufacturers Association, EFMA), Florida Industrial and Phosphate Research Institute formerly Florida Institute of Phosphate Research (FIPR), the AleffGroup, London, and International Atomic Energy Agency, Vienna (IAEA). Other stakeholders include national, regional, and local governments as well as the public at large. Details about fertilizer-processing technologies are published in various books and papers, e.g., in the Fertilizer Manual, edited by IFDC/UNIDO (UNIDO/IFDC 1998) and in a series of eight Best Available Technology (BAT) booklets [edited by the former European Fertilizer Manufacturers Association (EFMA 2000)].

Process optimization can be viewed from various perspectives. One issue to consider is the efficiency and environmental impact of the production lines. The separation/recovery of heavy metals, uranium, and members of the uranium family is a known challenge as is the processing/disposal of by-products and waste. A second issue involves adjusting the products to the real needs of soils in different geographical and climatic zones. A third issue focuses on transport and logistics. Whereas large centralized processing plants are usually located near the source of raw materials, secondary phosphate sources (recovered from sewage, animal wastes, biomass, and industry) will be processed in much smaller, local manufacturing plants, providing an opportunity for continuing or revitalizing local processing in absence of primary resources.

The scope of the processing node's activities encompasses mechanical, chemical, and thermal processing of *rock phosphate concentrate* as received from mining and beneficiation and *mineral, phosphate (and potash and magnesium) containing material* as received from conversion of biomass to energy or precipitation of P from liquids via the struvite route.

The normative reference point of the transdisciplinary approach is *sustainable development* largely following the interpretation of the Brundtland Report "Our Common Future (World Commission on Environment and Development; Brundtland 1987).

The goal of phosphate processing to fertilizers is to achieve a safe, environment-friendly material with a high plant nutrition value directly applicable to soils and plants. Fertilizers are primarily addressed because of their dominant position in the phosphate market.

1.1 Wet Chemical Processing of Phosphate to Fertilizer

With the exception of a few rocks classified as reactive, phosphate minerals as they exist in the ground are not soluble and are difficult for the plant to access. To provide the plant with the phosphate it needs, in a form it can take up through its roots, about 96 % of mined phosphate rocks are processed. More than 90 % are acidulated by the wet chemical route: converted with sulfuric acid to phosphoric acid (72–78 %), treated with sulfuric acid to single superphosphate and partially acidulated rock phosphate (10–14 %), or converted with nitric acid to nitrophosphates (2–4 %). A few small plants treat rock with hydrochloric acid with the largest one in Europe scheduled for shutdown by 2014. Less than 5 % is converted to elemental phosphorus by a thermal process which is further transformed into chlorides, oxides, and sulfides acting as the entry point to produce a multitude of (organo-)phosphorus compounds as well as to food and clean room-grade phosphoric acid.

More than 82 % of the phosphoric acid produced by the wet chemical route is used to make fertilizers, about 6–8 % to produce animal feed supplements and a small percent goes to detergents. The remainder is employed in a wide variety of products and applications (Prud'homme 2010; Jung 2012; Shinh 2012). There is phosphate in fire extinguishers, camera film, and indoor light bulbs. It also helps make steel harder and water softer. It plays a part in making and dyeing cloth as well as in washing dishes. Phosphate is in the fluids used to drill for oil and gas and in cementing the drilling holes. It helps to polish aluminum and to protect steel from corrosion. Most high-purity acids (produced by the thermal route or by liquid/liquid extraction) are used in the food, beverage, and electronics industries.

Wet phosphoric acid is usually produced in a fertilizer manufacturing facility (sometimes called a chemical processing plant) which is not necessarily connected to the mining operations, though new plants are conceived as integrated systems where mining, beneficiation, and processing are in one location. If mining operations follow the geographic extension of commercially viable deposits, such as in Florida's mining district, processing plants will not relocate. It is less costly to ship the rock back to existing plants for processing than to move the processing operations and the phosphogypsum stacks associated with the process.

Once the phosphate rock has been separated from the sand and clay at the beneficiation plant, it goes to the processing plant. At this point, the phosphate concentrate contains 25–40 % P_2O_5. The average grades have been slowly declining over the last 20 years (Prud'homme 2010).

In the processing plant, the phosphate concentrate is reacted with sulfuric acid to produce the phosphoric acid needed to make the most widely used high-analysis phosphate fertilizers diammonium phosphate (DAP) with a market share of 38 % and monoammonium phosphate (MAP) with a market share of 27 %. Both products are made by reacting ammonia with phosphoric acid. The third high-analysis phosphate fertilizer made from phosphoric acid is triple superphosphate (TSP) with a market share of 7 % (IFA 2012). DAP, MAP, and TSP fertilizers are water soluble and available for plants to absorb through their roots.

The sulfuric acid needed to convert the phosphate rock into phosphoric acid is frequently produced at the chemical processing plant using liquid (molten) sulfur or pyrite (a sulfur-containing ore), most of which is shipped to a port and then trucked to processing plants. Since the 1970s' energy crisis, most phosphate companies capture the heat released during sulfur burning and sulfuric acid production and use it to produce steam. The steam is used to produce the heat required to concentrate the phosphoric acid and to generate electricity to run the plant. Because of the exothermic process when converting sulfur to sulfuric acid, integrated plants produce most of the energy they need and some sell excess energy to a commercial provider.

When sulfuric acid is reacted with phosphate rock to produce phosphoric acid, the by-product *calcium sulfate known as phosphogypsum* is also produced. There are approximately five tonnes of phosphogypsum produced for every tonne of P_2O_5 in phosphoric acid. Phosphogypsum, like natural gypsum, is calcium sulfate, but it frequently contains a relevant amount of radioactivity due to the radium that naturally occurs in most phosphate rocks (IAEA 2006). Because of this radioactivity, a 1992 US Environmental Protection Agency (EPA) rule bans most uses of phosphogypsum in the country (Lloyd 2004).

Numerous applications for phosphogypsum, such as in agriculture as a soil conditioner, or in construction as plasterboards, have been developed in all phosphate-producing countries. However, estimates suggest that currently some 3–4 billion tonnes of phosphogypsum are disposed of in stacks in more than 50 countries. These stacks are growing by 150–200 million tonnes each year (Hilton 2010). Moreover, a small number of phosphoric acid plants still discharge phosphogypsum to the sea, entailing environmental hazards and eutrophication, particularly in the absence of strong currents, such as in the Baltic Sea (Wissa 2003).

Apart from the by-product phosphogypsum, the chemical process produces gaseous emissions in the form of hydrofluoric acid (HF) and silicon tetrafluoride (SiF_4), released during the digestion of phosphate rock, which typically contains 2–4 % fluorine. In case the energy generated from the exothermic reactions in the process is not effectively recovered (IFC 2007), modest amounts of CO_2 are released to the air. Large amounts of CO_2 are released, however, from carbonates as part of the crystal structure of sedimentary apatite (francolite) or as impurities in the form of calcite or dolomite, the latter being a challenge to remove even with the latest beneficiation technologies available.

Whereas most NP fertilizers are processed from phosphoric acid, roughly 15 % of acidulated phosphate fertilizers do not use phosphoric acid as a starting product.

Phosphate rock can be partly or fully acidulated with sulfuric acid, the latter marketed as single superphosphate with 16–22 % P_2O_5. If phosphate rock is reacted with nitric acid, nitrophosphates are produced, which cover a wide range of NP and NPK grades. The presence of ammonium nitrate and the hygroscopic nature of the product impose special manufacturing, handling, and storing conditions for nitrophosphates.

1.2 Thermal Processing of Phosphate Rock

In principle, there are two thermal routes for mineral phosphate processing: (1) production of elemental phosphorus at 1,500 °C in an electric arc furnace as an intermediary for technical phosphate compounds and phosphoric acid for applications requiring high-purity acid and (2) reacting mineral phosphates at temperatures at or above 1,000 °C with alkaline compounds to relatively low-grade calcined or fused phosphates.

The traditional electric arc process, which is used in the Netherlands, the USA, China, Vietnam, and Kazakhstan, consists of two steps. The first step is an agglomeration of the rock at 1,500 °C producing chunks up to 10 cm in size or a wet granulation with a clay binder, followed by a heating step at 800° which produces pellets of 1–2 cm in diameter. This step also serves to remove carbonates and sulfates that are detrimental to energy use in the subsequent arc furnace process. The rock pellets are mixed with cokes as a reducing agent and—as an option depending on the rock's silica content—pebbles (SiO_2) for slag formation. The mix is fed to a furnace heated to 1,500 °C by means of electric resistance. Under these conditions, phosphate is reduced to P_4, which leaves the furnace as a gas, together with the by-product CO and some dust. The dust is removed in an electrostatic precipitator and—after calcination—landfilled or recycled into the process. The P_4 is condensed to a liquid. This is further processed (oxidized) to phosphorus chlorides, sulfides, and oxides which serve as building blocks for a multitude of bulk and fine chemicals, often in the form of organophosphorus compounds. Typical examples include flame retardants, herbicides, lubricant additives, or lithium–ion battery electrolytes. Part of P_4 is converted to high-purity phosphoric acid (25 % of P_4 production) which is used in the food and electronic industries. The resulting CO gas stream is used as fuel for sintering plants and other on-site processes. The calcium oxide, which is left in the furnace after the phosphate has reacted, combines with the SiO_2 to form a liquid slag, which is tapped and either quenched directly with water or cooled and then crushed. It may be landfilled or used for road construction.

Iron, present as an impurity in the rock, is also reduced in the furnace. It forms a separate, ferrophosphorus slag which contains roughly 75 % Fe and 25 % P, with small amounts of other metals. It is used as a steel additive (Schipper et al. 2001).

Advantages of the furnace process are the ability to use low-grade phosphate rock, the production of a relatively high-value building block with a diverse and attractive application spectrum, as well as a higher tolerance to a number of impurities such as silica, magnesium, and aluminum. Disadvantages are the high energy consumption and investment cost.

The emissions typically associated with the electric arc process for elemental phosphorus and thermal phosphoric acid include phosphate, fluoride, dust, cadmium (Cd), lead (Pb), zinc (Zn), and radionuclides (Po-210 and Pb-210; see IFC 2007).

The Tennessee Valley Authority (TVA) has developed a thermal process where a mixture of phosphate rock and magnesium silicates (olivine or serpentine) is fused in an electric or fuel-fired furnace. Several hundred thousand tonnes of the resulting calcium–magnesium–phosphate glass are produced in Japan, Korea, Taiwan, China, Brazil, and South Africa. It contains about 20 % P_2O_5 soluble to over 90 % in citric acid and plant-available MgO in the order of 15 %. The product reportedly produces better yields than acid-based fertilizers on acidic soils (UNIDO/IFDC 1998).

Other products from thermo-chemical processes such as Thomas Slag or Rhenania Phosphate have disappeared from the—predominantly European—market because they were by-products of outdated steel production processes (Thomas converter) or because of the increased energy cost following the first global oil crisis, when the last Rhenania phosphate plant in Germany was shut down in 1982.

1.3 Phosphate Processing to Feed Supplements and Detergents

About 3.1 million tonnes of P_2O_5 per year or 6–8 % of processed phosphates is supplied to the (livestock) feed supplement industry, chiefly in the form of mono- or dicalcium phosphates (MCP/DCP). Despite significantly increased livestock production, this market has been growing slowly because of the addition of phytase to animal feed, an enzyme improving the digestibility of naturally occurring phosphates in plant-based feeds (phytate) by monogastric animals (Jung 2012), thus decreasing the need to add digestible feed phosphates.

About 800,000 tonnes of P_2O_5 per year or close to 2 % of processed phosphates are used in the detergent industry, largely as sodium tripolyphosphate (STPP). The demand for STPP has fallen by about 1,000,000 tonnes of P_2O_5 since 2007, and the outlook is a potentially stabilizing production capacity at 1,200,000 tonnes of product per year with about 58 % P_2O_5 (Shinh 2012).

Detergent phosphates perform various relevant functions, in particular as builders and disinfectants in dishwasher detergents. Eutrophication, potentially the result of phosphorus-rich wastewaters, can be effectively prevented by phosphate elimination in sewage treatment plants. Even if phosphates were replaced by other chemicals, phosphorus elimination from wastewater would still be mandatory to avoid eutrophication, as human excretions typically contribute at least 75 % or more to the sewage system's P inflow.

1.4 Current Status Summary

More than 90 % of the globally mined phosphates are processed using the wet chemical route. The best phosphoric acid processes (hemi-dihydrate process) achieve a P_2O_5 transfer efficiency of 98.5 %, which does not leave much room for further improvement (BAT No. 4, EFMA 2000). In addition, BAT processes for the production of the most common phosphate fertilizers (DAP/MAP and single superphosphate/triple superphosphate) release excess energy due to the huge surplus energy formation in modern sulfuric acid processes, making the wet chemical fertilizer process very energy efficient (Jenssen and Kongshaug 2003).

In contrast, P_4 production has a lower phosphate recovery rate of about 94 % and a significant energy consumption of 13–14 MWh/t of product (Schrödter 2008). The product has little, if any, overlap with the fertilizer market. The only real competition is found in the purified wet phosphoric acid sector and the equivalent pure phosphoric acid produced via the oxidation of P_4. If high-purity products are required for applications in the food and beverage industries and even more so in the electronics industries, the need for downstream purification of fertilizer-grade phosphoric acid by multi-step solvent extraction and—where necessary—re-crystallization put the electric arc process on equal footing.

Gaseous emissions from both process routes are minimized by the latest air pollution control systems.

In conclusion, a wide variety of mature BAT techniques, which are highly efficient in terms of energy and material use, are available at large scale on all five continents. Technology innovation is not a priority issue for the industry, at least if high-grade/low-impurity concentrates can be made available by mining and beneficiation operations.

However, there are an unknown number of older processing plants which are not retrofitted with BAT techniques and which continue to use unsustainable practices, some of them giving rise to serious concern about their environmental impact. A number of phosphoric acid plants continue to discharge phosphogypsum to the sea, adding point loads of thousands of tons of P to the diffuse inflows from runoff and erosion, regardless of the sensitivity and eutrophication state of the specific aquatic environment. These practices usually do not draw much public attention and foil widespread efforts to prevent P losses to aquatic bodies. In these cases, technology innovation is not the solution to protect aquatic environments; instead, transdisciplinary actions by regulators, shareholders, managers, engineers, and the public-at-large are all that is required to gradually retrofit or close polluting facilities.

As a result of the continuously employed efficiency improvement strategies, a sensitive balance between feedstock qualities and process parameters has been established. Lower-grade/higher-impurity phosphate ores and secondary materials entail lower efficiencies or require additional beneficiation steps. At this end, technology innovation may help to avoid higher losses and/or higher energy consumption. In a more system-oriented approach, the first step is a review of real nutrient requirements for regional crops and soils as well as an assessment of the common practices for their compliance with a changing environment.

Increasing efficiency has also been a driver for high water solubility and concentration of phosphates in fertilizers because it reduces the volumes of material as well as transport and handling costs. The high water solubility requirement will be called into question with the possible consequence of more local and smaller-scale processing.

A reason for environmental concern and economic burden is the large amount of phosphogypsum left behind as a by-product, and, to a large extent, as a waste needing long-term management. Phosphogypsum, its characteristics and potential uses have been the subject of research for over 30 years and stacks are still growing.

A certain pressure on heavy metal concentrations in phosphate fertilizers, in particular on cadmium, is expected from the European Commission which is currently working on a new fertilizer regulation to replace Regulation (EC) 2003/2003, the document currently in force.

Though desirable from an environmental perspective and feasible from a technical standpoint (20,000 tonnes of U_3O_8 were recovered between 1978 and 1998, for instance at Prayon in Belgium) and possibly profitable from an economic point of view (the current price of >USD 100/Kg should allow a payback time of <10 years for a plant producing >200,000 kg U_3O_8 per year), uranium recovery from phosphoric acid no longer happens (Stana 2009). Potentially revisiting the entire phosphorus process chain with one of the leading phosphoric acid suppliers could give a stimulus to restart uranium recovery.

Another stimulus could come from reviewing the potential benefits of extracting rare earth elements, often associated with phosphate deposits. Rare earth recovery from phosphoric acid has been a subject of research for years (Koopman 1999) with little practical impact.

2 Work in Global TraPs

2.1 Knowledge Gaps and Critical Questions

In a perfect world, phosphate ores are abundant and universally accessible, pollutants removed by technologies available at pilot or industrial scale, uranium recovered by industrially proven technologies, phosphogypsum safely used, and the phosphate cycles closed. However, in real life, most of these issues are controversially discussed and do not happen. In addition, an increasing number of phosphate mine operators have to cope with lower-grade ores and higher impurities—at least in the long run—resulting in inferior process efficiency (Burnside 2012). Moreover, due to plants and processes which are not being sustainably managed—still an issue at certain manufacturing locations—large amounts of phosphates are lost to aquatic bodies, leading to eutrophication and loss of biodiversity. Even if the phosphate industry is only marginally accountable for the eutrophication problem, occasional disregard of widely accepted practices is affecting its activities and reputation.

For several decades, efficiency and pollution prevention were the drivers for research (Scholz 2011) within the phosphate industry. The decrease in phosphate grades has become a rather recent issue (Prud'homme 2010), and efforts have been made to cope with the situation by improving traditional technologies. In response to environmental pressures, the phosphate industry assumed its responsibility by recommending fertilizing rates corresponding to good agricultural practice, largely limiting fertilizer application to phosphate uptake by crops.

Pollutant transfer from phosphate fertilizers to cropland is still an issue, though technologies for their removal have been developed, however, not on a large-scale level and not yet for universal application.

Removing cadmium, for instance, could be implemented during the calcination of rock or phosphoric acid processing by transferring existing technologies to the phosphate industries. However, industrial-scale implementation may still need some research work. Relevant barriers to overcome include commercial considerations and the current lack of legal limits values.

Long-term security of food and energy supplies may motivate uranium recovery from phosphoric acid where processes have been successfully performed in large-scale industrial plants for over ten years. By the end of the twentieth century, plants were closed for economic reasons after uranium prices had significantly dropped. There are some knowledge gaps related to the capital and operational costs of a second-generation plant built after lessons were learned and applied from first-generation plants (Stana 2009). Process and plant design optimization will be the technical tasks within a potential transdisciplinary case study which focuses on a concerted plan to manage critical resources in a strategic and sustainable manner.

An initial review of current phosphate processing indicates that technologies have been developed for separating and recovering uranium, heavy metals, and fluorosilicic acid as well as for different uses of phosphogypsum and for efficient production of phosphate fertilizers with high plant availability (Zhang et al. 2009). With a few exceptions in China, all but the last technology are not in operation, allegedly due to economic factors.

To address the knowledge gaps, the processing node team in Global TraPs has agreed on the following guiding question: How can the *energy balance*, *water balance,* and *material flows* of P processing be improved when producing fertilizer and other P-based products? More specifically detailed, critical questions for the processing node include the following:

- What are the actual P *recovery rates* by process and by other factors? What is the impact of lower-grade/higher-impurity rocks on *P-losses* in BAT processes? What *innovations and/or technologies* may help to support high processing efficiency in response to changing rock and concentrate elemental composition?

- How is *phosphogypsum* currently being disposed of, used, or stored? What are the current estimated quantities of phosphogypsum in storage around the world? What is the state of utilization globally, and, in particular, what barriers limit the use of phosphogypsum?
- What technological means and strategies are meaningful to cope with the *heavy metal, uranium,* and *other radionuclide* problems and what may be the costs and ramifications of removing/recovering these elements? What are the risk management options and their advantages and disadvantages?
- What may be the upside of separating (removing) elements of concern in terms of market value and sustainable management? Which elements potentially have a value that could sustain removal/separation technologies?
- What *non-apparent knowledge gaps* can be identified with respect to fertilizer processing?
- Is there a *future for* local processing *primary and recovered (secondary) phosphates*, or is the trend to large, centralized, and vertically integrated rock processing inevitable?
- How can we support the creation of *regulatory frameworks* for processing different P resources to marketable products?
- How can we promote the quality and acceptance by farmers of locally produced, potentially lower-analysis, and not water-soluble fertilizers?

2.2 Role, Function and Type of Transdisciplinary Process

Role, function, and the type of transdisciplinary process strongly depend on the issue at hand, such as case studies and the respective guiding question (Stauffacher et al. 2008).

With regard to processing, we can identify three types of potential case studies (see Sect. 3) which demand different transdisciplinary processes. While "re-thinking and innovating the overall processing of P to fertilizers" focus on technical issues, the "valorization of phosphogypsum" begins with a situation where technical questions have been almost completely answered, yet no agreement seems to be achievable among the key players due to differing risk assessment and risk management positions. The third type of case study, "enable the local processing of primary and secondary resources," includes both technical–scientific aspects and non-technical questions.

The *first type* of case study (see Table 1) involves optimizing phosphate production processing. The predominant function of the transdisciplinary process is *capacity building,* such as production of new knowledge. Depending on the topical focus, a number of scientific disciplines and expertise from various practices must be included. As new knowledge might directly affect a fertilizer manufacturer's performance, a clear commitment on how to use and share new knowledge between potential partners from science and industry will be necessary. The case facets might be about chemical and physical properties of fertilizers, the removal

Table 1 Illustrating type of case study, function/role of transdisciplinary process and potential guiding question

Type of case study (focus)	Primary function/role of transdisciplinary process	Potential guiding question
Rethinking and innovating the overall processing of P to fertilizers *(technical focus)*	Capacity building	How can phosphate production in terms of sustainability be optimized while meeting requirements of regional characteristics such as crop type, soil conditions, and climate?
Valorization of phosphogypsum *(non-technical focus)*	Mediation, consensus building	Are there options and ways of using phosphogypsum which can meet all stakeholder group needs?
Enable the local processing of primary and secondary resources *(technical and non-technical focus)*	Capacity building, consensus building	What are the requirements to match secondary resources with processing facilities in terms of stakeholder interests, material properties, logistics, and process parameters?

of impurities/contaminants from products, and different aspects of process optimization with regard to crop preferences and soil conditions in target regions.

The *second type* of case study involves the potential further use of phosphogypsum and the related environmental concerns and regulatory obstacles. A potential case study might bring together key stakeholder groups to further analyze and evaluate options for phosphogypsum use. A neutral institution such as ETH Zürich co-leading with the respective industry and regulators might initiate and establish a transdisciplinary process focusing on *mediation* and *consensus building* among stakeholder groups. Stakeholders from a number of fields such as the industry, potential users, administration/regulatory bodies, and NGOs together with leaders from the science community would participate in every step of the process.

The *third case study type* deals with locally processing primary and secondary P resources. In a first step, node team members must identify hot spot areas with high livestock density. Then, they need to invite potential co-leaders among manure producers and potential users of products from manure processing. In addition, relevant scientific disciplines and industrial expertise as well as key stakeholders need to be involved in the study. Dominant functions of the transdisciplinary process include *capacity building* and *consensus building*. Faceting can be done along the supply chain including *manure production, processing, distribution, use, and regulation*.

2.3 Suggested Case Studies

In response to the controversial topics of phosphogypsum management, the gradually degrading natural feedstock, the need for closing the phosphate loop, and preventing eutrophication of aquatic bodies, the node team identified three relevant areas for transdisciplinary case studies.

2.3.1 Innovating P Processing: Rethinking and Innovating the Overall Processing of P to Fertilizers

Primary resources are frequently located far away from regions of high phosphorus use, requiring high nutrient concentrations in fertilizers to avoid disproportionately high logistics costs. Thus, fertilizer types such as DAP, MAP, and TSP with nutrient concentrations ranging from 45 to 64 % were developed and have been dominating the phosphate fertilizer market ever since.

If large secondary raw material flows and decentralized processing plants were available in regions with high phosphorus consumption, lower nutrient concentrations might be acceptable due to the reduced transport distances. In addition, plant uptake from soils with medium and high phosphate concentrations could be replenished with fertilizers of lower nutrient concentrations and with no instant (water) solubility yet with a predictable release pattern as guaranteed by EU fertilizer types.

Changing framework conditions advocate reviewing the overall process chain.

- Low-grade primary and secondary phosphates are not easily processed to high-analysis fertilizers—many products from secondary resources come with P_2O_5 concentrations in the order of 15–25 %, largely comparable to single superphosphate and other low-analysis but high-fertilizing value products. They generally contain additional nutrients such as sulfur, calcium, magnesium, potash, manganese, and others.
- The growing awareness of dependency on limited resources and the increasing acceptance of political regulation may open a window of opportunity for reviewing and re-evaluating the potential for uranium recovery from phosphoric acid. If this opportunity is missed, the expertise gained during the engineering and operating of U recovery plants in the 1960s, 1970s, and 1980s may be irrecoverably lost with experts slowly retiring from the business.
- Legitimate decision makers such as the European Commission are currently reviewing fertilizer regulations aiming at a par market access of primary and secondary resources while limiting pollutant concentrations in fertilizers, regardless of their raw material basis. If the currently proposed limits are enforced, many feedstock materials will need to be processed by thermo- or wet chemical metal separation processes.
- Some phosphate rocks contain relevant mass fractions of rare earth elements that may be commercially extracted during rock processing. Extracting rare earth elements may improve the economic viability of processing low-grade rocks and removing undesirable elements from the fertilizers.

This is not a complete list of arguments for revisiting the phosphate production processes and trying to accommodate the needs of different stakeholders. Whereas most critical questions are raised from a global perspective, different answers may be developed from a regional or local perspective. Thus, similar transdisciplinary case studies should be performed in selected regions with different economic, societal, and agricultural framework conditions.

These critical questions need to be addressed:

- What is the most efficient and sustainable way to supply nutrients to crops in a designated target region? What are the characteristics (chemical and physical properties) of phosphate fertilizers that best comply with the criteria set by answering the first question? How could these criteria be translated into process and product specifications?
- What is the required nutrient solubility on various soils and under different climatic conditions to achieve a high fertilizing value and how is the determined solubility reproduced by chemical (solubility) tests to enable manufacturers to control the quality of their products online?
- How can the removal of impurities and pollutants, in particular cadmium and uranium, be integrated into phosphate processing without creating unacceptable cost hikes? Can the removal of impurities/pollutants be financed by the exploitation of additional high-value products such as rare earth elements? Which commercial, technical, political, and legal framework conditions are needed to facilitate the sustainable management of impurities and pollutants?

Case studies reviewing the fertilizer manufacturing process require fertilizer producers as practice stakeholders. Partnerships are sought with integrated phosphate fertilizer manufacturers who have the resources to actively participate and transfer the beneficial results to the daily processes. Scientific stakeholders can be selected from the Global TraPs team and from NGOs who have conducted a comprehensive amount of research work during the last half century.

2.3.2 P Gypsum: Valorization of Phosphogypsum

Phosphogypsum has been a major research subject for many decades. The FIPR has conducted targeted projects since 1979 and partnered in 2005 with the Aleff Group in the project "Stack Free by 53" (www.stackfree.com). The FIPR and Aleff Group have published a large amount of evidence on phosphogypsum, both on research and applications, much of it in the public domain. A phosphogypsum working group has been set up with members from industry, academia, and regulatory bodies. After the research phase was completed in 2009, the current action plan (2011–2014) aims at using the entire current and future production of phosphogypsum, and this target must be supported by both countries and international agencies.

After preliminary considerations, the goal seems to be feasible (Wissa 2003) with a focus on agricultural uses (fertilizer, soil conditioner, soil remediation, and increased water efficiency) and construction, including materials for the construction of wallboards and roads. The IAEA has drawn up "The Phosphate Industry Safety Report," to be published in the near future.

However, phosphogypsum stacks continue to grow by the order of 150–200 million tonnes per year (Hilton 2010), apparently due to the controversial perception of risks related to the various use options. A relevant percentage is still discharged to the sea, even in sensitive maritime environments.

With regard to the management of phosphogypsum, these critical questions have been raised:

- What are the barriers preventing more sustainable phosphogypsum management practices?
- What impact do reviewed and potentially modified processes have on the risks related to phosphogypsum use?
- How can we attain a less controversial risk assessment and develop a risk management which is acceptable to a majority of stakeholders?

A transdisciplinary approach to these questions will primarily focus on consensus building upon risk assessment and management. Technical issues will be addressed within the field of revisiting phosphate processing.

The problem strongly begs for knowledge sharing between science and society. If a case study is performed, scientific stakeholders who have been developing options for phosphogypsum reuse in the past need to play an important role. However, without a profound methodological change, no breakthrough can be expected as long as stacks are accepted as a management option. Among the studies proposed in this chapter, this case may be the best example of how technology-oriented science fails to abate environmental problems if stakeholders cannot come to a consensus. Thus, a truly transdisciplinary process involving scientists, risk assessors and managers as well as stakeholders from environmental activist groups, legitimate decision makers (political bodies), and industry should be involved.

Two scenarios are conceivable: (1) the review of general phosphate processing leads to disruptive progress in terms of eliminating radionuclides from fertilizers and phosphogypsum. In this case, the evidence of a pollutant-free material may clear up any concerns and open the pathway to the use of phosphogypsum without further dedicated action. (2) In the absence of technical process improvements, only the transdisciplinary approach may bring about a new and different assessment of risks and benefits concerning phosphogypsum reuse options.

2.3.3 Local Processing: Enable the Local Processing of Primary and Secondary Resources

The prevailing paradigm in fertilizer production is to concentrate every processing step in the vicinity of raw material resources. If this trend continues, local processing would gradually disappear, cutting jobs, idling production facilities, and increasing dependency on a limited number of vertically integrated suppliers.

Secondary resources largely derived from human or animal excreta could compensate for the limited supply of rock phosphate. They are mainly produced in densely populated areas or in regions with high livestock density. In contrast to phosphate rock mines covering an area of a few square kilometers, secondary resources must be collected from a number of widespread livestock farms and wastewater treatment facilities to achieve relevant volumes.

Thus, processing secondary resources preferably takes place in decentralized, regional, and comparatively small facilities. Existing local phosphate and fertilizer-processing plants may be appropriate and could replace imported phosphate rocks by utilizing locally available secondary resources, offering new business opportunities for industries which have either no or limited access to phosphate rock. However, having industrial processing plants in operation is not a precondition for a region to qualify for a transdisciplinary case study dealing with local processing. Relevant factors are as follows:

- Collectable regional resources which are currently not being sustainably utilized,
- suppliers and stakeholders willing to cooperate,
- industry or distributors to use or sell the product from the processing facilities,
- end-users who believe in the commercial value of the product,
- the implementation of legislation to stimulate or enforce the sustainable use of nutrients which are currently lost to sea, diluted beyond recovery, or landfilled, and
- the public at large which accepts industrial processing plants in their neighborhoods.

A preliminary investigation has produced evidence that livestock manure is the largest secondary phosphate resource requiring alternative management (Ott and Rechberger 2012). At present, manure is spread on cropland in the vicinity of livestock farms, largely contributing to phosphate losses to aquatic bodies and eutrophication. Regulatory bodies of some European countries, such as the Netherlands, Denmark, and Belgium, are starting to impose restrictions on current management practices.

The efficient exploitation of manure as a secondary phosphate resource is hampered by its (a) high water content and (b) pathogens and pollutants. In highly industrialized countries, incineration or gasification of biomass destroys pathogens and concentrates nutrients in the residues, thus providing an inorganic phosphate feedstock as required by the industry. However, there are only a handful of manure incinerators actually in operation, mainly because spreading manure on cropland is easier and cheaper than implementing sophisticated energy and nutrient recovery process chains. The current practice will only change if regulatory bodies enforce and control nutrient load limits on cropland. In less developed or less densely populated countries, incineration may not be the first choice and other processing methods could be preferred, provided that they produce a safe and plant-available product.

The task is to match secondary resources with processing facilities in terms of stakeholder interests, material properties, logistics, and process parameters. Apparently, this is a new research topic which was not addressed prior to the recent phosphate recovery initiatives. The knowledge gap is evident.

A project dealing with these issues faces several major challenges:

- Identifying regions with large phosphorus-containing waste flows with the potential to be transferred to a relevant feedstock for processing facilities;
- getting access to resources controlled by various stakeholders who are either in favor of or against the intended use;
- getting industrial processes politically supported, implemented, and licensed by regulatory bodies;
- achieving (relatively) high phosphorus concentrations at low cost in industrial processes upstream of the fertilizer and phosphorus manufacturing, necessitating close collaboration of the various stakeholders.

Thus, conflicting interests meet with uncertainty regarding the availability of appropriate upstream process technologies and require a transdisciplinary approach.

The case study begins with an initial screening to assess relevant P flows from agriculture in a spatially explicit manner. Based on this knowledge, up to three case study areas are selected to identify the hot spot regions with massive livestock waste flows where processing plants make sense. Following the initial phase, the methodology of "Area Development Negotiation" (Scholz and Tietje 2002; Scholz et al. 2006; Loukopoulos and Scholz 2004) will be performed.

Additional insight into complex management practices and waste flows can be gained by performing the case studies on different continents, preferably in countries with different agricultural practices in accordance with their state of development and development targets. Hot spots are determined by their importance within the geographical context. Preliminary investigations have produced evidence that northwestern Europe—the Netherlands, Belgium, Germany, and Denmark, the Russian province Leningrad, as well as the central and southern provinces of Brazil may host such hot spots.

Practice stakeholders are livestock farmers, agricultural cooperatives, the fertilizer and phosphorus industry, investors, and the public at large. Scientific stakeholders are members of the Global TraPs team who perform the case studies and selected experts with a specific local or specialist knowledge. Processing technologies (mechanical, wet chemical, and thermochemical) are selected in cooperation with the processing industry.

Assessment criteria (value trees) are determined along with the identified stakeholders, and the various technologies are evaluated using two approaches: (1) referring to data from existing industrial practice and research and by employing scientific research methods assessing aspects such as investment and operational costs, environmental impacts (CO_2 emissions), the overall environmental footprint of different process approaches, the impact on regional GDP; and (2) assessing preferences from various stakeholder groups ("Exploration Parcours").

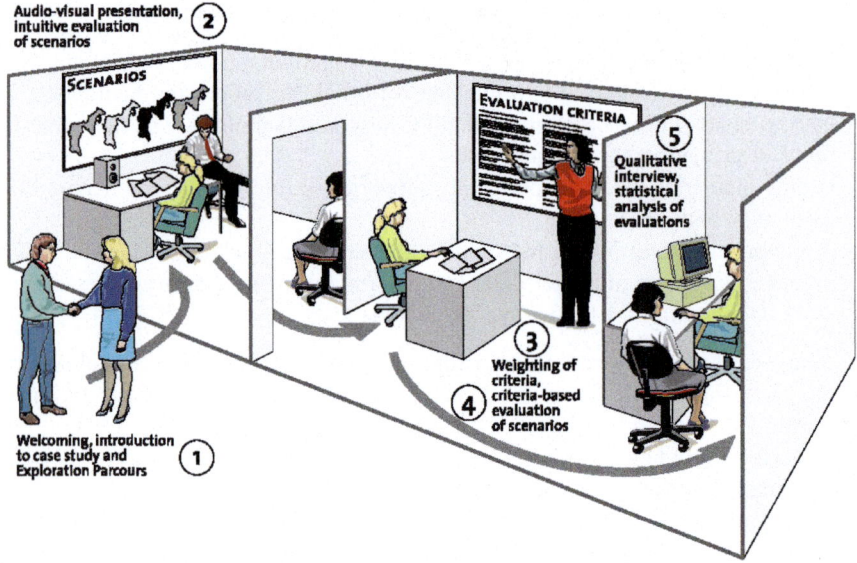

Fig. 1 Five stations of an exploration parcours (according to Scholz and Tietje 2002, p. 214)

Box
Exploration Parcours

The five-step procedure (see Fig. 1) aiming at the evaluation of stakeholder interests is called exploration parcours (EP). An EP generally starts from a set of scenarios identified in preceding steps and it facilitates gathering quantitative and qualitative data.
Step 1: The study team welcomes the participants and informs them about the goals of the study, the EP procedure and the rules. Step 2: The team presents the scenarios for thorough evaluation by the participants in a ranking procedure (first assessment). Step 3: Participants weigh evaluation criteria presented by the study team. Step 4: Participants evaluate the scenarios on the basis of the criteria (second assessment). Step 5: A qualitative interview supporting the interpretation of the stakeholders' evaluation patterns concludes the procedure (Scholz and Tietje 2002).

The EP setting is very flexible and can be performed in many ways. The study team may conduct single interviews, parallel sessions and group sessions with the various stakeholder groups. Group sessions facilitate conclusive group discussions which may initiate a reflection process to be followed by a negotiation process among the stakeholder groups.

> Universities or other institutions can collect the data. EP procedures are not limited to scenario assessments. Study teams have successfully performed in-depth interviews on stakeholder preferences with regard to key aspects of a repository site selection process for nuclear waste and safety issues (Krütli et al. 2010a, b).

Based on these steps, robust orientations can be developed; if and how the P resources in the selected hot spots can be processed with an acceptable environmental footprint and general lessons can be derived both on substantive and process level for similar hot spots.

2.4 Expected Outcome

The expected outcome follows the specific function of the transdisciplinary process in each area of its application.

In the area of revisiting phosphate processing, capacity building and incremental technical improvements in terms of the energy, water, and material balances are expected. The case studies should produce evidence that the environmental footprint of fertilizer production may be reduced without excessive cost hikes if all aspects of the benefits from potential by-products are taken into consideration.

With regard to the valorization of phosphogypsum, the immediate outcome of a transdisciplinary case study could be a risk management strategy in compliance with the needs of all stakeholders. Down the road, phosphogypsum may be used in a variety of safe applications.

As for the local processing of primary and secondary resources, the waste nutrient accumulation hot spots will be identified and secondary phosphorus resources processed to feedstock for safe and effective fertilizers and other products. The case study demonstrates the most appropriate techniques to accommodate local societal and environmental needs.

There is reason enough to believe that the market alone will not provide enough incentives to overcome the current barriers to more sustainable practices. Consequently, case studies must provide a decision support tool for an appropriate regulatory framework for the sustainable use of primary and secondary phosphate resources.

Acknowledgments We thank Kathy Mathers, Rosemarie Overstreet, Amit Roy, and Roland W. Scholz for important comments on earlier drafts of the paper.

References

Brundtland GH (1987) Our common future. World Commission on Environment and Development, Oxford

Burnside P (2012) Phosphate rock market outlook. Phosphates. CRU Group, El Jadida

EFMA (2000) Best available technologies for pollution prevention and control in the European fertilizer industry. EFMA European Fertilizer Manufacturers' Association (Fertilizers Europe), Brussels

Hilton J (2010) Phosphogypsum (PG): uses and current handling practices worldwide. In: Proceedings of 25th annual lakeland regional phosphate conference, Lakeland

IAEA (2006) Assessing the need for radiation protection—measures in work involving minerals and raw materials. Safety reports series no. 49, International Atomic Energy Agency, Vienna

IFA (2012) International fertilizer industry association. http://www.fertilizer.org/ifa/HomePage/STATISTICS/Production-and-trade (accessed 2012)

IFC (2007) Environmental, Health and Safety Guidelines for Phosphate Fertilizer Manufacturing. International Finance Corporation, Washington

Jenssen TK, Kongshaug G (2003) Energy consumption and greenhouse gas emissions in fertilizer production. International Fertiliser Society Meeting, London

Jung AJ (2012) Global market for inorganic feed phosphates. Phosphates. CRU Group, El Jadida

Koopman C (1999) Removal of heavy metals and lanthanides from industrial phosphoric acid process Liquors. Sep Sci Technol 34(15). doi:10.1081/SS-100100818

Krütli P, Stauffacher M, Flüeler T, Scholz RW (2010a) Functional-dynamic public participation in technological decision making: Site selection processes of nuclear waste repositories. J Risk Res 13(7):861–875

Krütli P, Flüeler T, Stauffacher M, Wiek A, Scholz RW (2010b) Technical safety vs. public involvement? A case study on the unrealized project for the disposal of nuclear waste at Wellenberg (Switzerland). J Integrative. Environ Sci 7(3):229–244

Lloyd GM Jr (2004) Phosphogypsum: should we just let it go to waste?—Part 1. FIPR Phosphogypsum Res, Bartow

Loukopoulos P, Scholz RW (2004) Sustainable future urban mobility: using "area development negotiations" for scenario assessment and participatory strategic planning. Environ Plann A 36(12):2203–2226

Ott C, Rechberger H (2012) The European phosphorus balance. Resour Conserv Recycl 60:159–172

Prud'homme M (2010) World phosphate rock flows, losses and uses. In: Proceedings of the phosphate 2010 conference and exhibition. IFA, Brussels

Schipper WJ, Klapwijk A, Potjer B, Rulkens WH, Temmink BG, Kiestra FD, Lijmbach AC (2001) Phosphate recycling in the phosphorus industry. Environ Technol 22(11):2001:1337

Scholz RW (2011) Environmental literacy in science and society: from knowledge to decisions. Cambridge University Press, Cambridge

Scholz RW, Tietje O (2002) Embedded case study methods: Integrating quantitative and qualitative knowledge. Sage, Thousand Oaks

Scholz RW, Lang DJ, Wiek A, Walter AI, Stauffacher M (2006) Transdisciplinay case studies as a means of sustainability learning: historical framework and theory. Int J Sustain Higher Educ 7(3):226–251

Schrödter K (2008) Phosphoric acid and phosphates. Ullmann's Encycl Ind Chem. doi:10.1002/14356007.a19_465.pub3

Shinh A (2012) The outlook for industrial and food phosphates. Phosphates. CRU Group, El Jadida

Sinden J (2012) Who eats what and why. Phosphates. CRU Group, El Jadida

Stana R (2009) Uranium and phosphorus—a "cooperative game" for critical elements in energy and food security. Technical meeting, International Atomic Energy Agency. IAEA, Vienna

Stauffacher M, Lang DJ, Scholz RW (2008) Problem framing in transdisciplinary case studies on sustainable development. Transdisciplinarity. In: Proceedings of conference inter-and transdisciplinary problem framing, Federal Institute of Technology, Zürich

UNIDO, IFDC (1998) Fertilizer Manual. Kluwer Academic Publishers, Dordrecht

Van Kauwenbergh SJ (2010) World phosphate rock reserves and resources. IFDC, Muscle Shoals

Wissa AEZ (2003) Phosphogypsum disposal and the environment. RTD report, Florida Institute of Phosphate Research, Bartow

Zhang F, Zhang W, Ma W et al (2009) The chemical fertilizer industry in China—a review and its outlook. Paris, France: original Chinese version by chemical industry Press, China; English Version by IFA

Appendix: Spotlight 5

Options in Processing Manure from a Phosphorus Use Perspective

Diane F. Malley

Of the livestock manures, hog manure is one of the most challenging for the management and recycling of nutrients. It is generally liquid, being a mixture of faeces, urine, feed residues, and wash water from the barns. Manitoba is one of the largest producers of hogs among Canadian provinces with a population of 8 to 10 million head in recent years (Honey 2011). The manure from these hogs is collected in open lagoons, or less commonly, in concrete storage tanks. The digestion of feed by the hogs is not complete, resulting in a considerable range of solids among samples of manure. Although the feeds may contain sufficient total P for nutritional purposes, the incomplete digestion leads to the need for supplemental P to be added to the diet. This adds to the P concentration in the manure. In recent years, the addition of the enzyme phytase to feed enhances digestion and somewhat reduces the use of supplemental P. The solids in the manure settle upon standing, such that manure stores are typically mechanically agitated, though incompletely, whenever manure is to be handled.

In Manitoba, periodically the manure is pumped from the manure stores through hoses up to 4 or 5 km distance and injected into the soil. This both disposes of the manure and provides nutrients for future crops. Despite the agitation, manure withdrawn from the stores is highly variable at the time of injection, varying over the pump-out from 0.4 to nearly 15 % solids, 0.5–6.5 g/L total N, and 0.03–5.7 g/L total P. Thus, solids may vary more than 37-fold, N 13-fold, and P 190-fold. The N/P ratios vary widely and on the average differ widely from the 15:1 needed to support agricultural productivity. Consequently, the application of manure to agricultural fields is more a form of disposal of the manure than an effective application of the nutrients as the valuable fertilizer that they are. Fertilizers are generally applied to these fields in addition to the manure to ensure the agronomic needs for N and P are met since the contribution of nutrients by the manure is unknown.

D. F. Malley
PDK Projects, 5072 Vista View Crescent, Nanaimo, B.C. V9V 1L6, Canada
e-mail: dmalley@pdkprojects.com

Beginning in 1999, a global technology used commercially first 30 years ago for the measurement of protein in wheat, now analyzing 85 % of marketed wheat globally, was applied to the analysis of nutrients in liquid manure. This technology, near-infrared spectroscopy (NIRS), analyzes samples on the spot in real time without the use of chemicals, when the instrument has been appropriately calibrated. Early results showed that NIRS accurately predicted total solids, total N, and total P in hog manure (Malley et al. 2002). This technology has been demonstrated in the laboratory with flowing manure. It has the potential to measure hog manure being applied to fields and to permit GPS mapping of the nutrients applied. In this way, a second pass with commercial fertilizers can result in the accurate application of nutrients for agronomic needs. Moreover, the nutrients in the hog manure can return financial value to the hog producers and manure applicators. The P and N can be accurately managed to avoid unintended losses to the surrounding environment.

The Netherlands

In the Netherlands, with a population of 12.2 million hogs in 2012, there is insufficient land base upon which to apply untreated liquid manure. Nutrients are highly managed under the economic instrument of MINAS (OECD 2005). Incoming nutrients onto farms are highly tracked and accounted for against nutrients outgoing from farms as products or manure. This reduces accumulation of nutrients on the land and loss of nutrients from the land to air and water. This is a shift in manure policy from regulations to economic incentives for managing nutrients. Yet, the oversupply of P in manure may amount to 60 million kg by 2015 (Schoumans et al. 2010). Manure processing is being seriously considered. Among the techniques are anaerobic digestion, manure separation into solid and liquid fractions, followed by composting or incineration of the solid fraction, reverse osmosis anaerobic treatment of the liquid fraction, and acidification (EC 2010). Not all manure treatments contribute to a better utilization of the N and P resources, nor are all focused on recycling P to agricultural soil. One option for reducing/recycling P is manure separation on livestock farms producing a liquid fraction containing N to be recycled back to their land along with a portion of the solids fraction containing the majority of the P. The second option is the recovery of P from manure as P fertilizer, biochar, or elementary P for export (Schoumans et al. 2010). These options require significant financial investments and institutional arrangements. Operational costs can be significantly reduced by employing near-infrared spectroscopy strategically in the processing stream for continuous, real-time monitoring of the process, the incoming raw materials, and the final products by batch. Furthermore, the technology has been demonstrated to measure the P in sediments of lakes and can be used as an indicator of unintended runoff of P from surrounding land.

References

European Commission (2010) DG Environment and Ministry of Economic Affairs, Agriculture and Innovation of the Netherlands. In: Workshop on managing livestock manure for sustainable agriculture, 24–25 Nov 2010, Wageningen, The Netherlands, p 68. Available at http://ec.europa.eu/environment/water/pdf/manure/report/Report.pdf

Honey J (2011) Manitoba pig and pork industry 2010. Department of agribusiness and agricultural Economics, University of Manitoba

Malley DF, Yesmin L, Eilers RG (2002) Rapid analysis of hog manure and manure-amended soils using near-infrared spectroscopy. Soil Sci Soc J 66:1677–1686

Organisation for economic co-operation and development (2005) Manure policy and MINAS: regulating nitrogen and phosphorus surpluses in agriculture of the Netherlands, p 47

Schoumans O, Oenema O, Ehlert PAI, Rulken WH (2010) Managing phosphorus cycling in agriculture: a review of options for the Netherlands. In: European commission, DG environment and ministry of economic affairs, agriculture and innovation of the Netherlands. Workshop on managing livestock manure for sustainable agriculture, 24–25 Nov 2010, Wageningen, The Netherlands, p 31

Chapter 5
Use: What is Needed to Support Sustainability?

Robert L. Mikkelsen, Claudia R. Binder, Emmanuel Frossard, Fridolin S. Brand, Roland W. Scholz, and Ulli Vilsmaier

Abstract Increased demands for agricultural output per unit of land area must be met in a way that encourages improved efficiency and better stewardship of natural resources, including phosphate rock. Modern crops remove between 5 and 35 kg P/ha, with P removal exceeding 45 kg P/ha for high-yielding maize. In situations such as Sub-Saharan Africa, where soil fertility is low and P removal exceeds average inputs of 2 kg P/ha/year, the resulting nutrient depletion severely restricts yields (e.g., maize yields < 1,000 kg/ha/year) and accelerates soil

---

R. L. Mikkelsen (✉)
International Plant Nutrition Institute (IPNI), 4125 Sattui Court, Merced, CA 95348, USA
e-mail: RMikkelsen@ipni.net

C. R. Binder
Ludwig-Maximilians University Munich, Lehrstuhl für Anthropogeographie mit Schwerpunkt Mensch-Umwelt Beziehungen, Luisenstraße 37, 80333 Munich, Germany
e-mail: Claudia.Binder@geographie.uni-muenchen.de

E. Frossard
ETH Zürich, Group of Plant Nutrition, FMG C 17.2, Eschikon 33,
8315 Lindau, ZH, Switzerland
e-mail: emmanuel.frossard@usys.ethz.ch

F. S. Brand · R. W. Scholz
ETH Zürich, Natural and Social Science Interface (NSSI), Universitaetsstrasse 22,
CHN J74.2, 8092 Zürich, Switzerland
e-mail: Frido_brand@hotmail.com

R. W. Scholz
Fraunhofer Project Group Materials Recycling and Resource Strategies IWKS,
Brentanostrasse 2, 63755 Alzenau, Germany
e-mail: roland.scholz@isc.fraunhofer.de

U. Vilsmaier
Leuphana University of Lüneburg, Methodenzentrum, Scharnhorststr. 1,
C4.009c, 21335 Lüneburg, Germany
e-mail: vilsmaier@leuphana.de

degradation. In other regions, excessive P inputs produce economic inefficiencies and increase the risk of P loss, with negative environmental consequences. During the year of application, plants recover 15–25 % of the added P, with the remaining fraction converting to less soluble forms or residual P which becomes plant available over time. Improving P efficiency requires a balance between the imperatives to produce more food while minimizing P losses. Utilizing transdisciplinary approaches, a number of social, economic, and environmental goals can be simultaneously achieved if progress is made toward short- and long-term food security and global P sustainability. This chapter provides an overview of efforts to improve P use efficiency in agriculture ranging from promising germplasm, improved crop, and soil management scenarios, additives in animal diets to reduce P inputs and surplus P in the manure, and opportunities for P recycling in food and household waste. Challenges and opportunities associated with each option are discussed and transdisciplinary case studies outlined.

Keywords Phosphorus and the food chain · Integrated nutrient management · Phosphorus recovery · Phosphorus losses from use · Improving access to phosphorus

Contents

1	Introduction	209
	1.1 Phosphorus and the Food Chain	209
	1.2 The Use of Chemically Processed Phosphorus Fertilizer	210
	1.3 Phosphorus and Forestry	212
	1.4 Phosphorus in the Food Chain	213
	1.5 Non-fertilizer Phosphorus Uses	214
2	Opportunities to Improve Phosphorus Use	214
	2.1 Integrated Nutrient Management	214
	2.2 Soil Testing and Phosphorus Recommendations	215
	2.3 Fertilizer Placement and Residual Phosphorus	216
	2.4 Fertilizer Products	217
	2.5 Plant Recovery	217
	2.6 Bioenergy Crops	217
	2.7 Increasing Phosphorus Use Efficiency at the Cropping System	218
	2.8 Decreasing Phosphorus Loss from Soil	218
	2.9 The Use of Phosphorus for Livestock Production	219
3	Opportunities to Improve Phosphorus Use at the Societal Level	221
	3.1 Postharvest, Retail, and Household Level	221
	3.2 Issues of Scale	222
4	Work in Global TraPs	223
	4.1 Knowledge Gaps and Critical Questions	223
	4.2 Role, Function, and Kind of Transdisciplinary Processes	223
	4.3 Potential Case Studies	224
References		226

1 Introduction

1.1 Phosphorus and the Food Chain

Phosphorus (P) is essential in every cell of living organisms. As a limited earth mineral, it is well accepted that greater efforts must be made to improve the use of P in global ecosystems. In addition to avoiding unnecessary losses, the time is right to manage all aspects of the P cycle as a whole process, not just as single pieces that are poorly connected (Fig. 1). To ensure long-term sustainability, consideration needs to be made on how to use P as efficiently as possible, minimize losses and waste, and promote recycling as much as economically feasible. Transdisciplinary approaches are especially suited to this venture, as they allow integration of scientific knowledge with insights gained in everyday practices.

Phosphorus does not become "lost" as it dissipates and cycles through geochemical processes, but the concentration becomes diluted to a point where it is too difficult or expensive to recover using current technology. Phosphorus loss to the ocean is a "natural" and unavoidable process. For example, an uncultivated soil from the Canadian prairies has lost up to 40 % of its total P within 10,000 years of pedogenesis (St Arnaud et al. 1988). However, human activities (including phosphate rock mining, soil tillage, animal production, and industrial/urban discharges) have greatly accelerated the rate and the quantity of P lost to water (Fig. 1). It is clear that excessive P loss to water bodies can have a strong negative impact on water quality, whereas P enrichment in terrestrial systems can cause a shift in native plant species.

Improving P recovery and efficiency while reducing negative environmental effects are the urgent short-term goals, but reducing irrecoverable losses of this finite natural resource will not keep pace with the geological concentration of P supplies without a shift in current approaches to P stewardship. Although P scarcity is not a pressing issue at this time, improving the efficiency of P use would extend the lifetime of global mineral reserves. However, long-range efforts need to converge with the ultimate goal of economic efficient P recovery and recycling.

Projections of global population growth anticipate the need for far more food, feed, and fuel to be produced on a limited amount of productive land. This pressure to increase agricultural output must be done in a way that encourages improved efficiency and better stewardship of natural resources. Appropriate applications of P fertilizer are an essential part of maintaining crop yields by minimally replacing the nutrients removed during harvest. It is reasonable to expect the requirement for P additions to grow in importance as crop intensification increases.

Improving the efficiency of P use in the human food chain will require cooperation of the P mining and fertilizer industry, the agricultural and forest production sectors, the food-processing industry, industrial P users, urban wastewater

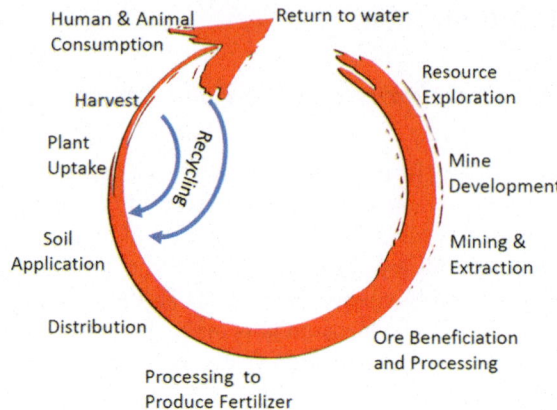

Fig. 1 The primary pathways of P extraction, use, and loss. Opportunities exist at each step for improved P use efficiency

treatment operators, and individual consumers (Hilton et al. 2010). There are significant economic and biological barriers to improving P use efficiency, thus this endeavor will require cooperation of many stakeholders.

The human recommended daily intake is 1 g P/person for maintaining proper health. This translates to an annual global requirement of over 2.5 Mt P (or nearly 6 Mt of P_2O_5 fertilizer equivalents). The actual amount of P consumed in daily diets is often much greater than this, so providing a reliable assessment on the world level is difficult. But since the average consumption is likely 3 g P/person/day (Smil 2000; Smit et al. 2009; Scholz and Wellmer 2013), the conversion of mined phosphate rock to edible human food is low and amounts only to 15–20 % (Schröder et al. 2011; Suh and Yee 2011), leaving ample room for improved efficiency. The unutilized P most commonly ends up stored in soil for subsequent crops or is transferred to water bodies through runoff, erosion and, to a lesser extent, leaching (Granger et al. 2010).

1.2 The Use of Chemically Processed Phosphorus Fertilizer

The majority of mined phosphate rock (>80 %) is processed to produce soluble fertilizers for plant nutrition. The insoluble phosphate rock (apatite) is first reacted with a strong acid to dissolve the raw mineral and form water-soluble compounds that will be available for plant uptake in the soil. Without the reaction with acid, apatite is very slow to dissolve in most agricultural soils and the rate of soluble P release is too slow to be of significant value for plant nutrition. There are, however, some sources of phosphate rock that dissolve sufficiently rapid in acidic soils to be a valuable source of P plant nutrition without first treating it with acid (Smalberger et al. 2006).

The discovery of this acid-treatment process in the 1840s marked the beginning of the modern P fertilizer industry and was a major advance in understanding plant nutrition. Since that time, considerable progress has been made to better understand P chemical reactions and plant nutrition. During the 20th century, the P fertilizer industry emerged as a fundamental partner in supporting the global food production system. As the demand for crop production increases, the global demand for additional P fertilizer also increases.

Since an inadequate concentration of P limits plant growth in many parts of the world, the chemistry of P fertilizer is one of the most studied topics in soil and plant nutrition (Hedley and McLaughlin 2005). Numerous global studies have repeatedly demonstrated that only a relatively small proportion of P annually added to agricultural soils ends up in the harvested product in the year of application (Syers et al. 2008).

Efficiency is commonly determined by measuring the amount of material "going in" (a field, watershed, or a nation) and comparing that with the quantity removed (such as a harvested crop). The quantities of soluble P added in excess to plant needs and remaining in the soil profile can react with Al and Fe oxides or with Ca and Mg carbonates depending on the soil pH and mineralogical properties of these sorbents. These reactions result in the conversion of soluble P to less soluble residual P forms that are not immediately available to the annual crop. The value of this accumulated residual soil P is difficult to quantify, but it makes an important contribution to plant nutrition for many years following the initial fertilizer application.

The high crop yields desired for food security, economic profitability, and farm sustainability place a significant demand on soil nutrient reserves. Modern crops remove between 15 and 35 kg P/ha for cereals, 15–25 kg P/ha for many leguminous and root crops, and 5–15 kg P/ha for vegetables and fruits. Phosphorus removal of over 45 kg P/ha can be seen in high-yielding maize and sugar crops. Modern high-yielding cultivars remove more P during harvest than during the mid-twentieth century. For example, Edwards et al. (1997) showed that English wheat removed about 7 kg P/ha in 1950, 13 kg P/ha in 1975, and about 20 kg P/ha in 1995.

Many regions of the world have soils that are P deficient. For example, there are many smallholder farms in East, Central, and Southeast Asia and Sub-Saharan Africa, where the lack of access to P fertilizers (MacDonald et al. 2011) and prolonged nutrient mining has led to severe land degradation (Craswell et al. 2010). Moreover, unsustainable agroecosystem management in these areas, such as burning or feeding of crop residues and shifts to more nutrient-demanding crops, results in severe soil erosion and drains on soil nutrient capital (Quinton et al. 2010; Vitousek et al. 2009). Large areas of once-productive soils have been abandoned due to soil degradation. At the same time, many farming areas in East Asian floodplains, North America, Europe, and parts of Latin America have a history of repeated P use, resulting in a buildup of the soil P reserve (IPNI 2010; MacDonald et al. 2011).

Calculating a P balance is a common way to estimate whether soils are becoming enriched or depleted in nutrients over time. All the nutrient inputs added to a field, watershed, or country are compared with the nutrients removed in the harvested portion of the crop and the P lost through erosion and surface runoff. Vitousek et al. (2009) conducted a P balance for three regions and found that they ranged to positive in North China (>50 kg P/ha/year), to neutral (Western Kenya) to negative balances in the Midwest USA (−9 kg/ha/year). The negative balance in the USA reflects current nutrient removals exceeding inputs following a history of P fertilizer and manure applications during 1970–1995 where additions were in excess of crop removal. Differences in current P balances in these three regions clearly demonstrate the need for targeted management strategies to sustain productivity.

A study by Richards and Dawson (2008) examined the P balance of the 27 European Union countries. They estimated that net imports to agricultural soil in these countries were 2 Mt P/year. This input is equivalent to an addition of 18 kg P/ha, with an average removal of 10 kg P/ha/year, resulting in an average surplus of 8 kg P/ha/year. This surplus does not mean that all soils in the EU are receiving an annual excess of P, but raises awareness that there are areas where P is accumulating and that those soils should be monitored to avoid excess P accumulation and potential negative environmental impacts. Positive P balances of agricultural systems are necessary during the soil-building phase, but the inputs of P will likely decrease over time as soil P concentrations reach the optimum level for crop production. Most farmers would not invest money in P fertilizer if the soil already contained adequate P and the likelihood of a return on investment was low. The surplus P situations most frequently occur as a result of livestock production and the difficulty of transporting manure to areas that would benefit from the nutrients. In these surplus-manure regions, government policies are often helpful to provide incentives for improved P management. For example, the introduction of nutrient subsidy schemes in Switzerland contributed to a decrease of the P surplus in agriculture from about 27 kg P/ha/year in 1980 to about 5 kg P/ha/year in 2008 (Spiess 2011; Lamprecht et al. 2011).

1.3 Phosphorus and Forestry

Forests cover about one-third of the total land surface of earth (FAO 2009). Besides providing fuelwood, which is still the largest use of wood worldwide, forests provide a number of timber and non-timber products. Forests also provide a range of vital environmental services (such as a reserve of biodiversity, C sink, protection against natural hazards, water fluxes regulations, and water purification). In the next decades, the demand for forest products will grow, and the provision of environmental services from forests will still be crucial.

Forest trees, like all organisms, need P to grow. For instance, the annual P needs of *Pinus* plantations have been estimated to be 10 kg P/ha to maintain leaf area index for optimal growth (Fox et al. 2011). However, apart from forest plantations where fertilization is common, forests rarely receive any fertilizers (Fox et al. 2011). In Switzerland, it is almost forbidden to apply any kind of fertilizers to forests. Although forest vegetation has developed very sophisticated mechanisms to survive and in some instances thrive on low P soils (Reed et al. 2011), there are signs that non-fertilized forests might become P limited in the next future. For instance, a significant decrease in P concentration has been observed in Eastern France in the leaves of 118 beech stands between 1970 and 1996 (Duquesnay et al. 2000). This can be related to the accelerated rates of P depletion from forest soils due to an increased use of wood for energy, to excessive N loads from atmospheric deposition (Genenger et al. 2003) leading to P limitation of tree growth (imbalanced nutrition), and to soil acidification which can increase P sorption on soil and decrease root growth. For example, in 1999, Switzerland was producing 25,000 t/yr of wood ash that could not be applied back to forest soil. Finally, climate change may also lead to an increased P requirement, as trees need higher P concentration to resist to drought (Duquesnay et al. 2000). In conclusion, the need for P fertilization might also increase in forest systems in the future.

1.4 Phosphorus in the Food Chain

When studying the whole global food chain, inefficiencies of P occur both during and after the production process. These apparent inefficiencies include accumulation in soil (4.5 Mt/year), erosion (8 Mt P/year); crop losses due to pests, diseases, and natural destruction (3 Mt P/year); post-harvest losses (0.9 Mt P/year); and losses at distribution, retail, and household level (1.2 Mt P/year) (Cordell et al. 2009). It is important to distinguish between true P losses (at a defined scale, such as erosion) and temporary factors that influence short-term P efficiency (such as drought-induced crop failure). Worldwide, the total amount of P consumed directly by humans has been estimated to be about 3 Mt P/year (Cordell et al. 2009).

Crop losses due to pests, diseases, and natural destruction account for 12 % of food in developed and for 22 % of the food in the developing world (Kader 2005). In some developing countries, this loss increases up to 40 % at the farm itself and 15 % during processing and storage. A close examination of P recovery of these post-harvest losses is needed in order to improve overall efficiency. In particular, appropriate mechanisms relevant for developed and developing countries should be considered.

At the retail and household level, food losses account for about 20–30 % of total production in developed and 10 % of total production in developing countries

(Kader 2005). A study in the UK showed that of the food (7 Mt/year) and drink (1.3 Mt/year) waste generated yearly in the UK, about 4.5 Mt (food) and 0.8 Mt (drink) would be potentially avoidable, and an extra 1.5 Mt (food only) would be possibly avoidable (Parfitt et al. 2010). In addition to the P savings that the reduction in food losses would confer, significant economic savings would also be expected.

1.5 Non-fertilizer Phosphorus Uses

About 10 % of the 24.5 Mt/yr of elemental P, which is currently mined, is used for non-fertilizer purposes (Jasinski 2011; Prud'homme 2010). This non-fertilizer P is used (a) as animal feed additive (about 7 %), which becomes part of the agricultural food chain, (b) as an additive of processed food and beverages such as dark sodas (1–2 %) [e.g., Coca Cola making up 0.01–0.02 % of total P consumed in Switzerland, Binder et al. (2008)], or (c) as an element of pharmaceutics and industrial uses such as detergents, metal treatment, and other industrial applications 9 % (Prud'homme 2010; Schröder et al. 2010).

2 Opportunities to Improve Phosphorus Use

With the complexities and challenges associated with improving the use of the global P supplies, three concurrent strategies can be proposed. The first approach suggests that the mining of phosphate rock and the processing of fertilizer can be improved to extend the known resources. The second path entails improving the efficiency of P use in agricultural production, food processing, industrial applications, and recovery from waste streams. The third approach is to change the demand for P use in various applications. Of these options, this chapter highlights some of the bottlenecks to improving P use in agriculture. We look first at different components of agricultural systems (mineral fertilizer, manure, plant) and then at the system in its entirety. This is done in the frame of integrated nutrient management (INM), which is presented thereafter. Some of the barriers for improving P use efficiency are constrained by scientific knowledge, by economics, or by social and political barriers—once again highlighting the need for transdisciplinary solutions.

2.1 Integrated Nutrient Management

INM is a comprehensive approach taken for improving nutrient use efficiency by crops and animals while decreasing nutrient losses to the environment (Frossard et al. 2009). To achieve these goals, INM must consider all the components involved in nutrient cycling (climate, soil, plants, animals, all inorganic and organic nutrient sources) as well as the relevant socioeconomic factors such as the production preferences of farmers, the food preferences of consumers, markets, and trade. From a biophysical standpoint, INM can include the input of exogenous nutrients that deliver available nutrients to plants when and where necessary, the use of crops with a high nutrient acquisition efficiency, nutrient recycling through the introduction of improved green manures, proper reuse of animal manure and urban wastes, and the decrease in nutrient losses by minimizing erosion, leaching, and runoff. Most of these points are discussed below for improving P use efficiency.

2.2 Soil Testing and Phosphorus Recommendations

It was estimated that crop growth on about two-thirds of farmland in the world is limited by insufficient P concentrations in soils (Cakmak 2002). To overcome this limitation, farmers have adapted many ways to provide the needed P, including using mineral fertilizer, animal wastes, composts, and various recycled materials to compensate for the P removed in the harvested crops and animals.

A fundamental issue is how to determine the appropriate amount of P to add to soil to meet crop production goals? Insufficient P fertilization risks a loss of crop yield and quality, while excessive fertilization poses economic inefficiency and a heightened risk of P losses with negative environmental impacts.

Chemical analysis of agricultural soils is widely accepted in developed countries as the primary technique to predict the need for additional P. There are many techniques and interpretations, which are adapted to local crop and soil properties. Unfortunately, costs associated with soil sampling and laboratory analysis, as well as access to laboratories limits this important tool in many parts of the world. Where chemical tests are not feasible, local research using the "omission plot" technique can identify the probability that P will improve plant growth.

Without access to information on the need for P on specific fields, farmers are left to guess the likelihood of P response based on factors such as historic fertilization practices, the amount of P required to compensate for crop removal, the quantity of P fertilizer that is affordable, and generalized soil fertility recommendations for the region. All of these factors can lead to either under- or over-fertilization. Modern fertilization practices should not rely on "insurance" doses of P, but should be based as much as possible on site-specific information of the local needs.

Soil testing is helpful to estimate the amount of P that will likely be available for plant uptake during the coming growing season, but less useful as a predictor of the long-term value of P held in relatively insoluble soil compounds.

2.3 Fertilizer Placement and Residual Phosphorus

Improving the placement of P fertilizer can benefit short-term P recovery by plants. For example, in the Netherlands, according to fertilizer recommendations, for a similar yield response twice as much P is needed if the fertilizer is broadcasted on the soil surface rather than positioned at the subsurface close to seed rows (Schröder et al. 2010). Since P is relatively immobile in most upland soils, it should be placed as close to the roots as practical. Specialized equipment that allows P to be injected (or knifed) into the soil increases the short-term recovery of P; placing P on the soil surface separates it from the active root zone and leaves it vulnerable to loss with runoff water. Tilling the soil to incorporate P deeper into the root zone carries the heightened risk of erosion if sediment is carried off the field. Compromises are sometimes made in deciding where P placement is optimal. Repeated P application onto the soil surface produces a stratified zone of nutrients, which may not be accessible to the deeper roots of many plants and pose a greater risk of P loss in runoff.

Plants only assimilate a small proportion of the fertilizer P in the first year following application. Most of the fertilizer remains in the soil in association with the mineral and the organic fractions of soil, largely unavailable for short-term plant uptake. Traditional soil testing only measures the P associated with the inorganic fraction, while soil organic matter can be an important reservoir for a large proportion of the total soil P.

During the year of application, plants may only recover 15–25 % of the added P. This low apparent recovery has caused considerable concern about P efficiency. Recovery is typically calculated as follows (Chien et al. 2012):

$$\% \text{ P recovery} = [\text{P in the fertilized plant} - \text{P in the unfertilized plant}] / \text{P applied} \times 100$$

However, the remaining fraction of freshly added P (75–85 %) remains in the soil, where it is slowly released over time. Once adequate P has accumulated in the soil, high-yielding crop production can often be maintained for several years without requiring further P additions. Many examples are available that demonstrate maintenance of high yields with reduced P inputs once an adequate supply of P has been accumulated (Gallet et al. 2003).

Results from long-term research suggest that P efficiency can exceed 90 % when a "balance method" is used for the calculation (Syers et al. 2008). The balance method considers the ratio of P uptake from a fertilized soil divided by the amount of P applied. This approach yields valuable information regarding P accumulation or depletion, but does not consider the amount P mined from soil without the addition of fertilizer (Chien et al. 2012).

2.4 Fertilizer Products

Most of the P added for plant nutrition comes from highly soluble fertilizer sources such as diammonium phosphate (DAP), monoammonium phosphate (MAP), and triple superphosphate (TSP). The high degree of water solubility that is provided by these fertilizers is not always justified from an agronomic viewpoint (Engelstad and Hellums 1992). It would be feasible in many conditions to use P fertilizer materials with a lower degree of solubility (measured with water and citric acid) and still achieve the same agronomic results. This would allow more of the lower-grade phosphate rock to be used for plant nutrition and result in less P discarded in the processing waste.

Unprocessed phosphate rock can be a useful source of nutrients in certain conditions (IFA 2013). The best predictor of agronomic performance for phosphate rock is its solubility (usually measured in a dilute solution of ammonium citrate or citric acid). Direct use of phosphate rock is best in acid soils (pH < 5.5) that have a high capacity to retain P. A low calcium concentration and high soil organic matter will enhance the agronomic performance of phosphate rock. Unlike water-soluble P fertilizers, there are specific factors (such as the rock reactivity, soil properties, management practices, and crop species) that must be considered before using phosphate rock (Smalberger et al. 2006).

2.5 Plant Recovery

Efforts to modify crops to be more efficient in recovering soil P also hold promise. Young plants and some plant species have limited root length and surface area that hinders the acquisition of P. Genetic research is underway to select plants that have more effective root architecture and abundant root hairs. The organic exudates excreted from roots play an essential role in solubilizing soil P. The release of organic acids solubilizes a portion of the inorganic P compounds for uptake. Root-excreted phosphatase enzymes are thought to hydrolyze organic P-containing compounds to inorganic phosphate.

Continuing work is underway with microbial additives that can assist with P nutrition, including promoting the association of mycorrhizal fungi with plant roots. This fungal symbiosis is an important mechanism for supplying P to plants, but limited progress has been made on improving this process for major food crops. This topic was thoroughly reviewed by Richardson et al. (2011).

2.6 Bioenergy Crops

The increased cultivation of bioenergy crops puts an additional drain on soil P reserves (CAST 2013). When crop biomass and residues are removed from the field in addition to grain or oilseed to produce bioenergy, there is less opportunity for nutrient recycling. The P removed in plant biomass should be returned to the cropland where the feedstocks were produced whenever possible. If bioenergy crops are grown on poor or marginal land in order to avoid competition with food production, it is probable that there will be a need for additional P fertilization to meet the crop requirements.

When grain or oilseeds are used for ethanol or biodiesel production, most of the P is concentrated in the coproducts (such as dried distiller's grain with solubles or canola meal). These P-rich by-products make excellent feed materials for animals. When this P is excreted in manure, it should be properly managed, like all P sources.

2.7 Increasing Phosphorus Use Efficiency at the Cropping System

Recent reviews have summarized how different components could be combined to improve P use efficiency at the cropping/grassland system level (Oberson et al. 2011; Simpson et al. 2011). This includes using adapted plant germplasm, better crop rotations to improve soil structure and plant health, appropriate use of both mineral and organic fertilizers, and appropriate soil preparation techniques to reduce erosion.

For example, Johnston and Dawson (2010) showed that organic matter additions improved soil structure, which promoted root health and greater P availability. In their Rothamsted field experiments, a 60 % increase in soil organic matter content (raised from 15 to 24 g kg^{-1} reduced the critical P concentration (from 46 to 17 µg P kg^{-1}) needed to achieve maximum yields. The soil organic matter improved the soil structure, so that root growth improved and acquired additional plant-available P.

Balanced nutrition, where the soil's nutrient-supplying capacities are maintained in relative balance to crop needs, is also essential for getting maximum P use efficiency. Phosphorus is just one of the many essential mineral nutrients that are supplied from soil. There are many examples where a shortage of one nutrient limits the uptake of several other nutrients and stunts overall growth. Phosphorus cannot be viewed in isolation from other factors that stimulate growth including maintenance of soil organic matter content and soil pH. Soil pH affects the availability of all plant nutrients, but has the greatest effect on P and several micronutrients.

2.8 Decreasing Phosphorus Loss from Soil

Soil erosion is the greatest source of P loss from cropped soils, while grassland soils lose more P through surface water runoff. Management practices that limit the loss of sediment-bound P will help avoid P-induced water quality problems. There are many cultural practices that can reduce soil and P losses including retention of crop residues on the surface, reduced tillage practices, subsoil tillage to improve water infiltration, terracing and contour tillage, use of cover crops, and conversion to perennial crops. Chemical additives, such as polyacrylamide, are effective at reducing sediment loss from fields by binding soil particles together.

Changing land use and management practices can limit P losses to water. Grassland and forests have the benefit of a permanent canopy that helps reduce runoff and enhance infiltration. Various management practices, such as planting field-edge vegetative buffer strips (riparian or grassland), have been shown to reduce P loss in runoff from agricultural land.

2.9 The Use of Phosphorus for Livestock Production

When crops and nutritional supplements are fed to animals and converted into animal products, there is a large inefficiency in P recovery. Only a fraction of the P present in the animal feed is converted into economic products such as milk, meat, eggs, or wool. The remaining P is excreted in manure and urine where it can be used as a valuable resource if it collected and applied back to cropland (Tarkalson and Mikkelsen 2003; Nelson and Mikkelsen 2005).

To avoid potential limitations to growth, animal producers in developed nations routinely add mineral P supplements to the feed ration to insure against deficiencies. Globally, between 5 and 8 % of the P use goes toward animal nutrition products (Schröder et al. 2010). Adjusting the amount of mineral P added to animal diets to reflect the actual P present in the feedstuffs can minimize the P excreted in manure and urine, but requires an advanced understanding of animal nutrition. Decreasing the quantity of P added to the diets as animals mature (phased feeding) also limits P excretion. The addition of phytase to the diets of nonruminant animals can enhance the nutritional availability of relatively indigestible organic P compounds present in plants (especially phytic acid present in seeds and grains). New plant varieties have been developed that contain less phytic acid in the seed. Such dietary adjustments can greatly reduce P inputs in the animal feed and the surplus of P in the manure (Cromwell 2005).

A major hurdle with intensive animal husbandry is to effectively use manure to sustain all plant nutrients in the sufficiency range for plant growth (Mikkelsen 2000a). The relatively low nitrogen to P ratio in most manure is the reverse of the plant requirement, which frequently leads to P accumulation in soils when manure is applied to meet the N needs of crops. Manure application rates need to be adjusted to meet the P requirement to avoid this accumulation. Uniform application of bulky and nonhomogenous material can also pose challenges.

Recycling organic manures in agricultural operations should be a fundamental principle in sustaining productive agriculture in mixed crop–animal production systems; however, this can be difficult to achieve in practice (Mikkelsen 2000b). The major problem with P for intensive animal production in North America is the separation of the feed-production areas from the animal-production areas. Since the challenge and expense associated with transporting manure over significant distances is difficult to overcome even with government regulatory or financial incentives, manure is too frequently applied to the most convenient fields.

Animal manures serve as a large pool of potentially recoverable P. Manure and animal waste products annually generate over 20 Mt P. When the animals are raised in confined areas, there is reasonable expectation to collect the manure and uniformly distribute it back on to crop production fields when possible, which is currently the case for about 50 % of the manure produced (Cordell et al. 2009). For the other 50 %, mostly free ranging or grazing animals, it is not generally feasible to collect manure, which leads to uneven distribution of manure and P. Additionally, grazing animals tend to congregate near water or shelter, leaving manure poorly distributed across the field. The fertilizer industry has repeatedly looked at how animal-derived nutrient sources can be incorporated into wide-scale crop production, but a suitable economic model has not yet been developed.

Various approaches have been proposed or tested to reduce P losses from manure-amended soils. For example, chemical treatments such as aluminum chloride (alum) added to manure have been shown to reduce soluble P and losses in runoff. Chemical additives might, however, have only a short-lasting effect (Schärer et al. 2007). Regardless, their additions cannot solve the problem of P inputs added in excess to crops' needs observed in systems with a high density of animals.

Experimental pigs have been genetically modified to produce more salivary phytase, allowing them to use the grain P much more efficiently ("enviropigs," Golovan et al. 2001). The public acceptance of these animals is not yet known, and whereas this genetic modification indeed significantly reduces the concentration of P in the manure, it does not address the emission of other nutrients which can be problematic in intensive swine production.

The P in animal wastes and composts is present in variable proportions of soluble P and scarcely soluble inorganic and organic P compounds (Oberson et al. 2010). The organic P compounds require microbial transformation for conversion to plant-available forms. The time required for the mineralization can range from days to years, depending on environmental factors and the type of organic compounds present.

A number of techniques have been developed to extract and recover P from animal manure. Several companies currently treat animal wastes and produce inorganic P compounds (such as calcium phosphates and magnesium ammonium phosphate or struvite) that can be used as commercial fertilizer.

From an efficiency view, the key issue is that P that is not retained in the animal should be recovered and returned to the field where the feed originated. There are many global examples where this is not being properly done, resulting in excessive P accumulation and potential water quality problems near the animal farms. Nutrient depletion is inevitable in the crop-producing areas unless the soils are supplied with supplemental P to replace the harvested P. There are a number of issues including economic constraints (such as expense and energy required to transport relatively dilute manure or to concentrate or dry it) that challenge the full recycling potential of animal wastes and pose a serious hindrance to improving global P efficiency.

3 Opportunities to Improve Phosphorus Use at the Societal Level

3.1 Postharvest, Retail, and Household Level

Large-scale food processing in developed countries is frequently contained in a closed process, making it easier to recover P. In less developed countries, it may be more challenging to recover P from food-processing residues and utilize them in a productive way. However, in both scenarios, there are multiple opportunities to reduce nutrient losses during food storage, processing, and transportation.

After harvest, crops are processed into final products for food, feed, fiber, and fuel. Losses occurring during storage from pests and disease can often be reduced by better management practices. Parts of food that are discarded during processing (such as husks or damaged and under-grade products) also represent a "loss" of P unless they are returned to the fields. Improving this may necessitate additional infrastructure, better management by the processor, and more thoughtful food handling by consumers.

With long and sophisticated food delivery systems that frequently stretch across the globe, care should be taken to minimize food losses. The challenge of dealing with wastes from a global food chain could be simplified by producing food as closely as possible to where it is consumed and then returning the waste products to the agricultural production fields. However, there are serious economic, social, technological, and logistic issues that need to be resolved to implement this ideal.

Huge quantities of food are regularly disposed at the retail level due to spoilage or elapsed expiration dates. To avoid unnecessary wastage, a balance between excessive disposal and food quality and safety must be maintained. Food that is slightly beyond its peak freshness or contains minor cosmetic blemishes should not be automatically disposed.

The loss of food and nutrients in the home is very high in some countries. It is estimated that over 50 % of food waste in some countries is still edible and could be used with better meal preparation and planning. Reasons for food being wasted in the UK are: (1) poor pre-shop planning; (2) not sticking to the shopping list; (3) not understanding the "use by" date and "best before" date; (4) food not stored correctly; (5) no meal planning; (6) cooking too large portions; and (7) poor skills in combining left-overs with fresh food (Parfitt et al. 2010). Thus, a cultural shift to create awareness among consumers is needed at different levels and might start with educational programs in school.

Projections of human population growth estimate that much of this increase will occur in peri-urban areas of developing countries where recycling could still be improved (Kiba et al. 2012a, b). Recycling urban wastes is easier where a large concentration of people is living. By reducing the quantity of food and nutrients going to landfills and sewers, compost or other organic products could be developed to beneficially reuse this resource (Kader 2005).

Recovery of P in household waste is not simple. The food wastes need to be collected and processed at a collection area. The P in human waste can be separated in specialized toilets prior to entering the sewer, but this technology is not widely adopted. Biosolids from waste treatment plants can provide a valuable P source for crops, but issues with public acceptance, sanitation, and the residual constituents must be considered. Phosphorus in sewage sludge ash can be recycled to cropland, but technologies to recycle this P are not widely adopted. In addition to the nutrients contained in biosolids, the organic matter can also be useful for soil improvement. While technologies to treat and recycle waste exist, socioeconomic constraints limit widespread adoption.

The ideal P cycle involves removal of nutrients from the soil by plants, harvest, and ultimate replacement of nutrients back to the soil without any losses. However, a perfectly closed P cycle has never existed (even in unfertilized ecosystems), and there will be some inevitable loss from the relatively leaky biological process of food production. However, the current system of moving large quantities of P from farms to the city without consideration of returning the P and other nutrients to the crop-producing area is short sighted and inefficient. Restoring this linkage is essential for making significant progress in improving P use efficiency.

3.2 Issues of Scale

As the long-term implications of poor utilization of P become better appreciated (such as water quality impacts, land degradation, and potential rock P depletion), a fresh examination of how P is used is timely, but the potential for improving P use efficiency in agriculture and food systems varies significantly by location. Countries that export significant quantities of agricultural products will necessarily have a different P balance than countries that rely on significant quantities of imported food and feed.

Improving P efficiency will involve careful examination of the entire biogeochemical process of this essential nutrient. Solutions to this issue require a balance between the imperatives to produce more food each year while minimizing losses that have wasteful and adverse impacts. The urgency to improve management practices is spurred by the recognition that P is a limited natural resource that plays an irreplaceable role in sustaining plant, animal, and human life. The social acceptance of changes that are made to improve P recycling, recovery, and efficiency will require a significant educational effort and knowledge of transition processes. A number of social, economic, and environmental goals can be simultaneously achieved as progress is made toward short-term and long-term food security and global sustainability. The complexity and regional specificity of the different practices influencing P use can benefit from a transdisciplinary approach to better understand and promote sustainable P use practices.

4 Work in Global TraPs

4.1 Knowledge Gaps and Critical Questions

Although a great deal of technical knowledge exists on ways to efficiently use and recycle P in agriculture, too few studies seek to understand how agricultural policy, financial services, farming technologies, and local capabilities interactively affect decisions about nutrient use and management (Hazell et al. 2007). Additionally, how such decisions affect soil fertility, food productivity, and profitability of the whole farm have not been sufficiently investigated. Human decision making at multiple levels, interactions between agroecosystem's components, and their dynamics over time and space is key to understand sustainable P use. An important first step toward promoting sustainable agriculture is incorporating region-specific parameters, such as how inorganic P fertilizers are being introduced to world regions at different times and with significantly different economic impacts (Elsner 2008).

The challenge of improving P use efficiency highlights important questions that need to be addressed:

- What potential does INM have for supporting soil fertility, improved food production, preservation of water quality, and eventually better livelihoods?
- What strategies can support more effective use of P resources (including manure and fertilizer) in market-oriented farms, so as to decrease vulnerability to fluctuating fertilizer prices and reduce negative environmental impacts?

4.2 Role, Function, and Kind of Transdisciplinary Processes

In order to comprehensively tackle knowledge gaps associated with improving P use, a collaboration of various disciplines (e.g., agricultural and soil sciences, biology, hydrology, environmental science, industrial ecology, economics, water engineering, sociology) and different societal fields/stakeholders (including administration, individual farmers and farmer organizations, fertilizer industry, NGOs, society/consumers) is essential.

Given the diverse tasks/targets of these groups, it is inevitable that decisions on P management can neither be understood nor managed by academia, industry, government, or international organizations alone. There is a need for close collaboration between these groups to create comprehensive knowledge and develop robust strategies. Transdisciplinary case studies allow for addressing highly complex societal problems such as P use, which need multi-perspective approaches and require the ability to deal with different knowledge cores, diverse interests, values, and norms. Through transdisciplinary case studies, optimizations related to closing different P cycles can be achieved, considering critical feedbacks with other cycles and P-related issues. They are especially suited on local/regional level, where the production of place-specific and application-oriented knowledge and transition abilities can be linked. The following examples provide illustrations of transdisciplinary case studies that could help with P sustainability issues at various scales.

4.3 Potential Case Studies

Bavarian P strategies: *Sustainable use of P in Bavaria and integrative transdisciplinary analysis.* Sustainable P management at a regional level implies a balance between processing, agricultural production, food consumption, and waste management. Focusing on key agricultural production and food consumption issues of relevance include: (a) the interrelation of P used for food and energy production, (b) the need for P in food production, (c) the regulatory framework with respect to the use of by-product-generated P, and (d) lifestyles affecting consumer behavior, such as diet (meat versus vegetarian) and the demand for certified food products.

This transdisciplinary case study focuses on the Federal State of Bavaria, Germany. Key research issues are: at which spatial scale of P balancing and trading among farmers should be fostered to reduce the risk of decreasing fertility of the soils, and what is the level of acceptance of "by-product fertilizers" among farmers? Further, the role which various food production systems could play in closing the P loop by linking consumption, waste management, and sustainable production will be examined.

P from animal manure: *Social and institutional constraints for integrating P from intensive animal production into crop production.* A large amount of research has been done to understand the behavior of P derived from animal manure in diverse cropping systems. When utilized properly, manure P can be an excellent nutrient source for sustainable agriculture. However, there are multiple examples where mismanagement of animal manure has resulted in excessive P loss to surface water, with an accompanying degradation of water quality and environmental services.

This case study will compare and contrast various approaches to incentivize more efficient P management in animal manure. Management approaches that involve institutional and government controls will be compared with technological remediation approaches. This will be done through a multi-stakeholder discourse by which the appropriate role of various private and public institutions in successful P management strategies will be assessed.

Modifying cropping systems: *Modifying cropping systems to improve P efficiency.* A number of practical on-farm agronomic techniques have been shown to improve the efficiency of P use. For example, the use of intercropping, cover crops, proper rotations, and integrated soil fertility management have all been demonstrated to have the potential of improving P use efficiency. In addition to the P benefits, the environmental value of these techniques has not been well quantified yet. These cropping system modifications are well suited for small-scale farmers who could immediately benefit from improved P management. However, there are a number of social, cultural, economic, and agronomic factors that can limit the widespread adoption of these practices. There are significant regional differences in farming and social conditions that can also pose significant barriers to adoption.

Appropriate P sources: *Identifying and delivering the most appropriate P source.* Many soils in the world contain insufficient P to sustain high crop yields. When a source of P is available to farmers in these P-deficient regions, it may not be the most appropriate material since they are often forced to rely on whatever P fertilizer is available in the market or is subsidized by the government. Market availability of P does not always equate to the most agronomically desirable material to meet economic, environmental, or production goals.

An analysis of what P fertilizer materials are currently available in the market along with a report on the most agronomically favorable P fertilizers would allow a better match in supply and demand. This would involve a transdisciplinary exchange between fertilizer companies, brokers, traders, and farm groups in developing countries where P deficiencies are common. However, other factors including cost per unit of nutrient, associated transport costs, storage and handling characteristics, availability, total tonnage all have to be considered. Involving market players will allow for better understanding and insight into factors that affect fertilizer demand beyond agronomic performance.

East African P management: *Phosphorus management in East African wheat, potato, and coffee smallholder farms.* Phosphorus management will be evaluated on smallholder farms with three different commodities (wheat, potato, and coffee) in three different agroecological and sociocultural regions of East Africa. Each commodity presents a unique challenge with P use. The case study aims at joint stakeholder research to understand and develop sustainable P management strategies at the farm level considering local conditions (such as local environmental, economic, market, social, and cultural conditions). Farm-specific learning processes and stakeholder interactions will be emphasized.

In the case study, three farmer groups will be engaged to develop insights to current P management in wheat, potato, and coffee production and then explore options for improvement. Based on best practices, recommendations for improved P efficiency may be extrapolated for a commodity within a broader regional context where constraints are similar. However, extrapolation to other commodities may be limited. For example, potato and coffee farmers typically have strong market linkages, but they may face labor constraints while wheat producers may have limited access to adequate markets.

Appropriate P in Vietnam: *Multi-actor strategies for avoiding P overuse (peri-urban) and P underuse (remote uplands) in Vietnamese smallholder farms.* Phosphorus fertilizer management challenges for Vietnamese farmers fall into two categories. (1) Many farmers engage in intensified production to meet market demand for food by applying P fertilizer at adequate or excessive rates to maintain production. However, these farmers frequently fail to use management practices to minimize P losses through runoff and erosion. This group includes smallholder farmers engaged in intensified agricultural production of cereals, fruits, and vegetables and who often produce two to three crops (e.g., rice, vegetables) per year on the same land area. (2) A second group includes poor subsistence farmers who cannot gain access to P fertilizer, leading to poor crop yields, soil degradation, and a cycle of poverty. Therefore, viable options for economically and environmentally efficient P use and recycling in these two agroecosystems need attention. The study considers Vietnam's smallholder systems in the Red River delta (available P fertilizer, market-oriented) and in the Northwest Mountain Region (no P fertilizer access, subsistence) as cases for the two contrasting P use regimes.

The case study will follow a transdisciplinary process focusing on P use in smallholder agroecosystems to form a better integration of scientific and stakeholders' knowledge. This will facilitate stakeholders' understanding about achieving sustainable P use. Stakeholders involved with P management (e.g., national and policy-makers, fertilizer companies, rural development donors and farmer groups) can benefit by improving their understanding of the role of P in long-term approaches to sustainable food production.

Acknowledgments We thank Patrick Heffer, Deborah T. Hellums, Quang B. Le, and Amit H. Roy for important comments on earlier drafts of the paper.

References

Binder CR, Wittmer D, Mouron P, Bieler P, Herren M (2008) Phosphorflüsse in der Schweiz: Stand, Risiken, Optionen, Schlusspräsentation Modul 1, BAFU

Cakmak I (2002) Plant nutrition research: priorities to meet human needs for food in sustainable ways. Plant Soil 247:3–24

Council for Agricultural Science and Technology (CAST) (2013) Food, fuel, and plant nutrient use in the future. Issue paper 51. CAST, Ames, Iowa, USA

Chien SH, Sikora FJ, Gilkes RJ, McLaughlin MJ (2012) Comparing of the difference and balance methods to calculate percent recovery of fertilizer phosphorus applied to soils: a critical discussion. Nutr Cycl Agroecosyst 92:1–8

Cordell D, Drangert JO, White S (2009) The story of phosphorus: global food security and food for thought. Glob Environ Change 19:292–305

Craswell ET, Vlek PLG, Tiessen H (2010) Peak phosphorous—implications for productivity and global food security. In: Proceedings of 19th world conference for soil solutions for a changing world, Brisbane, Australia

Cromwell GL (2005) Phosphorus: agriculture and the environment, Agronomy monograph. In: Sharpley AN et al (eds) Phosphorus and swine nutrition, vol 46. American Society of Agronomy, Madison, pp 607–634

Duquesnay A, Dupouey JL, Clement A, Ulrich E, Le Tacon F (2000) Tree Physiol 20:13–22

Edwards AC, Withers PJA, Sims TJ (1997) Are current fertiliser recommendation systems for phosphorus adequate? Int Fertiliser Soc 404:1–23

Elsner H (2008) Stand der Phosphat-Reserven weltweit (Status of phosphate reserves worldwide). Braunschweiter Nährstofftage. Julius Kühn-Institut, Hannover, Germany, Electronic Presentation, Bundesanstalt für Geowissenschaften und Rohstoffe

Engelstad OP, Hellums DT (1992) Water solubility of phosphate fertilizers: agronomic aspects a literature review. International Fertilizer Development Center (IFDC) paper series P-17, p 27

FAO (2009) State of the world's forest 2009. Food and Agriculture Organization of the United Nations, Rome, Italy

Fox TR, Miller BW, Rubilar R, Stape JL, Albaugh TJ (2011) Phosphorus nutrition of forest plantations: the role of inorganic and organic phosphorus. In: Bünemann E et al (ed) Phosphorus in action. Soil biology 26, Springer, Berlin Heidelberg, pp 317–338

Frossard E, Bünemann E, Jansa J, Oberson A, Feller C (2009) Concepts and practices of nutrient management in agro-ecosystems: can we draw lessons from history to design future sustainable agricultural production systems? Die Bodenkultur 60:43–60

Gallet A, Flisch R, Ryser JP, Frossard E, Sinaj S (2003) Effect of phosphate fertilization on crop yield and soil phosphorus status. J Plant Nutr Soil Sci 166:568–578

Genenger M, Zimmermann S, Hallenbarter D, Landolt W, Frossard E, Brunner I (2003) Fine root growth and element concentrations of Norway spruce as affected by wood ash and liquid fertilisation. Plant Soil 255:253–264

Golovan SP, Meidinger RG, Ajakaiye A, Cottrill M, Wiederkehr MZ, Barney DJ, Plante C, Pollard JW, Fan MZ, Hayes MA, Laursen J, Hjorth JP, Hacker RR, Phillips JR, Forsberg CW (2001) Pigs expressing salivary phytase produce low-phosphorus manure. Nat Biotechnol 19:741–745

Granger SJ, Bol R, Anthony S, Owens PN, White SM, Haygarth PM (2010) Towards a holistic classification of diffuse agricultural water pollution from intensively managed grasslands on heavy soils. Adv Agron 105:83–115

Hazell P, Poulton C, Wiggins S, Dorward A (2007) The future of small farms for poverty reduction and growth. International Food Policy Research Institute Washington, DC

Hedley M, McLaughlin M (2005) Reaction of phosphate by products in soils. In: Sharpley AN et al (eds) Phosphorus: agriculture and the environment, Agronomy monograph, vol 46. Madison, Madison, pp 181–252

Hilton J, Johnston AE, Dawson CJ (2010) The phosphate life-cycle: rethinking the options for a finite resource. Int Fertiliser Soc 668:1–42

International Fertilizer Association (IFA) (2013) Direct application of phosphate rock. http://www.fertilizer.org/ifacontent/download/97688/1430905/version/3/file/2013_ifa_darp.pdf

International Plant Nutrition Institute (IPNI) (2010) Soil test levels in North America. 30-3110. Norcross, GA, USA

Jasinski SM (2011) 2010 minerals yearbook—phosphate rock. U.S. Geological Survey, Washington DC

Johnston AE, Dawson CJ (2010) Physical, chemical and biological attributes of agricultural soils. Proc Int Fertiliser Soc 675:1–40

Kader AA (2005) Increasing food availability by reducing postharvest losses of fresh produce. In: Mencarelli F, Tonutti P (eds) Fifth international Postharvest symposium, Acta Horticulturae

Kiba DI, Lompo F, Compaore E, Randriamanantsoa L, Sedogo PM, Frossard E (2012a) A decade of non-sorted solid urban wastes inputs safely increases sorghum yield in periurban areas of Burkina Faso. Acta Agric Scand Sect B: Plant Soil Sci 62:59–69

Kiba DI, Zongo NA, Lompo F, Jansa J, Compaore E, Sedogo PM, Frossard E (2012b) The diversity of fertilization practices affects soil and crop quality in urban vegetable sites of Burkina Faso. Eur J Agron 38:12–21

Lamprecht H, Lang DJ, Binder CR, Scholz RW (2011) The trade-off between phosphorus recycling and health protection during the BSE crisis, in Switzerland, A "Disposal Dilemma", GAIA 20, pp 112–121

MacDonald GK, Bennett EM, Potter PA, Ramankutty N (2011) Agronomic phosphorus imbalances across the world's croplands. PNAS 108:3086–3091

Mikkelsen RL (2000a) Beneficial use of swine by-products: opportunities for the future. In: Proceedings of beneficial uses of agricultural, industrial, and municipal by—products, Soil Science and Society of American Specialists Publication, pp 451–480

Mikkelsen RL (2000b) Nutrient management for organic farming: a case study. J Nat Resour Life Sci Educ 29:88–92

Nelson NO, Mikkelsen RL (2005) Balancing the phosphorus budget of a swine farm: a case study. J Nat Resour Life Sci Educ 34:90–95

Oberson A, Bünemann E, Frossard E, Pypers P (2011) Management impacts on biological P cycling in cropped soils. In: Bunemann E (ed) Phosphorus in action. Soil biology, vol 26. Springer, Berlin Heidelberg, pp 431–458

Oberson A, Tagmann HU, Langmeier M, Dubois D, Mäder P, Frossard E (2010) Fresh and residual phosphorus uptake by ryegrass from soils having different fertilization histories. Plant Soil 334:391–407

Parfitt J, Barthel M, MacNaughton S (2010) Food waste within food supply chains: quantification and potential for change to 2050 Phil. Trans R Soc B 365:3065–3081

Prud'homme M (2010) Peak phosphorus: an issue to be addressed. Fertilizers and agriculture, International Fertilizer Industry Association (IFA), Feb 2010

Quinton JN, Govers G, van Oost K, Bardgett RD (2010) The impact of agricultural soil erosion on biogeochemical cycling. Geo Nat 3:311–314

Reed SC, Townsend AR, Taylor PG, Cleveland CC (2011) Phosphorus cycling in tropical forests growing on highly weathered soils. In: Bünemann E (ed) Phosphorus in action. Soil biology, vol 26. Springer, Berlin Heidelberg, pp 339–369

Richards IR, Dawson DJ (2008) Phosphorus imports, exports, fluxes and sinks in Europe. Int Fertiliser Soc 638:1–28

Richardson AE, Lynch JP, Ryan PR, Delhaize E, Smith FA, Smith SE, Harvey PR, Ryan MH, Veneklaas EJ, Lambers H, Oberson A, Culvenor RA, Simpson RJ (2011) Plant and microbial strategies to improve the phosphorus efficiency of agriculture. Plant Soil 349:121–156

Schärer M, Stamm C, Vollmer T, Frossard E, Oberson A, Flühler H, Sinaj S (2007) Reducing phosphorus losses from over-fertilized grassland soils proves difficult in the short term. Soil Use Manage 23:154–164

Scholz RW, Wellmer FW (2013) Approaching a dynamic view on the availability of mineral resources: what we may learn from the case of phosphorus? Glob Environ Change 23:11–27

Schröder JJ, Smit AL, Cordell D, Rosemarin A (2010) Sustainable use of phosphorous. Report No. 357. Plant Research International, Wageningen, NL

Schröder JJ, Smit AL, Cordell D, Rosemarin A (2011) Improved phosphorus use efficiency in agriculture: a key requirement for its sustainable use. Chemosphere 84:822–831

Simpson RJ, Oberson A, Culvenor RA, Ryan MH, Veneklaas EJ, Lambers H, Lynch JP, Ryan PR, Delhaize E, Smith FA, Smith SE, Harvey PR, Richardson AE (2011) Strategies and agronomic interventions to improve the phosphorus-use efficiency of farming systems. Plant Soil 349:89–120

Smalberger SA, Singh U, Chien SH, Henao J, Wilkens PW (2006) Development and validation of a phosphate rock decision support system. Agron J 98:471–483

Smil V (2000) Phosphorus in the environment: natural flows and human interferences. Annu Rev Energy Environ 25:53–88

Smit AL, Bindraban PS, Schroder JJ, Conijn JG, van der Meer HG (2009) Phosphorus in agriculture: global resources, trends and developments. Plant Research International B. V, Wageningen

Spiess E (2011) Nitrogen, phosphorus and potassium balances and cycles of Swiss agriculture from 1975 to 2008. Nutr Cycl Agroecosyst 91:351–365

St Arnaud RJ, Stewart JWB, Frossard E (1988) Application of the "pedogenic index" to soil fertility studies in Saskatchewan. Geoderma 43:21–32

Suh S, Yee S (2011) Phosphorus use-efficiency of agriculture and food system in the US. Chemosphere 84:806–813

Syers JK, Johnston AE, Curtin D (2008) Efficiency of soil and fertilizer phosphorus use. FAO fertilizer and plant nutrition bulletin 18. Food and Agriculture Organization of the United Nations, Rome, p 108

Tarkalson DD, Mikkelsen RL (2003) A phosphorus budget of a poultry farm and a dairy farm in the Southeastern U.S. and the potential impacts of diet alterations. Nutr Cycl Agroecosyst 66:295–303

Vitousek PM, Naylor R, Crews T, David MB, Drinkwater LE, Holland E, Johnes PJ, Katzenberger J, Martinelli LA, Matson PA, Nziguheba G, Ojima D, Palm CA, Robertson GP, Sanchez PA, Townsend AR, Zhang FS (2009) Nutrient imbalances in agricultural development. Science 324:1519–1520

Appendix: Spotlight 6

Health Dimensions of Phosphorus

James J. Elser

Phosphorus is an essential element for all living things, needed (in the form of phosphate, PO_4) in cells for construction and renewal of DNA & RNA, of phospholipids, and of energy transduction molecules such as ATP. In vertebrates, PO_4 (P_i, hereafter) is a main component of the mineral apatite (a form of calcium phosphate) in bones. Thus, human health depends on an adequate dietary supply of P every day (DACH 2008, Reference Values for Nutrient Intake): 700 mg for an adult, 500–1,250 mg for children and youth). The body tightly regulates P_i homeostasis, primarily by modulating P_i excretion in the kidney, closely in concert with calcium (Ca) due to their joint role in bone formation.

Deficiency

Deficiencies of P are well studied in association with Vitamin D metabolism, as Vitamin D regulates levels of Ca and P in the bloodstream and thus controls bone growth and remodeling (Perwad and Portale 2011). Vitamin D deficiency leads primarily to bone fragilization ("rickets"), a relatively rare condition in the modern world. Direct deficiency of P ("hypophosphatemia") is also relatively rare for humans nowadays, as P_i is relatively abundant in many foods and easily assimilated in the gut. However, it can occur independent of Vitamin D deficiency in, for example, cases of general malnutrition, alcoholism, damage to the gastrointestinal tract, or tumors that produce the P_i-regulating hormone FGF23.

J. J. Elser
Arizona State University, Downtown Phoenix campus, 411 N. Central Avenue, Phoenix, AZ 85004, USA
e-mail: j.elser@asu.edu

Excess

Elemental P: Nearly all P on Earth is in the oxidized form, as phosphate (PO_4). However, various industrial processes involve production and use of two molecular forms of elemental P: white phosphorus (P_4) and red phosphorus (a polymer based on the P_4 unit). These forms of P are used in production of matches, munitions, and illicitly methamphetamines. Elemental P is highly unstable and reactive, readily reacting with oxygen. Thus, exposure to elemental P is damaging to tissues, and use and exposure in industrial settings are usually subjected to close regulation.

Dietary PO_4 and kidney dialysis: A significant challenge facing kidney disease patients is proper regulation of body PO_4 because existing dialysis methods are relatively inefficient in removing PO_4 (Kuhlmann 2007). Thus, patients with kidney disease often exhibit excess serum phosphate ("hyperphosphatemia") that must be addressed using PO_4-binding medications and/or by reducing levels of dietary P intake. The latter can be challenging because foods with high P content are often the same as those with high protein content (thus patients can become protein deficient) and because many foods contain PO_4 additives used as preservatives (e.g., deli meats, many processed foods) or flavorants (e.g., phosphoric acid in some soft drinks).

Dietary PO_4, cardiovascular disease, aging, and cancer: The health risks of P are well established for elemental forms and for patients with kidney disease. However, tentative epidemiological and animal data also suggest link between P_i intake (both high and low) and certain degenerative diseases and cancers. For example, P_i has been called the "new cholesterol" (Ellam and Chico 2012) due to evidence linking high P_i intake to cardiovascular disease via a mechanism in which P_i reacts with Ca in formation of mineral plaques that contribute to atherosclerosis ("hardening of the arteries"). The reader should bear in mind, however, that such epidemiological studies cannot conclusively establish such links and considerably more work is needed. Some data connecting elevated P_i intake to accelerated aging have recently appeared, in the form of studies of mice-bearing mutations in the gene Klotho (John et al. 2011), which acts with FGF23 to regulate kidney P_i transport. Mice lacking either FGF23 or Klotho show hyperphosphatemia and develop multiple aging-like symptoms. This work is in a preliminary stage and has not been extensively evaluated for its relevance to humans. Finally, it has recently been proposed that dietary P_i has a mechanistic link to tumor progression because of the high P demands needed to construct ribosomal RNA in rapidly growing tumor cells (Elser et al. 2003). Tentative support for this hypothesis has appeared from comparative study of human tumors (Elser et al. 2007) and experimental dietary P manipulations in mice (Wulaningsih et al. 2013). As with P_i's possible association with aging, much further work remains to confirm or reject a putative connection to human cancer.

Acknowledgements The author is grateful for the helpful comments of Haley C. Steven, Roland W. Scholz, Marc Vermeulen and two anonymous reviewers, whose input helped clarify important issues in the text.

References

Camalier CE, Young MR, Bobe G, Perella CM, Colburn NH, Beck GR, Jr. (2010) Elevated Phosphate Activates N-ras and Promotes Cell Transformation and Skin Tumorigenesis. Cancer Prevention Research 3 (3):359-370. doi:10.1158/1940-6207.capr-09-0068

DACH (2008) Referenzwerte für Nährstoffzufuhr, 3rd edn. http://www.sge-ssn.ch/de/ich-und-du/rund-um-lebensmittel/Referenzwerte-fuer-die-Naehrstoffzufuhr/

Ellam TJ, Chico TJA (2012) Phosphate: the new cholesterol? The role of the phosphate axis in non-uremic vascular disease. Atherosclerosis 220(2):310–318. doi:10.1016/j.atherosclerosis.2011.09.002

Elser JJ, Kuang Y, Nagy J (2003) Biological stoichiometry of tumor dynamics: an ecological perspective. BioScience 53:1112–1120

Elser JJ, Kyle M, Smith M, Nagy J (2007) Biological stoichiometry in human cancer. PLoS ONE 10:e1028. doi:10.1371/journal.pone.0001028

Jin H, Xu C-X, Lim H-T, Park S-J, Shin J-Y, Chung Y-S, Park S-C, Chang S-H, Youn H-J, Lee K-H, Lee Y-S, Ha Y-C, Chae C-H, Beck GR, Jr., Cho M-H (2009) High dietary inorganic phosphate increases lung zumorigenesis and alters Akt signaling. American Journal of Respiratory and Critical Care Medicine 179 (1):59-68. doi:10.1164/rccm.200802-306OC

John GB, Cheng CY, Kuro-o M (2011) Role of Klotho in aging, phosphate metabolism, and CKD. Am J Kidney Dis 58(1):127–134. doi:10.1053/j.ajkd.2010.12.027

Kuhlmann MK (2007) Practical approaches to management of hyperphosphatemia: can we improve the current situation? Blood Purif 25(1):120–124. doi:10.1159/000096410

Perwad F, Portale AA (2011) Vitamin D metabolism in the kidney: regulation by phosphorus and fibroblast growth factor 23. Mol Cell Endocrinol 347(1–2):17–24. doi:10.1016/j.mce.2011.08.030

Wulaningsih W, Michaelsson K, Garmo H, Hammar N, Jungner I, Walldius G, Holmberg L, Van Hemelrijck M (2013) Inorganic phosphate and the risk of cancer in the Swedish AMORIS study. Bmc Cancer Unsp 13:257. doi:10.1186/1471-2407-13-257

Appendix: Spotlight 7

Phosphorus in the Diet and Human Health

Rainer Schnee, Haley Curtis Stevens, and Marc Vermeulen

A minimum of phosphorus in human diets is essential for health, because the body needs phosphorus (P) for bones, teeth, DNA, energy metabolism and many other functions. The recommended daily requirement for health (DV) for P is 1,000 mg/day (Council for Responsible Nutrition 2013, based on US Food and Drug Administration data). However, modern Western diets often have higher levels of P, because of improved diets, increased meat and dairy product intake (rich in P, calcium and other minerals) and food phosphates used to ensure safety (bacteria free preservation) or processing of many modern foods, such as cakes, soft cheese, cold meats, pre-prepared meals. Today in Europe, dietary P intake is around 1.3–2.7 g P/day (Flynn et al. 2009). The balance between P and dietary calcium is important.

The dietary contribution of food phosphate additives is around 0.15 g P/day, based on an estimate by the Phosphoric Acid and Phosphates Producers Association of 25,000 t P/year in food additives for EU25 in 2006. This is well below the MTDI (Maximum Tolerable Daily Intake) for food phosphates considered safe (JECFA 1982). The large majority of P in diets comes from natural sources, such as milk and dairy products, meat and many other foods.

R. Schnee
Chemische Fabrik Budenheim KG, Rheinstrasse 27, 55257 Budenheim, Germany
e-mail: rainer.schnee@budenheim.com

H. C. Stevens
International Food Additives Council (IFAC), 1100 Johnson Ferry Road, Suite 300, Atlanta, GA 30342, USA
e-mail: hstevens@kellencompany.com

M. Vermeulen · R. Schnee
Phosphoric Acid & Phosphates (PAPA), European Chemical Industry Council,
Avenue E. van Nieuwenhuyse 4, 1160 Brussels, Belgium
e-mail: mve@cefic.be

For patients with kidney problems, it is generally recognized that P accumulation in the body can cause significant health damage. This is because P is normally balanced in the body by excretion of unneeded intake by the kidneys.

A number of studies suggest a statistical relationship between blood phosphorus concentrations (serum P) and indicators of cardiovascular disease (CVD) in the general population. Some recent studies (Westerberg et al. 2013; Itkonen et al. 2013), however, show no link after taking into account other factors. It is thus unclear whether P and/or other minerals (e.g., calcium) contribute to these risks, or whether both are consequences of other factors such as unhealthy diet, obesity, undeclared kidney insufficiency, or other body metabolism problems.

While it is recognized that deterioration of kidney function leads to increased blood P (serum P), there is no evidence that higher dietary P levels lead to increased serum P (except immediately after the P containing meal) in persons not suffering from kidney function deficiency. Indeed, correctly functioning kidneys normally maintain an optimal serum P level. Analysis of the numerous studies available comparing P intake to serum P show that many do not contain relevant data (diet not known, data only for very short term such as one dose of high dietary P), and that those containing useful data do not enable any conclusion because they reach conflicting conclusions.

Very few studies are available comparing dietary P levels directly with heart disease (as opposed to comparing P intake to serum P or serum P to heart disease), and the studies which do exist show contradictory results, some suggesting a correlation and others suggesting no correlation or an inverse correlation (dietary phosphorus inversely related to heart disease symptoms) (Alonso et al. 2010; Elliott et al. 2008; Joffres et al. 1987).

Thus scientific data does not at present indicate that the P content of Western diets increases heart disease risk. Further investigation is warranted to better understand the roles of diet and life style in the etiology of heart disease.

Although several publications suggest a possible link between dietary P and other health impacts these are based on metabolic hypotheses, with very little in vitro basis and virtually no experimental evidence. Because there are very few experimental data, it must be considered that these may be caused by artifacts. The experimental evidence regarding cancer concerns only genetically modified mice, artificially susceptible to cancer. It can be considered that these results cannot be reliably extrapolated to normal mice, or to humans (IFAC 2009). In the most cited case (Jin et al. 2009) a later similar experiment by the same authors with the same genetically modified mice produced the contrary result: a *low* phosphate diet increased lung cancer (Cheng-Xiong et al. 2010). We face the problem of extrapolation of experimental findings to human in vivo conditions also with a study suggesting that high dietary P in Drosphila flies reduced their lifespan (aging), which again cannot be reliably extrapolated to humans, has been completed by a further study suggesting that this is due to impairment of flies' equivalent to kidneys (Bergwitz et al. 2013).

A few epidemiological or in vitro studies with humans suggest a statistical relationship between dietary P and cancer occurrence, but others show no relationship (Berndt et al. 2002; Chan et al. 1998; Giovannucci et al. 1998; Tavani et al. 2005), and others an inverse relationship (increased dietary P related to lower cancer incidence; Brinkman et al. 2010; Chan et al. 2000; Kesse et al. 2005; Spina et al. 2012; Takata et al. 2013; van Lee et al. 2011) or positive/negative relationships for different forms of cancer (Wulaningsih et al. 2013). Other recent publications suggest a possible link between low dietary P and obesity (Celik and Andiran 2011; Lindegarde and Trell 1977; Haglin et al. 2001; Lind et al. 1993; Obeid 2013) or between dietary P and reduced cholesterol (Kim et al. 2013; Ditscheid et al. 2005; Lippi et al. 2009; Trautvetter et al. 2012). There are many other studies which indicate that P intake is safe at current levels (see e.g., JECFA 1982; Weiner et al. 2001).

Based on the current scientific evidence, it can be concluded that normal Western levels of dietary P intake are safe: no studies have reported a clear link to human health risks. This is to conform to the conclusions of the European EFSA scientific panel (EFSA 2005).

Acknowledgements Thanks to James J. Elser, Myra Weiner, Roland W. Scholz and two anonymous reviewer, whose input helped to improve the text.

References

Alonso A, Nettleton JA, Ix JH, de Boer IH, Folsom AR, Bidulescu A, Kestenbaum BR, Chambless LE, Jacobs DR Jr (2010) Dietary phosphorus, blood pressure, and incidence of hypertension in the atherosclerosis risk in communities study and the multi-ethnic study of atherosclerosis. Hypertension 55 (3):776–U109. doi:10.1161/hypertensionaha.109.143461

Bergwitz C, Wee MJ, Sinha S, Huang J, DeRobertis C, Mensah LB, Cohen J, Friedman A, Kulkarni M, Hu Y, Vinayagam A, Schnall-Levin M, Berger B, Perkins LA, Mohr SE, Perrimon N (2013) Genetic Determinants of Phosphate Response in Drosophila. Plos One 8(3): e56753. doi:10.1371/journal.pone.0056753

Berndt SI, Carter HB, Landis PK, Tucker KL, Hsieh LJ, Metter EJ, Platz EA (2002) Calcium intake and prostate cancer risk in a long-term aging study: the Baltimore longitudinal study of aging. Urology 60(6):1118–1123. doi:10.1016/s0090-4295(02)01991-x

Brinkman MT, Karagas MR, Zens MS, Schned A, Reulen RC, Zeegers MP (2010) Minerals and vitamins and the risk of bladder cancer: results from the New Hampshire study. Cancer Causes Control 21(4):609–619. doi:10.1007/s10552-009-9490-0

Celik N, Andiran N (2011) The relationship between serum phosphate levels with childhood obesity and insulin resistance. J Pediatr Endocrinol Metab 24(1–2):81–83. doi:10.1515/jpem.2011.116

Chan JM, Giovannucci E, Andersson SO, Yuen J, Adami HO, Wolk A (1998) Dairy products, calcium, phosphorous, vitamin D, and risk of prostate cancer (Sweden). Cancer Causes Control 9(6):559–566. doi:10.1023/a:1008823601897

Chan JM, Pietinen P, Virtanen M, Malila N, Tangrea J, Albanes D, Virtamo J (2000) Diet and prostate cancer risk in a cohort of smokers, with a specific focus on calcium and phosphorus (Finland). Cancer Causes Control 11(9):859–867. doi:10.1023/a:1008947201132

Cheng-Xiong X, Hua J, Hwang-Tae L, Yoon-Cheol H, Chan-Hee C, Gil-Hwan A, Kee-Ho L, Myung-Haing C (2010) Low dietary inorganic phosphate stimulates lung tumorigenesis through altering protein translation and cell cycle in K- rasLA1 mice. Nutr Cancer 62(4):525–532

Council for Responsible Nutrition (CRN) (2013) Vitamin and mineral recommendations. http://www.crnusa.org/about_recs.html

Ditscheid B, Keller S, Jahreis G (2005) Cholesterol metabolism is affected by calcium phosphate supplementation in humans. J Nutr 135(7):1678–1682

EFSA (2005) Opinion of the scientific panel on dietetic products, nutrition and allergies on a request from the commission related to the tolerable upper intake level of phosphorus. EFSA J 233:1–19 (http://www.efsa.europa.eu/de/efsajournal/doc/233.pdf)

Elliott P, Kesteloot H, Appel LJ, Dyer AR, Ueshima H, Chan Q, Brown IJ, Zhao L, Stamler J, Grp ICR (2008) Dietary phosphorus and blood pressure—international study of Macro- and Micro-nutrients and blood pressure. Hypertension 51(3):669–675. doi:10.1161/hypertensionaha.107.103747

Flynn A, Hirvonen T, Mensink GBM, Ocke MC, Serra-Majem L, Stos K, Szponar L, Tetens I, Turrini A, Fletcher R, Wildemann T (2009) Intake of selected nutrients from foods, from fortification and from supplements in various European countries. Food Nutr Res 53:20–20. doi:10.3402/fnr.v53i0.2038

Giovannucci E, Rimm EB, Wolk A, Ascherio A, Stampfer MJ, Colditz GA, Willett WC (1998) Calcium and fructose intake in relation to risk of prostate cancer. Cancer Res 58(3):442–447

Haglin L, Lindblad A, Bygren LO (2001) Hypophosphataemia in the metabolic syndrome. Gender differences in body weight and blood glucose. Eur J Clin Nutr 55(6):493–498. doi:10.1038/sj.ejcn.1601209

IFAC (2009) Statement of the international food additives council and the European chemical industry council. www.cefic.org/Documents/Other/Backgrounder_of_et_al_1-21-09.doc

Itkonen ST, Karp HJ, Kemi VE, Kokkonen EM, Saarnio EM, Pekkinen MH, Karkkainen MUM, Laitinen EKA, Turanlahti MI, Lamberg-Allardt CJE (2013) Associations among total and food additive phosphorus intake and carotid intima-media thickness—a cross-sectional study in a middle-aged population in Southern Finland. Nutrition J 12:94. doi:10.1186/1475-2891-12-94

JECFA (World Health Organisation Joint Expert Committee on Food Additives) (1982) International programme on chemical safety, WHO food additives series 17. http://www.inchem.org/documents/jecfa/jecmono/v17je22.htm

Jin H, Xu C-X, Lim H-T, Park S-J, Shin J-Y, Chung Y-S, Park S-C, Chang S-H, Youn H-J, Lee K-H, Lee Y-S, Ha Y-C, Chae C-H, Beck GR Jr, Cho M-H (2009) High dietary inorganic phosphate increases lung zumorigenesis and alters Akt signaling. Am J Respir Crit Care Med 179(1):59–68. doi:10.1164/rccm.200802-306OC

Joffres MR, Reed DM, Yano K (1987) Relationship of magnesium intake and other dietary factors to blood-pressure—the Honolulu heart-study. Am J Clin Nutr 45(2):469–475

Kesse E, Boutron-Ruault MC, Norat T, Riboli E, Clavel-Chapelon F, Grp EN (2005) Dietary calcium, phosphorus, vitamin D, dairy products and the risk of colorectal adenoma and cancer among French women of the E3N-EPIC prospective study. Int J Cancer 117(1):137–144. doi:10.1002/ijc.21148

Kim WS, Lee D-H, Youn H-J (2013) Calcium-phosphorus product concentration is a risk factor of coronary artery disease in metabolic syndrome. Atherosclerosis 229(1):253–257. doi:10.1016/j.atherosclerosis.2013.04.028

van Lee L, Heyworth J, McNaughton S, Iacopetta B, Clayforth C, Fritschi L (2011) Selected dietary micronutrients and the risk of right- and left-sided colorectal cancers: a case-control study in Western Australia. Ann Epidemiol 21(3):170–177. doi:10.1016/j.annepidem.2010.10.005

Lind L, Lithell H, Hvarfner A, Pollare T, Ljunghall S (1993) On the relationships between mineral metabolism, obesity and fat distribution. Eur J Clin Invest 23(5):307–310. doi:10.1111/j.1365-2362.1993.tb00779.x

Lindegarde F, Trell E (1977) Serum inorganic phosphate in middle-aged men I. Inverse relation to body weight. Acta Medica Scandinavica 202:307–311

Lippi G, Miôntagna M, Salvagno GL, Targher G, Guidi GC (2009) Relationship between serum phosphate and cardiovascular risk factors in a large cohort of adult outpatients. Diab Res Clin Pract 84:e3–e5

Obeid OA (2013) Low phosphorus status might contribute to the onset of obesity. Obes Rev 14(8):659–664

Spina A, Sapio L, Esposito A, Di Maiolo F, Sorvillo L, Naviglio S (2012) Inorganic phosphate as a novel signaling molecule with antiproliferative action in MDA-MB-231. BioRes doi:10.1089/biores.2012.0266

Takata Y, Shu X-O, Yang G, Li H, Dai Q, Gao J, Cai Q, Gao Y-T, Zheng W (2013) Calcium intake and lung cancer risk among female nonsmokers: a report from the Shanghai women's health study. Cancer Epidemiol Biomark Prev 22(1):50–57. doi:10.1158/1055-9965.epi-12-0915-t

Tavani A, Bertuccio P, Bosetti C, Talamini R, Negri E, Franceschi S, Montella M, La Vecchia C (2005) Dietary intake of calcium, vitamin D, phosphorus and the risk of prostate cancer. Eur Urol 48(1):27–33. doi:10.1016/j.eururo.2005.03.023

Trautvetter U, Ditscheid B, Kiehntopf M, Jahreis G (2012) A combination of calcium phosphate and probiotics beneficially influences intestinal lactobacilli and cholesterol metabolism in humans. Clin Nutr 31(2):230–237. doi:10.1016/j.clnu.2011.09.013

Weiner ML, Salminen WF, Larson PR, Barter RA, Kranetz JL, Simon GS (2001) Toxicological review of inorganic phosphates. Food Chem Toxicol 39:759–786

Westerberg P-A, Tivesten A, Karlsson MK, Mellstrom D, Orwoll E, Ohlsson C, Larsson TE, Linde T, Ljunggren O (2013) Fibroblast growth factor 23, mineral metabolism and mortality among elderly men (Swedish MrOs). Bmc Nephrol 14:85. doi:10.1186/1471-2369-14-85

Wulaningsih W, Michaelsson K, Garmo H, Hammar N, Jungner I, Walldius G, Holmberg L, Van Hemelrijck M (2013) Inorganic phosphate and the risk of cancer in the Swedish AMORIS study. Bmc Cancer 13:257. doi:10.1186/1471-2407-13-257

Appendix: Spotlight 8

Technological Use of Phosphorus: The Non-fertilizer, Non-feed and Non-detergent Domain

Oliver Gantner, Willem Schipper, and Jan J. Weigand

Out of a total of 191 million tons phosphate rock mined yearly (2011 figure; Jasinski 2012), a large part is used to make fertilizers, via merchant grade phosphoric acid (MGA), and only a small part, typically 10–15 %, is used for non-fertilizer applications (IFA 2011).

Accurate figures for non-fertilizer uses are not available. IFA (2008) estimates about 7 % is used to make detergents, 10 % to make feed supplements for livestock, and 3 % is used in other applications. Recent estimates (CRU 2013) put feed usage at 5 %, detergents at 2 % and all other uses at 3 % combined. Most of these other applications, as well as a fair part of the detergent phosphate production, involve the manufacturing of elemental, white phosphorus P_4 ("thermal route") which constitutes the only other relevant processing route for rock besides MGA/fertilizer ("wet acid route").

The authors estimate a very approximate breakdown for world use of phosphorus (P) as follows:

O. Gantner
Resource Strategy, University of Augsburg, Universitätsstr. 1a, 86159 Augsburg, Germany
e-mail: oliver.gantner@wzu.uni-augsburg.de

W. Schipper
Willem Schipper Consulting Middleburg, The Netherlands
e-mail: willemschipper@hotmail.com e-mail: wsconsulting@zeelandnet.nl

J. J. Weigand
Department of Chemistry and Food Chemistry, TU Dresden
Mommsenstr, 4, 01062 Dresden, Germany
e-mail: jan.weigand@tu-dresden.de

- 85–90 % fertilizer
- ±5 % animal feeds
- 1 % food
- 2 % detergents
- 1 % glyphosate
- 1 % other P_4 derivatives
- 1 % other technical phosphates

These numbers are estimates reflecting the uncertainties and differences between published data and are given to the best of knowledge of the authors for the years 2010–2012. They serve above all to demonstrate the relative importance of each use segment. This spotlight focuses on the non-fertilizer uses of P, which include both MGA derived and P_4- derived products, hereafter referred to as the "technological use of P." These include ortho- and polyphosphates, from either MGA (usually) or P_4, as well as a large catchall category of organophosphorus compounds and inorganic phosphorus derivatives that can only be manufactured from the element, P_4.

As MGA contains up to 5 % of sulfate and metallic impurities originating from the rock and manufacturing conditions, phosphates derived from MGA usually need some form of purification of the acid, by extraction, precipitation or crystallization, thus essentially forming a product that competes with high-purity acid and its derived phosphates as obtained through oxidation of P_4 and subsequent hydrolysis ("thermal acid"). Applications include technical fields (detergent, firefighting, flame retardants, water treatment, and many others), feed and food uses, and these determine the amount of purification needed.

Feed phosphates are added to livestock feed, typically as mono-, dicalcium, and several sodium phosphates. As purity requirements are not as strict as for human consumption, these are routinely made from partially purified MGA. This also includes a number of phosphates used in the pet food industry (pyro/polyphosphates).

Food phosphates and food grade phosphoric acid perform a host of functions, such as moisture retention, sequestering, and acidulation. These are made either from highly purified MGA or from thermal acid, with soda ash or caustic soda in most cases. Ammonium, calcium, magnesium, aluminum, and potassium salts are also commercially relevant.

Technical phosphates such as detergent sodium tripolyphosphate (STPP) or phosphates for water treatment, firefighting compositions, and ceramics can be made either from MGA (usual) or via the thermal route. The choice between the two routes is above all cost driven, as a higher quality requires more rigorous purification and hence additional cost (Table 1).

The remaining P compounds all necessarily need to be made through the most reactive allotrope of the element, i.e., white phosphorus (P_4). This is

Table 1 Overview of phosphate containing applications (non-exhaustive) *References* Budenheim (2013), Emsley (2000), Phosphate Facts (2013), Prayon; Villalba et al. (2008)

Phosphate containing applications	Chosen examples for a finished product	Ingredients	Phosphate function
Agrochemicals	Glyphosate	Glycine phosphonate	Herbicide
Beverage	Cola	Food grade phosphoric acid	Acidulant
Building/construction	Cement	Sodium tripolyphosphate, tetrasodium pyrophosphate, Sodium monofluorophosphate	Retarding agent for cement
Cleaning agents	Heavy duty cleaner	Trisodium phosphates, tetra sodium and potassium pyrophosphates, sodium potassium pyrophosphate	Remove oils and greases
Coloring and coating	Intumescent coating	Polyphosphates	Flame retardant
Flame protection	Fire extinguisher	Mono- and diammonium phosphates	Fire retardants
Flame retardants	Automotive polyurethane foam	Phosphate triesters (tricresyl phosphate, etc.)	Flame retardant
Food processing	Dairy, Seafood, Meat	Polyphosphates (sodium tripolyphosphate, potassium tripolyphosphate, etc.)	Moisture retention, emulsifier, shelf life
Glass and ceramic	Ceramic	Tricalcium phosphate	Impart opalescence
Lighting	Neon tubes	Halo phosphates	Phosphor (fluorescent substance)
Lubricant additives	Car engine protectors	Sulfur compounds (zinc dialkyl thiophosphates, etc.)	Protection
Medical engineering	Dental cement	Magnesium phosphate, ammonium phosphate	Dental investment, binding agent
Military use	Fire bomb	White phosphorus	Destruction by fire
Pharmaceutical articles and cosmetics	Toothpaste	Tetrasodium pyrophosphate, tetrapotassium pyrophosphate, etc	Cleaning agent, polishing agent
Phosphoric acid for industrial use	Circuit board	Phosphoric acid	Metal surface treatment
Polish	Polished aluminum	Phosphoric acid	Chemical polishing
Storage technology	Lithium-ion batteries	Lithium iron phosphate	High energy density, rapid charging, and discharging abilities
Textile and leather	Dying wool	Diammonium phosphate	Control of pH in dye bath to allow even penetration of dye through the wool

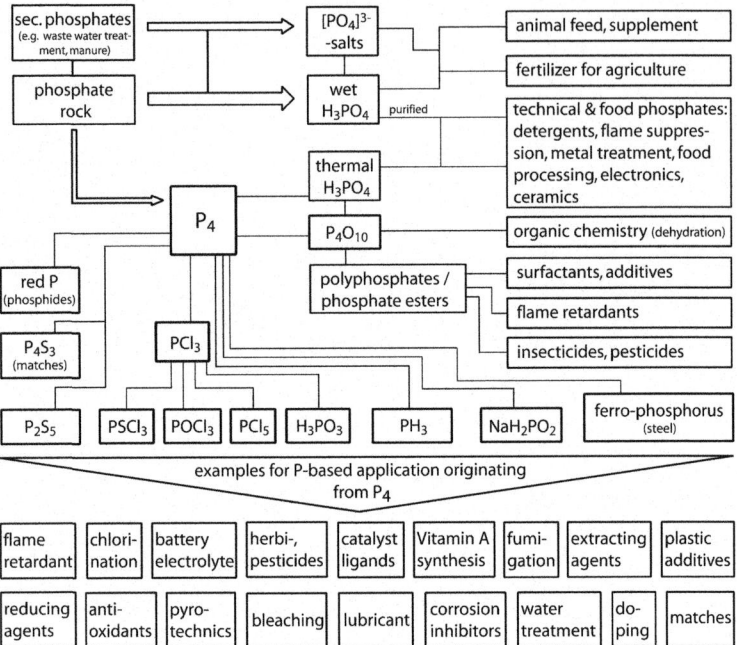

Fig. 2 Technical use of P

obtained through electrothermal reduction of phosphate rock (apatite) in an electric arc furnace at 1,600 °C with coke (reducing agent) and gravel (slag former). White phosphorus as such has limited applications in military incendiaries, but otherwise serves as the father compound to a large palette of (organo) phosphorus compounds (OPCs) through its first derivatives: P chlorides (PCl_3, $POCl_3$, and PCl_5), sulfides (P_2S_5 and P_4S_3) and oxides (P_2O_5 and polyphosphoric acid; see Fig. 2). These consist of mono-, di- and tri-esters of phosphoric and phosphonic acid (phosphates and phosphonates).

Applications include crop protection agents, flame retardants, lubricant additives, extracting agents (e.g., uranyl salts for the fuel production for nuclear power plants), pharmaceuticals, biocides, battery electrolytes and many more. Glyphosate, an organic P compound, the largest volume agrochemical worldwide is a derivative of PCl_3. For 2017 this application will use 250,000 t of P which is 1 % of total P usage (PRWeb 2011).

Apart from these derivatives involving oxygen, chlorine and sulfur, sodium hypophosphite and red phosphorus play a role in respectively nickel electroplating and flame retardant compositions.

Phosphine (PH_3 derived) chemistry is very minor volume-wise but has a huge field of application in such diverse fields as extraction agents, catalyst ligands, fumigation and fine chemistry.

The rationale to choose the energy intensive "thermal" route through P_4 is that it gives access to compounds that could otherwise not be produced. Reasons to make P_4 and its derivatives include:

- creating water soluble molecules that perform a function such as chelating, surface treatment or antiscale action, often as a variation on structurally related carboxylic acids (such as phosphonates and acid organophosphates),
- introducing P into a plastic or other flammable material to obtain flame retardancy, with functional groups around the P atom to create compatibility with the material to be flameproofed, or particles made compatible with the matrix material, (e.g., phosphate esters, DOPO, phosphinates),
- mimicking a molecule from nature to obtain a pesticide/herbicide (such as glyphosate),
- providing functionality to obtain a catalytic action, usually together with a transition metal ion, usually by providing a lone pair, i.e. acting as a Lewis base, such as in hydrocarbonylation (i.e., polymerization and modification of petrochemical building blocks),
- chemical reduction (such as sodium hypophosphite which is difficult to replace in electroless nickel plating),
- certain P compounds provide specific functions, including chlorination (P chlorides) or a strong dehydrative power in reactions (phosphorus pentoxide, polyphosphoric acid) which are difficult to replace with other reagents.

Academic research continues to develop our understanding on different branches of P chemistry, as shown in the recently updated 1,500 page handbook on Phosphorus (Corbridge 2012). This includes superconductivity, interesting thermo-chemistry, and magnetic behavior (Pöttgen et al. 2005). Some compounds are also important in homogeneous and heterogeneous catalysis (Peruzzini and Gonsalvi 2011). With respect to P_4 routes, given the drawbacks of chlorine-based and heavy salt-waste syntheses, stringent, environmental, and transportation regulations increasingly demand not only new ways of P_4 functionalization to useful molecules, but also new economic and ecological ways to meet current challenges also in the non-fertilizer P use. The need to bypass P chlorides, e.g., to obtain OPCs straight from P_4 is a continuing research topic (Caporali et al. 2010). Recycling phosphine oxides as a by-product of the Wittig synthesis, which are currently treated as waste in most cases, is a typical example of smart re-use of P compounds in a small, dedicated loop (Feldmann et al. 2011).

References

Budenheim (2013) Solutions for a better life. Accessed from https://www.budenheim.com/en/

Caporali M, Gonsalvi L, Rossin A, Peruzzini M (2010) P4 activation by late transition metal complexes. Chem Rev 110:4178–4235

Corbridge DEC (2012) Phosphorus: chemistry, biochemistry and technology, 6th edn. CRC Press/Taylor & Francis Group, Boca Raton

CRU (2013) Ashok Shinh (presentation), Phosphates 2013 conference, Monte Carlo

Emsley J (2000) The 13th element: the sordid tale of murder, fire, and phosphorus. Wiley, New York

Feldmann K-O, Schulz S, Klotter F, Weigand JJ (2011) A Versatile protocol for the quantitative and smooth conversion of phosphane oxides into synthetically useful pyrazolylphosphonium salts. ChemSusChem 4(12):1805–1812

IFA (2008) Accessed from http://www.fertilizer.org/

IFA (2011) Global phosphate rock production trends from 1961 to 2010—reasons for the temporary set-back in 1988–1994. Feeding the Earth Series, Oct 2011. http://www.fertilizer.org/ifa/HomePage/LIBRARY/Publication-database.html/Global-Phosphate-Rock-Production-Trends-from-1961-to-2010.-Reasons-for-the-Temporary-Set-Back-in-1988-1994.html

Jasinski SM (2012) Phosphate rock. In: Salazar K, McNutt MK (eds) Mineral commodity summaries 2012. USGS, Reston, p 118

Peruzzini M, Gonsalvi L (2011) Phosphorus compounds: advanced tools in catalysis and material science in catalysis by metal complexes. Springer, Heidelberg

Phosphate Facts (2013) Uses and applications. http://www.phosphatesfacts.org/uses_apps.asp

Pöttgen R, Hönle W, von Schnering HG (2005) Phosphides: solid state chemistry

Prayon (n.d.) Phosphates and phosphoric acid in everyday life. Accessed from http://www.prayon.com/media/pdf/publications/brochure-tech-en.pdf

PRweb (2011) Global glyphosate market to reach 1.35 million metric tons by 2017. http://www.prweb.com/releases/glyphosate_agrochemical/technical_glyphosate/prweb8857231.htm

Villalba G, Liu Y, Schroder H, Ayres RU (2008) Global phosphorus flows in the industrial economy from a production perspective. J Ind Ecol 12(4):557–569. doi:10.1111/j.1530-9290.2008.00050.x

Appendix: Spotlight 9

Phosphorus in Organic Agriculture

Bernhard Freyer

Today, worldwide more than 1.8 million farmers have a total 37.2 million ha of agricultural land that would meet the criteria for organic crop production (FAO 2009; Willer and Kilcher 2011). Organic Agriculture is defined by internationally accepted guidelines, standards, and certification systems (IFOAM-EU 2012). The organic system is built upon vision and understanding of the farm as an organism. With such an understanding, organic farmers seek to close their farm nutrient cycles, to reduce resource input from off farm sources and to increase the efficiency of resources used. These goals lead organic farmers (as well as other farmers) to consider nutrient balances as a management and decision making tool that leads to increased nutrient availability in soils, the efficiency of nutrient uptake in plants. In addition, organic farmers use mixed cropping systems along with the application of organic manures to promote microbial diversity that strengthens the antiphyto-pathogenic potential of the soil. Similarly, some fodder legumes (e.g., alfalfa, clover) improve overall soil fertility by biological fixing nitrogen (some of which is available to the subsequent crop), provide residues to maintain or improve soil physical and biological properties and promote nutrient re-cycling and access to water through deep rooting. The combination of rich root systems, residue production and humus production by legumes, and cropping systems with green manure mixtures, contributes to a permanent soil cover, minimizing soil erosion and thereby the loss of phosphorus (P) and other nutrients. Compost from plant residues and stable manure from livestock permits an efficient internally closed nutrient cycle on the farm.

B. Freyer
University of Natural Resources and Life Sciences BOKU Vienna,
Gregor Mendel Strasse 33, 1180 Wien Vienna, Austria
e-mail: bernhard.freyer@boku.ac.at

Types of Farm External P-Sources

According to the organic guidelines, the readily available mineral fertilizers (e.g., triple superphosphate, DAP, etc.) are excluded, while the application of low soluble phosphate rock (hypherphos) is accepted. The use of Thomas-phosphate is restricted in almost all countries because of Cadmium content. Low energy input for the provision of mineral P-fertilizer is one reason for the use of P-mineral fertilizers with low solubility. A second reason is that the main strategy for ensuring proper plant nutrition in organic farming is to improve the conditions for nutrient mobilization from slightly soluble sources by plant–soil-microorganism interactions instead of directly fertilizing plants with readily available mineral fertilizers. Specifically, in acid soils, liming improves the availability of many nutrients including P. Accepted farm external organic P-sources are: communal biowaste composts (sewage sludge is excluded because of the risk of contamination with heavy metals and organic compounds); mineral fodder including P; diverse organic industrial P-fertilizers (e.g., slaughterhouse waste); fodder and organic manure from organic farms or conventional low input farms, with production intensity limited to the site-specific production potential.

P-Cycles

The amount of external P-sources is strictly regulated through guidelines and control systems. Farmers are required to provide calculations on nutrient balances as a precondition for getting permission to apply farm external P-fertilizers. Negative balances could allow P use from mineral fertilizers limited to approximately 5–15 kg P ha^{-1} y^{-1}. The input of P through industrial organic P fertilizers or from other sources is limited due to the site-specific yield potential. P import through mineral fodder is accepted up to approximately 4 kg P LU^{-1} ha^{-1} y^{-1} (LU = Livestock Unit). P export with respect to farm types increases as follows: grassland farms with cattle < mixed arable farms with livestock < stockless cereal producing farms < stockless root crops producing farms < vegetable farms. Without P imports, P-balances decline by approximately 1 to 5 kg P ha^{-1} y^{-1} in the first farm type to 10–15 kg P ha^{-1} y^{-1} in the last farm type (Berner et al. 1999; Martin et al. 2007). Even under limited P-input conditions, P content in organic products is equal to those in conventional farms (Dangour et al. 2009). In smallholder farms under subtropical conditions, the P balances in low input/organic farms range from highly positive to negative demonstrating a lack of access to P inputs, and possibly awareness of P needs and farm management practices in general (Onwonga and Freyer 2006).

P-Dynamics

Sources of plant-available phosphate (PO_4^{3-}) in many soils are the reserves of labile organic and inorganic soil pools, which have been built up over time through the use of mineral P fertilizers, rock phosphate, or other "organic" amendments including green manure residues, and livestock manure. Species-specific root exudates, microorganisms, and fungal enzymes induce the mineralization processes of phytates (storage of P in organic matter), thereby increasing the phosphate levels in the soils (Hinsinger et al. 2011). The extension of the root surface is highly relevant for plant P uptake. Optimal soil structure and organic manure will increase root growth and mycorrhiza colonization (Muthukumar and Udaiyan 2000) and with that the volume of the soil where P is accessible for uptake by roots. Finally, solubility of phosphate rock is supported by specific legume root exudates (Vanlauwe et al. 2000). To summarize, technologies including—lime for pH-regulation, farmyard manure, compost, phosphate rock and diversified legume based crop rotations are key for sustainable use of phosphorous (Onwonga et al. 2008).

References

Berner A, Heller S, Mäder P (1999) Nährstoffbilanzen im biologischen Landbau. 5. Wissenschaftstagung zum Ökologischen Landbau. Berlin. Retrieved June 10 http://orgprints.org/00002908

Dangour AD, Dodhia SK, Hayter A, Allen E, Lock K, Uauy R (2009) Nutritional quality of organic foods: a systematic review. Am J Clin Nutr 90(3):680–685

FAO (2009) High level expert forum. FAO. Accessed 15 Aug 2013

Hinsinger P, Betencourt E, Bernard L, Brauman A, Plassard C, Shen J, Zhang F (2011) P for two, sharing a scare resources: soil phosphorus acquisition in the rhizosphere of intercropped species. Plant Physiol 156(3):1078–1086

IFOAM (2012) European organic guidelines. Retrieved 16 June 2012. http://www.ifoam-eu.org/positions/publications/regulation/

Martin R, Lynch D, Frick B, van Straaten P (2007) Phosphorus status on Canadian organic farms. J Sci Food Agric 87:2737–2740

Muthukumar T, Udaiyan K (2000) Influence of organic manures on arbuscular mycorrhizal fungi associated with Vigna unguiculata (L.) Walp. in relation to tissue nutrients and soluble carbohydrate in roots under field conditions. Biol Fertil Soils 31:114–120

Onwonga R, Freyer B (2006) Impact of traditional farming practices on nutrient balances in smallholder farming systems of Nakuru district, Kenya. In: Asch F, Becker M (eds) Conference on international agricultural research for development, Tropentag, pp 158–161

Onwonga RN, Lelei JJ, Freyer B, Friedel JK, Mwonga SM, Wandhawa P (2008) Low cost technologies for enhancing N and P availability and maize (Zea mays L.) performance on acid soils. World J Agric Sci 4(2):862–873

Vanlauwe B, Diels J, Sanginga N, Carsky RJ, Deckers J, Merckx R (2000) Utilization of rock phosphate by crops on non-acidic soils on a toposequence in the Northern Guinea savanna zone of Nigeria, response by maize to previous herbaceous legume cropping and rock response by maize to previous herbaceous legume cropping and rock phosphate treatments. Soil Biol Biochem 32:2079–2090

Willer H, Kilcher L (2011) The world of organic agriculture. Statistics and Emerging Trends 2011, IFOAM, Bonn and FiBL, Frick

Chapter 6
Dissipation and Recycling: What Losses, What Dissipation Impacts, and What Recycling Options?

Masaru Yarime, Cynthia Carliell-Marquet, Deborah T. Hellums,
Yuliya Kalmykova, Daniel J. Lang, Quang Bao Le, Dianne Malley,
Leo S. Morf, Kazuyo Matsubae, Makiko Matsuo, Hisao Ohtake,
Alan P. Omlin, Sebastian Petzet, Roland W. Scholz,
Hideaki Shiroyama, Andrea E. Ulrich, and Paul Watts

Abstract This chapter describes the activities in the Dissipation and Recycling Node of Global TraPs, a multistakeholder project on the sustainable management of the global phosphorus (P) cycle. Along the P supply and demand chain, substantial amounts are lost, notably in mining, processing, agriculture via soil

---

M. Yarime (✉) · A. P. Omlin
University of Tokyo, Graduate School of Public Policy, Hongo 7-3-1,
Bunkyo-ku, Tokyo 113-0033, Japan
e-mail: yarimemasa@gmail.com

C. Carliell-Marquet
University of Birmingham, Edgbaston, Birmingham B15 2TT, UK
e-mail: C.M.Carliell@bham.ac.uk

D. T. Hellums
International Fertilizer Development Center (IFDC),
P.O. Box 2040, Muscle Shoals, AL 35662, USA
e-mail: dhellums@ifdc.org

Y. Kalmykova
Civil and Environmental Engineering, Chalmers University of Technology,
412 96 Gothenburg, Sweden
e-mail: Yuliya.Kalmykova@chalmers.se

D. J. Lang
Leuphana University of Lüneburg, Scharnhorststr. 1, UC 1.314, 21335 Lüneburg, Germany
e-mail: daniel.lang@uni.leuphana.de

Q. B. Le · R. W. Scholz · A. E. Ulrich
ETH Zürich, Natural and Social Science Interface (NSSI), Universitaetsstrasse 22,
CHN J74.2, 8092 Zürich, Switzerland
e-mail: Quang.le@env.ethz.ch

A. E. Ulrich
e-mail: andrea.ulrich@env.ethz.ch

erosion, food waste, manure, and sewage sludge. They are not only critical with respect to wasting an essential resource, but also contribute to severe environmental impacts such as eutrophication of freshwater ecosystems or the development of dead zones in oceans. The Recycling and Dissipation Node covers the phosphorus system from those points where phosphate-containing waste or losses have occurred or been produced by human excreta, livestock, and industries. This chapter describes losses and recycling efforts, identifies knowledge implementation and dissemination gaps as well as critical questions, and outlines potential transdisciplinary case studies. Two pathways toward sustainable P management are in focus: To a major goal of sustainable P management therefore must be to (1) quantify P stocks and flows in order to (2) identify key areas for minimizing losses and realizing recycling opportunities. Several technologies already exist to recycle P from different sources, including manure, food waste, sewage, and steelmaking slag; however, due to various factors such as lacking economic incentives, insufficient regulations, technical obstacles, and missing anticipation of unintended impacts, only a minor part of potential secondary P resources has been utilized.

D. Malley · P. Watts
PDK Projects, 5072 Vista View Crescent, Nanaimo, BC V9V 1L6, Canada
e-mail: dmalley@pdkprojects.com

P. Watts
e-mail: paulwatts52@yahoo.com

L. S. Morf
Abfall, Wasser, Luft und Energie (AWEL), Sektion Abfallwirtschaft Weinbergstrasse 34,
Postfach, 8090 Zürich, Switzerland
e-mail: leo.morf@bd.zh.ch

K. Matsubae
Tohoku University, 6-6-11-1005, Aoba, Aramaki, Aoba-ku,
Sendai City, Miyagi 980–8579, Japan
e-mail: matsubae@m.tohoku.ac.jp

M. Matsuo
The University of Tokyo, 7-3-1 Hongo, Bunkyo-ku, Tokyo 113–0033, Japan
e-mail: matsuoma@j.u-tokyo.ac.jp

H. Ohtake
Osaka University, Yamadaoka 2-1, Suita, Osaka 565–0871, Japan
e-mail: hohtake@bio.eng.osaka-u.ac.jp

S. Petzet
Technische Universität Darmstadt, Petersenstr. 13, 64287 Darmstadt, Germany
e-mail: s.petzet@iwar.tu-darmstadt.de

R. W. Scholz
Fraunhofer Project Group Materials Recycling and Resource Strategies IWKS,
Brentanostrasse 2, 63755 Alzenau, Germany
e-mail: roland.scholz@isc.fraunhofer.de

H. Shiroyama
Graduate School of Public Policy, The University of Tokyo,
7-3-1 Hongo, Bunkyo-ku, Tokyo 113–0033, Japan
e-mail: siroyama@j.u-tokyo.ac.jp

Minimizing losses and increasing recycling rates as well as reducing unintended environmental impacts triggered by P dissipation require a better understanding of the social, technological, and economic rationale as well as the intrinsic interrelations between nutrient cycling and ecosystem stability. A useful approach will be to develop new social business models integrating innovative technologies, corporate strategies, and public policies. That requires intensive collaboration between different scientific disciplines and, most importantly, among a variety of key stakeholders, including industry, farmers, and government agencies.

Keywords Phosphorus and eutrophication · Environmental costs of phosphate reduction · Phosphorus recycling in industry · Phosphorus recycling in agriculture · Phosphorus recycling from sewage

Contents

1	Background	249
2	Flows, Stocks, and Balances of P	250
3	P Dissipation and Eutrophication	254
	3.1 Eutrophication and Dead Zones	254
	3.2 Processes Involving P in Soils	256
	3.3 Environmental Costs of P Dissipation	256
4	Recycling of P-Containing Wastes	257
	4.1 Manure	257
	4.2 Wastewater	258
	4.3 Solid Wastes	263
	4.4 Industrial Wastes	265
5	Work in Global TraPs	266
	5.1 Knowledge Gaps and Critical Questions	266
	5.2 Roles, Functions, and Varieties of the Transdisciplinary Process	268
	5.3 Suggested Case Studies	269
References		271

1 Background

Around 191 Mt of phosphate rock (PR), containing 83.3 Mt of phosphorus (P), was mined in 2011 (U.S. Geological Survey 2012), of which more than 80 % was used as agricultural fertilizer. Population growth and changing diets are increasing demands for PR, which is a finite resource, yet much of P is lost along its way through the supply chain, in mining, in processing, or in fertilizers as the most substantial form of use. Often P is lost into the natural environment, where it transforms from a resource into a pollutant of aquatic (Bennett et al. 2001; Ulrich et al. 2009) and terrestrial (Olde Venterink 2011) systems. Sustainable P management must map its stocks and flows to identify key points at which to minimize dissipation and increase recycling opportunities.

The Dissipation and Recycling Node covers the supply chain from the point where P-containing *waste* has been produced by *humans, livestock*, and *industries*. Recycling covers the *processing, marketing*, and *use of recycled waste products, including biosolids*, whereas *dissipation* refers to the intended or unintended *loss of P in mining, processing, use*, and, *to a minor extent, recycling*. The dissipated P is, depending on the sink, more or less accessible for subsequent recycling activities. P-recycling activities span diverse scales and contexts: from farms to households and to megacities, where food is consumed away from production; from simple household and livestock waste composting to complex recovery of P from sewage, food, and industrial waste products; and from large scale, advanced facilities in industrialized countries to small-scale efforts in developing countries, where direct local-level recycling would be carried out, e.g., urine diversion.

A very critical challenge for sustainable P management is to make P recovery economical, reliable, and predictable while ensuring that the use of recycled P products will not result in adverse health or environmental impacts. It is therefore crucial to manage P stocks and flows through a system-based approach, linking dissipation, eutrophication, and recycling. The primary focus is on current and potential markets, quality and price of products, costs of production processes, available and future technologies, institutional structures and public perception and behavior. Stakeholders involved are diverse, including industries creating waste streams and others focusing on P recovery such as farmers, governmental regulatory and specialized agencies, public environmental and health organizations, researchers and agricultural and health NGOs.

The overall aims of the chapter are to (1) give a comprehensive and structured overview of the current state of knowledge; (2) formulate major research gaps; and (3) outline potential case studies.

2 Flows, Stocks, and Balances of P

For sustainable management of P, an essential first step is mapping direct and indirect demands for P in an economy. For example, Fig. 1 shows the substance flows of P in the Japanese economy, which are estimated to be 618 kt (Matsubae et al. 2011). Approximately 284 kt of P is applied annually to farms and ranches in the form of fertilizer, one of the largest input flows in the entire domestic P flow. Input flows to food and feed sectors also have large values, mainly from world imports and marine resources (163.1 kt) and domestic crop production from farmlands (45.2 kt), with the P mainly consumed by humans and livestock (97.6 and 111.0 kt). Livestock grow by eating grass and feed on ranches, and the P in livestock manure ends up accumulating in the soil, the amount of which (285.3 kt) is nearly equal to input from fertilizer to farms and ranches. Another main output is the human waste that ends up flowing down rivers, in the ocean, or in landfill. In addition, 110.5 kt of P is associated with the steel industry as mineral resources, most of which is condensed in steelmaking slag.

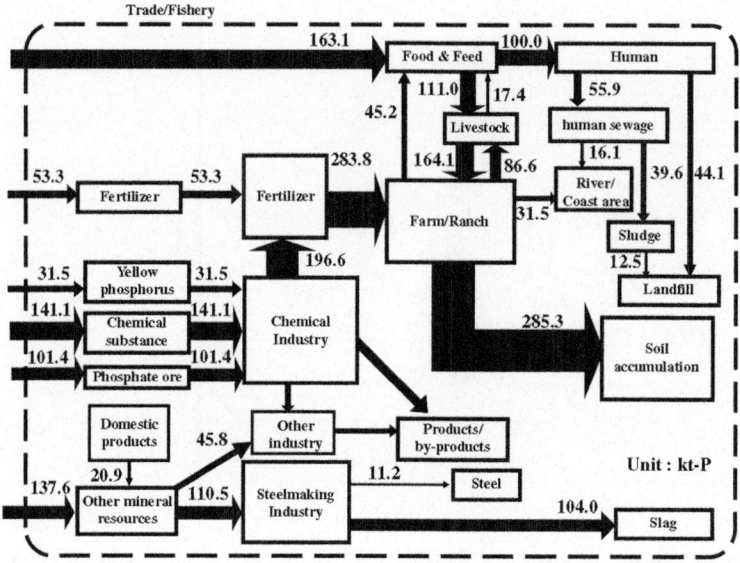

Fig. 1 Substance flows of P in the Japanese economy (2005). *Source* (Matsubae et al. 2011)

Figure 1 reveals a variety of potential P resources within the economy in Japan, including food waste, sewage sludge, steelmaking slag, and other industrial wastes, totaling approximately 240 kt per year, which is comparable to the P demand for fertilizer of approximately 284 kt per year. Hence, an appropriate nation-wide recycling strategy could potentially provide the majority of P required for agricultural production in the country. Sustainable P management is also of economic importance. While the size of the domestic fertilizer market is only five billion US dollars per year, it supports all the food-related industries and businesses in the country, whose total sales reach US$800 billion.

In the vegetation process, fertilizer is used for plant growth. Not all of P in fertilizer, however, is transformed into the harvested products, as loss is caused by absorption in the pedosphere, diffusion into the hydrosphere, and waste in residual portions of agricultural products. Substance flow analysis focusing on P contained in products tends to neglect such P flows. As a new indicator to consider the direct and indirect P requirements for our society, virtual P ore requirement (VPOR) is proposed (Matsubae et al. 2011). As in the case of virtual water (Hoekstra and Chapagain 2007), the estimation of VPOR requires to consider hidden P flows, which constitute the total P requirement excluding the amount contained in agricultural products, including the loss to the environment, non-edible parts, and feedstuff for livestock.

Figure 2 illustrates the example of VPOR for the Japanese economy in 2005. The economy consumed 3,662 kt of P ore in overseas countries to produce the agricultural products. While the amount of real P ore import was 774, 6,160 kt (=3,662 + 407 + 240 + 1,077 + 774) of P ore was required in total. The left side

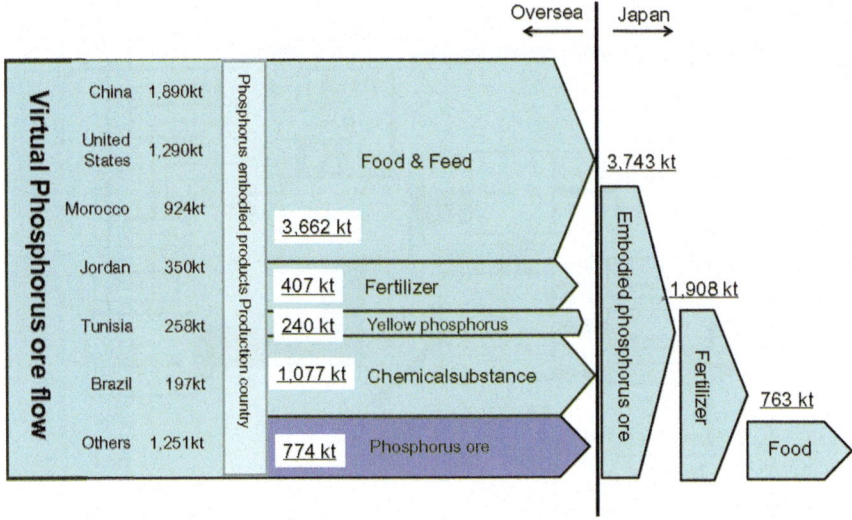

Fig. 2 Virtual P ore requirement of the Japanese economy (2005). *Source* (Matsubae et al. 2011)

of the figure shows where that amount of ore came from, and the value refers to the direct and indirect demand of P ore by country of production. The largest virtual P ore supplier for Japan is China, followed by the USA and Morocco. Although there was no actual P ore imported from the USA, 20 % of VPOR came from the USA through fertilizer and food imports.

VPOR indicates the direct and indirect demand of P ore transformed into agricultural products and fertilizer. The amount of embodied P ore flow associated with commodities imported to Japan was 3,743 kt, which was based on 6,160 kt of VPOR. Approximately half of the imported embodied phosphate ore was transformed into fertilizer and utilized to produce agricultural products. An amount of 763 kt of P ore was actually eaten, and the rest ends up being dissipated in soil and water. The results suggest that P consumed in agricultural products accounts for only 12 % (=763/6,160) of VPOR. As the sites of direct and indirect P consumption are different, VPOR is useful to analyze the global network of P ore requirement derived from the consumption of agricultural products. For sustainable P resource management, it is very important to recognize the virtual P ore demand for the agricultural product consumption through the international P demand and supply network.

In the case of EU 27, it is estimated that an average of 8 kg P/ha per year has been accumulated in agricultural lands (Richards and Dawson 2008). There are significant national and regional differences. For instance, the Netherlands, with a high density of intensive livestock farming, accumulates 20 kg P/ha per year on average (Smit et al. 2009), whereas regions with a low livestock density often show a P deficit, resulting in nutrient mining from agricultural land (Albert 2004).

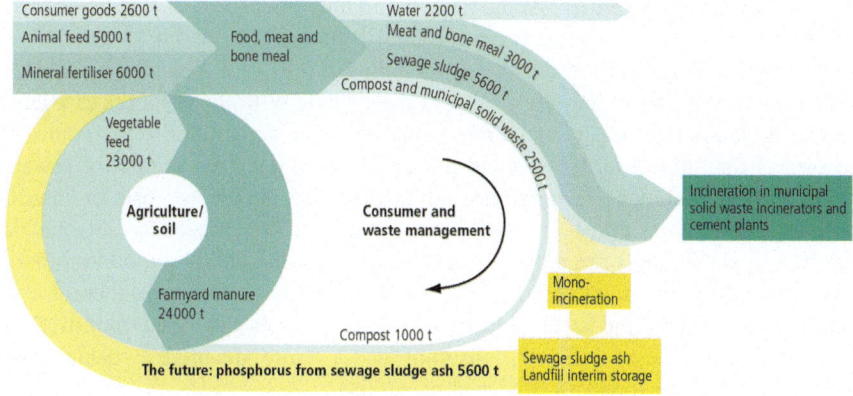

Fig. 3 Simplified illustration of P balance in Switzerland. *Source* (AWEL 2008)

P is mostly lost in crop waste, food spoilage, as well as livestock and human waste. As an indicator of the scale, it is estimated that one-third of the global food production is discarded during its production and consumption process (Gustavsson et al. 2011). This does not only mean that considerable amounts of fertilizer and other resources for food production are used in vain, but also that a significant part of the contained P is lost in various waste streams. It is estimated that 70–80 % of the P mined in PR for food production actually never reaches the plate of the end consumer (Cordell et al. 2009) (see Chap. 1). Solid organic wastes such as slaughterhouse waste (bone, blood, etc.) and inedible parts of agricultural products are frequently mixed with other types of solid waste and thus are dumped into landfills or lost in the form of disposed incinerator ashes. Similarly, P caught in liquid wastes such as livestock manure and human excrement often ends up in bodies of water, where it is very difficult to recover from.

Figure 3 shows simplified substance flows of P for Switzerland based on a detailed substance flow analysis for phosphorus. In Switzerland, like in the rest of Europe, there are no natural phosphorus deposits. The most important factor for the phosphorus yield of the country is the return of farmyard manure and harvest residue back into agriculture. For decades, however, large quantities of phosphorus have been imported for feeding humans and animals. In the meantime, investigations have shown that the amount of phosphorus bound in sewage sludge per year is approximately the same as the quantity imported into Switzerland, with around 7.5 million inhabitants, in mineral fertiliser—approximately 6 kt per year. This is roughly the same ratio compared to the Japanese case discussed above. The P balance for Switzerland also indicates that there still remain losses to hydrosphere, despite of very effective modern sewage treatment plants removing large amounts of phosphorus out of sewage water to avoid environmental problems, i.e., eutrophication. Also, it reveals the remaining accumulation of phosphorus into the soil during agriculture activities. Besides the losses into the hydrosphere, the substance flow analysis for Switzerland also demonstrates the dissipation rate of waste management activities. At the time, being only about 1 kt of P in the form of compost is recycled in a year. From the total P potential of about 11 kt of P per year in waste fractions for recycling, including 5.6 kt of P per year in sewage sludge, currently more than 90 % of it is dissipated. With the disposal to a large extent in municipal waste incinerators (MSW) and cement kilns, P could be recovered for future use. If we could manage to treat the total amount of sewage sludge in monosludge incinerators and landfill the incinerator ash in interim monocompartment storage or apply phosphorus retrieval procedures from ash directly, approximately the amount of P equivalent to the yearly imported mineral fertilizer into Switzerland could be substituted. Compared to Japan, there are no activities related to the steel industry or other phosphorus-based industries in Switzerland with associated P consumption. A more detailed P balance for Switzerland can be found in Binder et al. (2009).

The mapping of stocks, flows, and balances of P provides critical data for understanding where P accumulates and dissipates, based on which we will be able to identify where we could intervene for promoting recycling of P effectively and efficiently. This analysis also illustrates the importance of collecting and analyzing accurate data on the quantity of flows and stocks, with geographical distributions of P supply and demand and potential gaps between them, for designing and implementing sustainable P management.

3 P Dissipation and Eutrophication[1]

3.1 Eutrophication and Dead Zones

Until recently, P has been recognized as a nutrient that, in some circumstances, can cause nuisance bloom of microalgae known as the eutrophication problem. Excessive input of P to lakes, bays, and other surface waters causes algal bloom. In some cases, when algal bloom occurs, dissolved oxygen is consumed as the cells are decomposed. Economically important fishes and other aquatic organisms cannot survive under oxygen-depleted conditions. In addition, a toxic substance, called algal toxin, may be released from the bloom. This also causes a difficult problem in drinking water supply (Falconer 1993).

Eutrophication is attributed to such factors as the increase in human populations, lack of tertiary sewage treatment, intensive cropping, increased use of fertilizers, and increased cattle and hog production, all of which increase the loads of biowastes to the watershed. In particular, the use of phosphate fertilizers in agriculture has been associated with most cases of eutrophication (World Resources Institute 2012).

---

[1] Parts of this section are derived from an unpublished report, Food and Water Security in the Lake Winnipeg Basin—Transition to the Future (Malley et al. 2009).

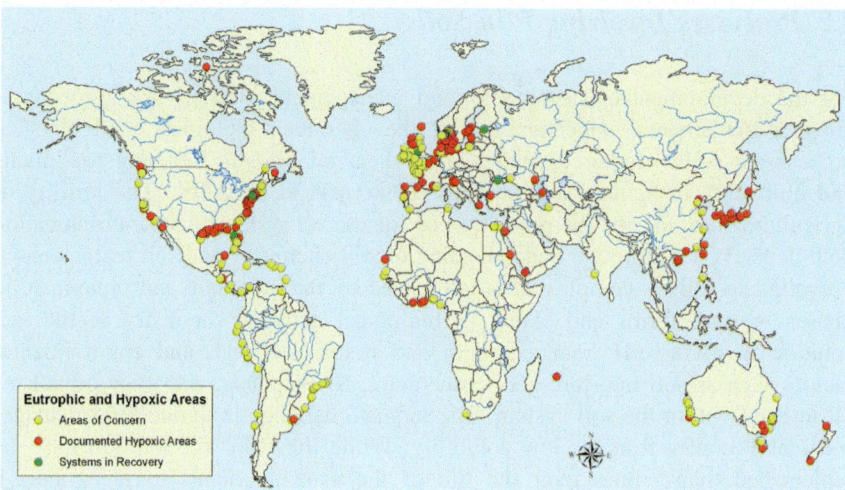

Fig. 4 Locations of hypoxic areas, areas of concern, and areas of recovery. *Source* (World Resources Institute 2012)

Large amounts of phosphate fertilizers are important in today's agricultural systems. At this stage of P use, particularly in industrial fertilizers, the goal of agriculture to increase crop yields and the goal of environmental quality of aquatic ecosystems are essentially working at cross-purposes, though not intentionally. Nevertheless, it is not the total P, but the water-soluble, available form of P, i.e., dissolved reactive P, that is the most effective form in causing eutrophication and should be the focus of management of P losses from agricultural land (Foy 2005).

Unlike the often-conspicuous responses of freshwater ecosystems to severe eutrophication, the signs in oceans are often not directly visible. Rather, the oceans respond to the wasteful addition of nutrients, including P, washed from the land with dead zones. *Dead zones* are areas of low oxygen in the world's oceans and large lakes that can no longer support most marine life. Incidences of dead zones have been increasing since oceanographers began noting them in the 1960s. These occur near inhabited coastlines, where aquatic life is most concentrated. In March 2004, the first Global Environment Outlook Yearbook reported 146 dead zones in the world oceans where marine life could not be supported. Some of these were as small as 1 km^2, but the largest dead zone covered 70,000 km^2. A total area of more than 245,000 km^2 is affected, functioning as a key stressor on marine ecosystems, which is shown in Fig. 4 (Selman et al. 2008). Their formation has been exacerbated by the increase in primary production and consequent worldwide coastal eutrophication fueled by riverine runoff of nitrogen (N) and P in fertilizers and the burning of fossil fuels. It is estimated that more than 0.212 Mt of food is lost to hypoxia in the Gulf of Mexico, an amount which would be enough to feed 75 % of the average brown shrimp harvest (Biello 2008).

3.2 Processes Involving P in Soils

For the development of sustainable land management practices for agroecosystems, a fundamental understanding of the chemical, biological, and physical processes in soils is required, as they affect the availability of P to terrestrial plants and ultimately to humans and animals (Pierzynski et al. 2005). The fertility of agricultural systems and the protection of aquatic ecosystems from eutrophication both depend on a thorough understanding of soil chemistry and soil management. P cycling in soils is complex. It is influenced by the inorganic and organic solid phases present, forms and extent of biological activity, chemistry of the soil solution involving pH, ionic strength and redox potential, and environmental factors such as soil moisture and temperature. Soils, plants, and microorganisms all interact within the soil system. The largest challenge in agricultural management of P comes from its low solubility. While the P in soil solution must be replenished many times over the life of the growing plants to meet their P requirements, it is also prone to be removed by erosion in runoff or to become part of the sediment load and delivered to the aquatic ecosystems. Managing the soil P concentrations involves dissolution–precipitation, adsorption–desorption, mineralization–immobilization, and oxidation–reduction processes. In solution, P moves within soils primarily by diffusion. Crop removal is the main route by which P is removed from soil, whereas erosion and surface runoff are the environmentally significant removal processes.

3.3 Environmental Costs of P Dissipation

Environmental costs related to P dissipation need to be taken into account in considering P recycling. There are difficulties in measuring them accurately, however, although many methodologies have been proposed to deal with the problem, for example, contingent valuation methods. This issue will have important implications for policy intervention for incorporating the costs into market and pricing mechanisms.

An economic analysis is conducted on the impacts of higher mineral P prices and externality taxation on the use of organic P sources in US agriculture (Shakhramanyan et al. 2012). This study examines alternative hypothetical scenarios concerning the prices for PR-based mineral fertilizers and the taxation of external damages from the application of the latter fertilizers. These scenarios reflect both an increasing scarcity of PR, which led to substantial price increases in recent years, and increasing political efforts to address and correct adverse externalities from land use. To adequately depict adaptation of producers as well as adjustments in agricultural commodity markets, the authors modified and applied a price-endogenous agricultural sector model of the USA. They considered alternative fertilizer options which substitute mineral phosphorous by manure-based organic phosphorus sources and link phosphorous supply from livestock manure to phosphorous demand from crop production. The results indicate that substantial reductions in the use of mineral phosphorus in US agriculture are possible, if the price and tax signals are strong enough.

The results do not indicate a severe physical scarcity of organic phosphorus sources. Low rates of organic phosphorous at a low cost for mineral sources reflect mainly the cost of manure application. The shadow prices on the regional manure supply–demand balances remain zero or at fairly low levels throughout all examined scenarios. Thus, the substitution of mineral by organic phosphorus is primarily an economic or regulatory issue. Furthermore, the overall impact of a higher cost of mineral phosphorous would have little impact on aggregate crop and livestock production, trade, and prices, because of sufficient supply of organic P sources.

4 Recycling of P-Containing Wastes

In this section, four major domains are selected to describe P management and recycling practices in more details, namely manure, wastewater, solid wastes, and industrial wastes.

4.1 Manure

One of the main sources of P for recycling is animal manure. Manure is a valuable source of plant nutrients and organic matter. In the case of intensive livestock farming, P is imported to the farm with the animal feed. In particular, monogastric animals, such as chicken and swine, take up only a small portion of the P contained in the feed and most of the P is excreted in the manure (see Table 1). P in manure is mostly present in the inorganic form and similar to commercial fertilizer in that it is readily available for plant uptake. Substitution of inorganic fertilizers by manure, however, is often not a preferred choice because of higher transportation cost, difficulty to define the appropriate manure application rate, the risk of transmitting pathogens, and undesirable odor effects. Currently, 0.9 % of the agricultural land is organic; by region, the highest shares are in Oceania (2.9 %) and in Europe (2.1 %), and within the European Union, 5.1 % of the farmland is organic (Willer and Kilcher 2012). In principle, manure could be applied far more on cropland, mitigating the risks that arise from excessive concentrations of manure and replacing high-priced commercial fertilizers (MacDonald et al. 2009). As a result of the transport problems, animal manure is repeatedly spread on fields in the vicinity of the livestock farm, resulting in P surplus and causing water contamination. It is estimated that to balance the areas of P surplus and deficit in England (UK), 4.7 kt of P (2.8 Mt of manure) must be exported annually from the areas of livestock farming in the west to the areas of arable farming in the east of the country (Bateman et al. 2011).

Table 1 P content in manure

	Liquid pig kg P/l	Liquid dairy	Liquid poultry	Solid dairy kg P/t	Solid beef	Solid poultry
Average	1.4	1.0	1.0	1.4	1	9
Minimum	0.1	0.1	0.3	0.5	0.3	0.5
Maximum	3.8	10.1	1.7	7.7	5.9	25.2

Source (Government of Manitoba 2008)

In contrast to sewage sludge biosolids, inorganic pollutants (heavy metals) are not an issue in the case of animal manure, although organic pollutants such as veterinary medicines, antibiotics, and biocides could be problematic as well as pathogens. Thus, the main technological issues to be resolved are how to increase the transportability of P within the manure matrix and how to extract P from the manure matrix for further processing, for example, through crystallization as struvite or calcium phosphate. The P-depleted manure may then be applied on fields with a reduced P load.

An approach to increasing the transportability of P in manure could be to use manure as feedstock for energy production, although manure-to-energy projects are not currently in widespread use (MacDonald et al. 2009). Current available technologies include combustion power plants (Hermann 2011) and anaerobic digestion systems designed to capture methane gas and burn it as fuel for electricity generation. Anaerobic digestion does not consume the nutrients in manure and leave them in residuals. Anaerobic digestion reduces pathogen counts and denatures weed seeds in raw manure, and the odors of raw manure are greatly reduced in the effluent, thereby easing the storage, movement, and application of manure nutrients. During combustion processes, most nitrogen nutrients are burned, whereas the ash residues from combustion retain P and potassium in concentrated form. The resulting ashes would be free of organic pollutants and could be easily processed into P fertilizers (Schoumans et al. 2010).

4.2 Wastewater

Wastewater is a significant source of P; globally, wastewater contains approximately 4.6 Mt P per year, corresponding to more than 2 % of the world PR production. In 1890, approximately 90 % of P in global wastewater was recycled, which decreased to 30 % by the end of the twentieth century (Liu et al. 2008). A prerequisite for P recycling from wastewater is an adequate sanitization infrastructure, including wastewater treatment. Nutrient recycling in fast-growing urban areas is needed to address nutrient imbalances and to return P to the places of food production. The urban population is expected to double from currently 3.5–6.5 billion in 2050. By then, 4.7 Mt of P per year will be emitted to wastewater in urban areas alone.

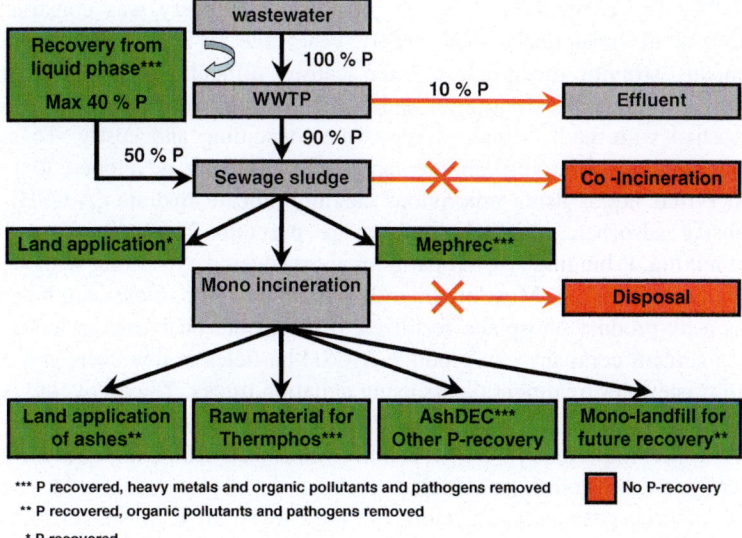

Fig. 5 Diagram of complete P recovery from sewage sludge and incineration ashes. *Source* (Petzet and Cornel 2010; Petzet and Cornel 2011)

4.2.1 Centralized Measures

Wastewater treatment plants may become important sources of P recovery and recycling. The removal of P from municipal wastewater is well established and widely applied; typically, 80–90 % of influent P is transferred from wastewater to sludge solids by chemical precipitation (Al^{3+} or Fe^{3+}) or biological uptake (Petzet and Cornel 2011). Chemical precipitation is less amenable to P extraction and recovery because it forms insoluble phosphate precipitates within the sludge matrix, whereas biologically removed P is readily released as soluble phosphate and can be further recovered as struvite or calcium phosphate (Hirota et al. 2010). Complete recovery of P from wastewater is technically feasible using current technologies (Petzet and Cornel 2011) (Fig. 5).

P recovery from concentrated wastewater in wastewater treatment plants by crystallization as struvite ($MgNH_4PO_4$) or calcium phosphate ($Ca_{10}(OH)_2(PO4)_6$) is a simple and proven technology, which in some cases is economically feasible. The overall recovery potential is limited, however, since it can only be applied to a certain type of wastewater treatment process, namely enhanced biological P removal. Even in this case, the recovery is limited to 30 % of the P contained in wastewater with existing sewage treatment systems; the rest remains in the sewage sludge and the effluent. While biological P removal is increasingly showing a higher performance of removing P, especially in places where P discharge limits are very low, chemical removal would be applied and crystallization technologies might not be feasible.

A full-scale struvite crystallization plant for P recovery was constructed at Matsue city in Japan more than twenty years ago. Without this P recovery, spontaneous struvite precipitation often causes difficult fouling problems in pipelines. To remove it, the reactor and downstream pipelines need to be repeatedly washed with acids, which is very time-consuming and costly. To address these problems, a relatively simple technology is being developed to recover P from P-rich liquor using amorphous calcium silicate hydrate (A-CSH) as an inexpensive adsorbent. A-CSH particles are put into P-rich liquor, and after 10 min mixing, P-binding A-CSH particles are recovered by settling without using any chemical coagulant. Most importantly, the recovered particles can be directly used as a by-product phosphate fertilizer. If a feasible business model is established, a cement company can produce A-CSH particles at low costs and deliver them to wastewater treatment plants using agitation trucks. The recovered A-CSH particles can be delivered to a fertilizer company and used as by-product phosphate fertilizer. Importantly, A-CSH can be synthesized from unlimitedly available resources such as calcium and silicate.

Many microorganisms can accumulate high levels of P in the form of polyphosphate (Hirota et al. 2010), a unique long-chain polymer of inorganic phosphate residues having a chain length of 1,000 or more. Enhanced biological P removal relies primarily on the ability of sludge microorganisms to accumulate polyphosphate. Anaerobic sludge digestion is a well-established process to stabilize waste sludge and to reduce its volume by methane production. If anaerobic sludge digestion process is available, P can be readily released from poly-P-rich sludge biomass to the liquid phase.

Recycling of processed (e.g., digested or composted) sewage sludge (biosolids) to land, the simplest method to recycle P from wastewater, is in some cases impeded by legal bans (e.g., Switzerland and Japan) due to concerns about potential organic and inorganic contaminants in the sludge, or by a lack of agricultural land in the vicinity of large urban agglomerations.

An increasing amount of sewage sludge (for example, 50 % in Germany) is incinerated in monoincineration plants or coincinerated in power plants, municipal waste incinerators, or cement kilns, where P is usually not recycled. In the case of coincineration of sewage sludge, P is permanently tied to the resulting ashes and products, and recovery is not economically and/or technically feasible. In the case of monoincineration, P and non-volatile metals are concentrated in the ashes, which are good raw material for P recycling.

Various options exist for P recovery from ashes, which depend on the chemical composition; in Germany, 30 % of the ashes with low heavy metal content can be directly recycled as fertilizers, although there is a debate about the plant availability of the P. In contrast to sewage sludge, organic contaminants (pharmaceuticals, endocrine disruptors, pathogens) are destroyed, and P can be solubilized by the addition of acid (Petzet and Cornel 2011). Some ashes can be directly recycled as a raw material in the production of yellow P. At least technically speaking, PR can be substituted with suitable sewage sludge ash. One requirement for both direct recycling options is a low iron content of the ashes, which can be achieved by using aluminum instead of iron as the precipitant in wastewater treatment or by using biological nutrient removal.

In cases where direct recycling of either sewage sludge or sewage sludge ashes is not possible, technologies are required to remove organic or inorganic pollutants and to transform P into a bioavailable form or into a raw product that can be used by the P industry. Many innovative approaches have been investigated, which can be divided into thermal and wet chemical processes; some of these are now ready for implementation: The ASH-DEC process removes heavy metals as chlorides by a thermochemical treatment of ashes. The Mephrec process is a smelting gasification technology for sewage sludge that simultaneously recovers energy and P as a P-rich slag. Both technologies have been successfully tested at pilot scales, and their full-scale implementation is envisaged. Wet chemical leaching procedures, including bioleaching, for sewage sludge and ashes have been extensively investigated. Usually, P is leached together with metals, followed by different separation steps that can be combined, such as ion exchange, liquid–liquid extraction, precipitation, and nanofiltration. Wet chemical processes have been tried on a large scale for sewage sludge in Germany; while they are technically feasible, they have high operating costs due to chemicals (Petzet et al. 2012).

Where P removal is operating in sewage works, sludge monoincineration ash contains P at concentrations similar to those of rock phosphate. A full-scale plant for P recovery from sludge incineration ash has recently started operation at Gifu city in Japan (Goto 2009). The full-scale plant is now making a great contribution to the sustainability of local agriculture, because the quality of recovered P matches well the local demand. There are critical challenges, however, including the high capital cost for plant construction and the difficulty in establishing stable channels for distribution and sale of recycled P, which might discourage expanded uses of this technology.

In Switzerland, similar to other countries in middle Europe, nutrients (P, N, etc.) had been used as a resource from sewage sludge through their direct application in agriculture since the construction of efficient sewage purification systems. But for the last 40 years, questions have been raised in connection with this use (heavy metals, persistent organic pollutants, BSE). Quality demands of consumers and wholesalers on agricultural products (e.g., organic farming, high environmental awareness, "no risk" strategy) have increased. These concerns led to the ban on the direct use of sewage sludge in agriculture from 2006 in Switzerland, based on the precautionary principle. From that time, thermal treatment of sludge to destroy the pollutants or to concentrate and store them safely (e.g., waste incineration plants) or to bind them in a mineral matrix (cement plant) was mandatory. This trend, however, was also associated with the fact that nutrients in sludge are no longer able to be used.

In 2006, it was recognized in the Canton Zürich that capacity bottlenecks are to be expected with this disposal concept from 2015. The Canton Zürich is the most populated canton of Switzerland, with roughly 1.4 million inhabitants and an area of about 1,600 km2. Also, the knowledge that phosphorus is severely limited as an important nutrient became increasingly evident. Both factors were then used as an opportunity to define a new sludge recycling strategy based on the goals of modern waste and resource management under consideration of the optimized conservation of resources regarding phosphorus (Morf 2012). With this new strategy, the three most relevant general conditions have been defined in a resolution in 2007 for the planning in the Canton Zürich: (1) the (later) retrieval of phosphorus is possible; (2) the renewable energy in sludge is used; and (3) regardless of the place, it is treated in the optimum economic manner.

In a long-term and holistic-oriented approach developed and defined during the last six years, the Canon Zürich managed to change from a decentralized resource-inefficient to a very efficient centralized system in less than eight years. The selected concept based on one single sewage sludge monoincinerator avoids further phosphorus dissipation to a large extent and secures this scarce resource starting in 2015. This strategy allows to roughly substitute the total phosphorus imported in the form of mineral fertilizer. It is planned to enable successful direct P recovery from incinerator ash soon, in order to avoid intermediate storage costs. Therefore, a project to evaluate in detail direct P recovery from monoincineration ash with the focus on wet chemical extraction similar to primary phosphate production (LEACHPHOS®-Process) has been started since 2012. The project incorporates (a) the technical evaluation with a first full-scale test at the end of 2012, (b) detailed investigations regarding product quality and management (phosphorus fertilizer or secondary raw material, e.g., for white phosphorus production), and (c) market and economical aspects. The wet chemical extraction process was compared with two alternatives, namely a thermochemical process (ASHDEC®-Process) and a phosphoric acid treatment (RECOPHOS®-Process) at the moment.

4.2.2 Decentralized Measures

Urine diversion is a relatively established technology and has been tested in several places in Germany and Sweden, with consequent nutrient recycling to agriculture (GTZ Deutscher Gesellschaft für Technische Zusammenarbeit 2005a; Tanum Kommun 2008; Sustainable Sanitation Alliance 2010). While urine constitutes no more than 1 % of the total volume of wastewater, it contains 50 % of the P (Vinneras and Jonsson 2002). Urine is almost free from heavy metals and pathogens and is easily hygienized by storage (Kvarnström et al. 2006), ozone, or UV light. Urine can also be evaporated or precipitated as struvite, as, for example, by local solar-driven systems in Nepal and Vietnam (Etter et al. 2011; Antonini et al. 2012). Through fertilization with separately collected urine, the input of heavy metals in general, and the disputed cadmium in particular, to agriculture could be remarkably decreased, compared with spreading of sewage sludge from combined systems (Remy and Jekel 2008). Although human urine contains ingested pharmaceuticals and hormones, the level of concentration is much lower than in animal manure, which is commonly used as crop fertilizer today (Lienert et al. 2007; Winker 2010).

In blackwater diversion, the whole toilet wastewater is transported either by gravity or by vacuum to a decentralized or semi-centralized treatment site and treated separately. Blackwater contains little pollutants as their main sources are the household greywater and urban stormwater (Vinnerås 2001; Lamprea and Ruban 2008; Hernandez Leal 2010). Addition of urea or ammonia reduces eventual pathogens in blackwater (Winker et al. 2009). Blackwater is then treated aerobically by liquid composting, storage, or ammonia treatment. Alternatively, it is treated anaerobically to produce biogas as an additional product. The anaerobic digestion process results in mineralization of nutrients in the digested sludge, and, in particular, nitrogen becomes more plant-available (Meinzinger 2010). Blackwater separation is being applied on a building scale as well as a district scale in Sweden (GTZ Deutscher Gesellschaft für Technische Zusammenarbeit 2005b; Karlsson et al. 2008).

4.3 Solid Wastes[2]

A study of P flows for the EU 27 and for a municipality in Sweden recently showed that solid waste contains as much P as does the sewage sludge (Kalmykova and Harder 2012; Ott and Rechberger 2012). For the EU, per capita discharge through wastewater is larger: 0.6 kg P/cap per year, compared to 0.45 kg P/cap per year, through biowaste from households, restaurants, and canteens. P contained in a range of other waste materials, however, is not included: wood (0.31 kg P/t TS), textiles (0.14 kg P/t TS), paper and cardboard (0.24 kg P/t TS), porcelain, and chemical products (variable). Moreover, 20 % of the sewage sludge in the EU is landfilled directly, while another 11 % is incinerated before being landfilled. Also, equally large stocks of P were measured in the municipal solid waste incineration (MSWI) residues and the sewage sludge in Sweden (Kalmykova and Harder 2012).

Solid waste represents an underestimated sink of P and needs to be taken into account for sustainable P management. Extraction of P from untreated solid waste has not been investigated thoroughly. Incineration is a commonly used method for treatment of waste before landfilling, to reduce volume and sometimes also to recover energy. MSWI residue offers a relatively homogeneous and concentrated stock for mineral recovery, and methods for P recovery have been developed recently (Kalmykova and Karlfeldt Fedje 2012).

An estimation of the P flows in municipal solid waste based on the generation rates, waste composition, and subtracting recycling (recycled fractions are only available for OECD) results in 94,400 t of P per year for 60 % of the world's population, that is, OECD, China, India, Brazil, Russia, and South Africa

---

[2] Parts of this section are derived from the report, Food and Water Security in the Lake Winnipeg Basin—Transition to the Future (Malley et al. 2009).

(Organisation for Economic Co-operation and Development 2008). Generation rates for 2005 have been used, and the landfilled sewage sludge, wood, and industrial wastes are not included. The biodegradable waste (biowaste) fraction is known to decrease with increasing affluence of the population. While the fraction of biowaste is 30 % for the EU, North America, and Australia, it is 60–80 % for China, India, Bangladesh, Latin America, and African countries. Therefore, potentially even larger quantities of P are landfilled in the developing countries.

4.3.1 Food Waste

Food and food-processing wastes are a major source of P in solid waste due to both the large quantities and high P content of 0.4 kg P/t TS. The extreme wastage of food is a unique modern phenomenon. A report by the UN FAO, Stockholm International Water Institute, and the International Water Management Institute indicates that close to half of all food produced worldwide is wasted (Lundqvist et al. 2008). This amounts to about 1.3 billion tonnes per year, even though calculations are still uncertain due to large data gaps (Gustavsson et al. 2011).

Developed and developing countries differ in their characteristics in food loss and waste. In developing countries, more than 40 % of the food losses occur at post-harvest and processing levels, while in industrialized countries, more than 40 % of the food losses occur at retail and consumer levels (Gustavsson et al. 2011). It is argued that the per capita food loss and waste by consumers in Europe and North America is 95–115 kg/year, whereas in sub-Saharan Africa and South/Southeast Asia, this amounts to 6–11 kg/year (Gustavsson et al. 2011). In the developing world, lack of infrastructure and technical and managerial skills in food production is a key driver in the creation of food waste. Consequently, the majority of uneaten food is lost, with P included in it also lost.

In the developed world, in contrast, the majority of the food waste is driven by the low price of food relative to the income, consumers' high expectations of food cosmetic standards, and the increasing disconnection between consumers and the place where food is produced (Parfitt et al. 2010). Astonishingly, much of the food wasted in the developed countries is in entirely edible condition. For instance, in the EU, around 90 million tonnes of wasted food includes losses from agricultural production due to quality standards, which discharges food items not perfect in shape and appearance (39 %), distribution and retail (5 %), food services and catering (14 %), and final household consumption (42 %), due to, e.g., inconsistency in date labels (Commission of the European Communities 2010). Britain, for example, throws away half of all the food produced on farms, amounting to about 20 Mt of food, which would be equivalent to half of the food import needs for the whole of Africa (Mesure 2008). Approximately 16 Mt of this is wasted in homes, shops, restaurants, hotels, and food manufacturing. Much of the rest is thought to be destroyed between the farm field and the shop shelf.

Separate collection and treatment of food waste enable recycling of nutrients through application of compost, while both energy and nutrients can be recycled via biogas production with consequent agricultural application of the residue. In addition, diverting of biowaste from landfills prevents production of the landfill gas methane, a greenhouse gas 21 times more potent than carbon dioxide. Separate collection of food waste is implemented in several countries worldwide. Effective separate collection, however, is difficult to achieve due to the low collection rate from households and large non-separated flows from food distribution and retail, restaurants, and public institutions (Kalmykova and Harder 2012). In Scandinavia and Canada, 10–15 % of biowaste is composted, with higher efficiency of 20–40 % in Austria, Germany, the Netherlands, France, Italy, and Spain (Organisation for Economic Co-operation and Development 2008).

4.3.2 Slaughterhouse Waste

Slaughterhouse waste is another P-rich waste, which contains up to 60 g P/kg TS. Bones contain even around 100 g P/kg DM (Lamprecht et al. 2011). During the slaughter process, 33–43 % of live animal weight is discarded as inedible waste, so-called animal by-products. This waste is processed by the rendering industry into high-quality fats used by the oleochemical industry and the meat and bone meal (MBM) used as protein and phosphorous supplements for animal feed. Every year, 16 million tonnes of animal by-products is processed by renderers and fat-melters in the EU, 25 million tonnes in North America, and 12 million tonnes in Argentina, Australia, Brazil, New Zealand, and India (60 % water content).

In the EU, USA, Australia, and New Zealand, the use of animal by-products is severely restricted, due to the fear of BSE, what is often called "mad cow disease" (Australian Government 2011; Commission of the European Communities 2002; United States Food and Drug Administration 2008). Animal by-products are divided into three risk categories; that is, class 1 must be incinerated, while categories 2–3 can be composted or digested for biogas production. There are no data available on the amounts of slaughterhouse waste entering different disposal routes; therefore, the fraction of P either recycled as compost or landfilled as ash cannot be estimated. Several techniques have been developed in a laboratory scale for P extraction from animal by-products or its ash, and carbonization into charcoal has been tested on a pilot scale in the EU (Someus 2009; Zalouk et al. 2009).

4.4 Industrial Wastes

One of the most economically important pathways would be P recycling in the manufacturing sector, including some of the high-tech industries. P is used in surface treatment chemicals, for example, such as the iron phosphate coating material in the automotive industry. P is also one of the crucial raw materials for the production of rechargeable batteries such as lithium ion batteries. Furthermore, P is used in etching agents for aluminum line-patterned substrates in the production of computer chips and liquid crystal panels and flame retardants for a wide variety of industrial products.

In terms of quantity, P recycling in the steelmaking industry is particularly important. P is present in iron ore and coal at concentrations as low as 0.12 % and is removed into steelmaking slags at concentrations of 2–3 % at the maximum. Since the steelmaking is a very large industry, the amount of P emitted as slag is considerable.

The manufacturing industrial sectors require high-quality phosphoric acid, which is derived from elemental P. There would be a strong demand for maintaining domestic elemental P production from the manufacturing industrial sectors. Although the consumption of elemental P is minor from the quantitative point of view, it is strategically important especially for high-tech industries.

Industrial chemical processes such as direct hydration of ethylene to ethanol also use large amounts of quality phosphoric acid as the catalysts. P recovery has been put into operation in the process of synthetic alcohol production. The recovered P is reused in fertilizers for agricultural purposes. Edible oil refining process also uses large amounts of phosphoric acid to remove impurities from crude vegetable oil. Since no harmful substance is used in the edible oil refining process, the P recovered from wastes and wastewater is well suited to the use for agricultural purposes. This is also the case for fermentation wastewater. Fermentation companies have also been recovering P from the fermentation wastewater using the HAP precipitation technique. Several companies of electronic equipment manufacturing have been recovering P from liquid wastes and are attempting to use the recovered P in liquid fertilizer for urban plant factories.

Pulp and paper production is another industry disposing large volumes of P-containing waste. Annually 11 million tonnes of the waste is produced in the EU and 8 and 3 million tonnes in the USA and Japan, respectively (Monte et al. 2009; Wajima et al. 2006). The waste is usually incinerated in order to reduce the volume. While a part of the resulting ash is used as a construction material, mainly for landfill covering layers, most of it is simply landfilled. Although the P content of the ash is only 0.1 %, because of the large waste volumes, considerable amounts of P are disposed of as a result. Assuming 60 % of water content of the waste, an amount equivalent to 20,000 t of P is disposed annually in the EU, USA, and Japan.

5 Work in Global TraPs

5.1 Knowledge Gaps and Critical Questions

The current P management practices and approaches in the different sectors show diverse characteristics with regard to temporal and spatial scales of the issue; technological measures, including types of technology, energy consumption, costs of investment, and operation; key stakeholders involved, such as farmers, industry,

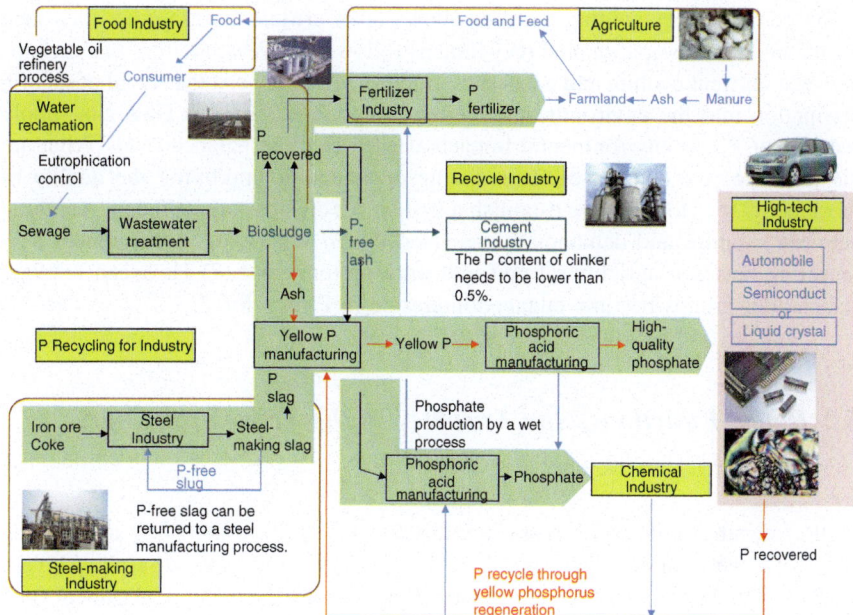

Fig. 6 Possibilities of P recycling in agricultural and industrial sectors. *Source* (Ohtake 2010)

consumers, and public sectors; and institutional conditions, including public policies and interventions. Solid understanding of the factors influencing P dissipation and recycling in different domains will be of critical importance for sustainable P management.

Potentially, there are three main areas for implementing P recycling as illustrated in Fig. 6:

1. Recycling of P contained in food and feed;
2. Recycling of P from wastewater; and
3. Recycling of P from industries using high-quality phosphate.

Recognizing the significance and potential of P recycling, experts and practitioners have started to pay attention to the development of P management and recycling as a new green industry. Active involvement of industry, however, is still limited, and there are not many cases in which P recycling is successfully implemented in practice. At the current stage of development, recycling of P is not a feasible business opportunity, as the conventional practice of buying normal fertilizers while wasting water and sludge would be much cheaper. Recycling of P, therefore, has not yet become a strategic issue for major companies in the industry. One of the critical issues which we need to tackle is *how to establish socially robust business models in a broader sense, integrating scientific understanding, technological development, corporate strategies, and public policies, for successful implementation of P recycling.*

Based on the key guiding question above, other critical questions are identified for the node on dissipation and recycling as follows: (1) What are the relationships between nutrient cycling and ecosystem stability? (2) How can the social costs of P dissipation into the environment be assessed appropriately? (3) How can supply security of P and environmental impacts caused by P be balanced? (4) What are the technological challenges for reducing costs and/or improving the quality of recycled P? (5) How can we establish a system for effective and efficient matching between supplies and demands for recycled P? (6) What are the differences in P recycling between agricultural and industrial sectors? and (7) How can relevant stakeholder groups become engaged/interested in P recycling?

5.2 Roles, Functions, and Varieties of the Transdisciplinary Process

In the transdisciplinary process, it is crucial to build partnerships among key stakeholders to conduct joint problem definition and joint transdisciplinary case studies (Scholz 2011). Relevant actors in academia, industry, government, and NGOs need to be actively engaged in identifying the critical challenges (Yarime et al. 2012; Trencher et al. 2013). While the need for P recovery has been identified as an important issue by various institutions, actions taken by stakeholders around the world have been still limited. In the design of a P-recycling system, it is necessary to approach this issue at multiple levels, including local, national, and global levels. While the recycling system would be very much conditioned by local characteristics and contexts, at the same time, it is also influenced by national resource management strategies and institutional frameworks. Generally speaking, the recycling of P is not yet considered to be a high priority issue at the national level, except for a few countries such as Sweden, where a target is set to recover 60 % of P from sewages by 2015. A full commitment to implementing P recycling is not yet dedicated by the industry, which is increasingly influenced by the fluid business environment in the global economy.

As we have seen in the previous sections, there are many actors in different sectors that have stakes in recycling P. P can be characterized as an essential, non-substitutable, but low-cost commodity that each person consumes, as well as a source of environmental pollution. It is crucial to identify who has what kind of stake within the system and to find out the best way to realize a situation in which a common solution would satisfy different interests and objectives (Shiroyama et al. 2012). For instance, there can be a potential of implementing P recycling through close collaboration between cement companies, fertilizer companies, and the local government. The sewage department of the local government has to extract P to meet the water quality standard to avoid environmental degradation. Fertilizer companies need P for producing fertilizer. And cement companies require a low level of P contained in sludge because sludge with high concentration of P can weaken the strength of the product (Ohtake 2011). Currently, one of the most serious challenges for the recycling of P is how to expand the market for recycled P. For that purpose, it would be important to maintain the stability of supply and the quality of products involving recycled P. Institutional measures to accelerate the closing of the P chain would include the implementation of P discharge criteria in waste stream regulations and the revision of lengthy and costly permission procedures and requirements for recovery technologies (Drizo 2012).

We then need to prepare for sustainable transitions by exploring feasible strategies for social business models, with the relevant stakeholders closely involved. For that, it would be possible to consider pursuing consensus at two levels. The first one is whether we should go for a soft or detransformation of sewage and wastes. While there has been a concern about heavy metals, harmful chemicals, pathogens, and other biological issues, it is not completely clear what kind of soft processing including organic matters might be a better option, in comparison with incineration. This type of consensus building could take place in a precompetitive arena, although it might also affect industrial activities from a long-term perspective. Based on that, we could consider what technical process (thermal or chemical extraction) should be applied, probably in a competitive arena.

5.3 Suggested Case Studies

To address the key guiding questions and the dimensions of transdisciplinary processes, we can suggest case studies to be conducted for further research. First, it is very important to explore how to make an appropriate assessment of the external costs associated with the dissipation of P into the environment. A case study in highly contaminated regions would be useful to find out how much is actually caused and affected by P in a freshwater region in a sea area, for example, the Manila Bay in the Philippines, where P-based detergents are still used, probably with significant environmental impacts.

We also need to conduct detailed case studies to examine some of the emerging cases of successful implementation of recycling P in different sectors and regions. For example, P recycling in the sewage treatment plant in Gifu, Japan, has been operating since 2010, and the fertilizer involving recycled P has been sold to farmers. In Europe, a couple of companies that previously have operated in the detergent field are now utilizing their extensive knowledge on P for different types of purification and reprocessing. Ostara has been running five plants for recycling P in Europe. Companies such as ICL have already joined the Global TraPs project. It could be possible to conduct case studies at the watershed level such as Lake Winnipeg in Canada as well as the city level, for example, the case of urban metabolism in Gothenburg, Sweden. Development and implementation of innovative technologies are currently explored in Germany, and recovery measures in the water sector and from manure have started to be introduced in the UK.

The knowledge generated in the transdisciplinary process is expected to be used by the practitioners in their business and policy decisions for realizing sustainable use of P, which demands cooperation and coordination across different sectors. Close collaboration among relevant stakeholders including academia, industry, and the public sector is urgently required to cope with this critical challenge. In an attempt to address that, the Phosphorus Recycling Promotion Council of Japan (PRPCJ 2008) was established in 2008 by inviting experts and practitioners from academia, industry, and the public sector. This nation-wide association is supported by the four relevant ministries of the Japanese government and currently has approximately 140 members, including more than 70 corporate members. Based on the PRPCJ activities, a national platform for industry–academia–government collaboration was also initiated in 2011 to discuss and implement national strategies for robust P-recycling systems.

In Europe, the Nutrient Platform was established in 2011 in the Netherlands, with more than 20 Dutch companies, knowledge institutes, government authorities, and NGOs signing the Phosphate Value Chain Agreement (Dutch Nutrient Platform 2011). The Nutrient Platform is a cross-sectoral network of Dutch organizations that share a common concern for the global impact of phosphorus depletion and the way the society is dealing with nutrients in general. Together with the Dutch government, the Nutrient Platform is aimed at facilitating the organizations throughout the value chain in closing the phosphorus cycle. They all share the ambition of creating a sustainable market within two years, where as many reusable phosphate streams as possible will be returned to the cycle in an environmentally friendly way and where the recycled phosphate will be exported to the fullest extent possible, as long as surplus exists in the Dutch market, in order to contribute elsewhere to soil improvement and food production. To achieve the vision and mission, the platform practices an approach of learning by doing within a framework of action learning and new types of partnerships.

The first European Sustainable Phosphorus Conference 2013 was held in Brussels in March 2013, with the purpose of raising awareness about the necessity for a more sustainable phosphate management within the context of a Resource Efficient Europe (European Phosphorus Platform 2013). It was aimed at facilitating support for a clear and coherent legislative framework to create an enabling environment for ecoinnovation, a sustainable European market for secondary phosphorus and more efficient phosphorus use. Different nutrient waste flows and market possibilities will be connected between stakeholders, including private sector throughout different sectors, knowledge institutes, government, and NGO's, for further development of sustainable nutrient chains within Europe. At the conference, participants reached consensus to launch the European Phosphorus Platform to continue dialogues, raise awareness, and trigger actions to address the phosphorus challenge, with significant implications for ensuring food security, geopolitical stability, and environmental sustainability.

In North America, a kickoff workshop was organized in May 2013 to launch Research Coordination Network (RCN) in Washington DC, USA (Sustainable P Initiative 2013). The workshop was mean to bring together some of the world's top scientists, engineers, and technical experts to spark an interdisciplinary synthesis of data, perspectives, and understanding about phosphorus to envision solutions for P sustainability. Key stakeholders from relevant sectors shared their knowledge and expertise on various dimensions of the global phosphorus system, including farmers and growers, food processors, fertilizer producers, waste managers, water quality managers, regulators, legislators, and others. Two challenges of phosphorus efficiency and phosphorus recycling have been identified. RCN on coordinating phosphorus research has been funded by the National Science Foundation to create a sustainable food system.

These experiences of establishing national/regional platforms involving key stakeholders will provide valuable lessons and implications for implementing P recycling successfully in different technological, economic, and institutional contexts.

References

Albert E (2004) Versuchsergebnisse zur Stickstoff- und Phosphorwirkung von Fleischknochenmehl. In: 93. Sitzung des DLG Ausschuss für Pflanzenschutz. Deutsche Landwirtschafts-Gesellschaft e.V., Derenburg, 25 Mai 2004. Sächsische Landesanstalt für Landwirtschaft

Antonini S, Nguyen PT, Arnold U, Eichert T, Clemens J (2012) Solar thermal evaporation of human urine for nitrogen and phosphorus recovery in Vietnam. Sci Total Environ 414:592–599

Australian Government (2011) Australia New Zealand Food Standards Code—Standard 2.2.1—Meat and Meat Products—F2011C00615

AWEL (2008) Phosphor im Klärschlamm—Informationen zur künftigen Rückgewinnung. Baudirektion Kanton Zürich, AWEL Amt für Abfall, Wasser, Energie und Luft [phosphorus in sewage sludge—Information about future retrieval, Building Department of the Canton of Zürich, AWEL Office for Waste, Water, Energy and Air] (http://www.klaerschlamm.zh.ch)

Bateman A, van der Horst D, Boardman D, Kansal A, Carliell-Marquet CM (2011) Closing the phosphorus loop in England: the spatio-temporal balance of phosphorus capture from anaerobically-digested manure versus crop demand for phosphorus. Resour Conserv Recycl 55:1146–1153

Bennett EM, Carpenter SR, Caraco NF (2001) Human impact on erodable phosphorus and eutrophication: A global perspective. Bioscience 51(3):227–234

Biello D (2008) Oceanic dead zones continue to spread. Scientific American. http://www.sciam.com/article.cfm?id=oceanic-dead-zones-spread. August 15

Binder CR, de Baan L, Wittmer D (2009) Phosphorflüsse in der Schweiz: Stand, Risiken und Handlungsoptionen. Abschlussbericht. Umwelt-Wissen Nr. 0928. Bundesamt für Umwelt, Bern

Commission of the European Communities (2002) Regulation (EC) No. 1774/2002 of the European parliament and of the Council of 3 October 2002 on management rules for animal by-products not intended for human consumption

Commission of the European Communities (2010) Preparatory study on food waste across EU 27. http://ec.europa.eu/environment/eussd/pdf/bio_foodwaste_report.pdf

Cordell D, Drangert J-O, White S (2009) The story of phosphorus: global food security and food for thought. Global Environ Change 19(2):292–305

Drizo A (2012) Innovative phosphorus removal technologies. Available at http://www.azocleantech.com/article.aspx?ArticleID=226-6. Accessed 7 Mar

Dutch Nutrient Platform (2011) Phosphate value Chain agreement. Dutch Nutrient Platform, 4 Oct
Etter B, Tilley E, Khadka R, Udert KM (2011) Low-cost struvite production using source-separated urine in Nepal. Water Res 45(2):852–862
European Phosphorus Platform (2013) Joint declaration for the launch of a European Phosphorus Platform. In: 1st European sustainable phosphorus conference 2013, Brussels, 6–7 March
Falconer I (ed) (1993) Algal Toxins in seafood and drinking water, 1st edn. Academic Press, London
Foy RH (2005) The return of the phosphorus paradigm: agricultural phosphorus and eutrophication. In: Sims JT, Sharpley AN (eds) Phosphorus agriculture and the environment, vol 46. American Society of Agronomy Monograph, Madison, pp 911–939
Goto K (2009) Advanced utilization of sludge incineration ashes (conversion to phosphorus fertilizers and use of incineration ashes). In: Ohtake H (ed) Recovery and effective utilization of phosphorus resources. Science & Technology, Tokyo, pp 365–382
Government of Manitoba (2008) MARC 2008 User's Manual: manure application rate calculator version 2.1.3. Government of Manitoba, represented by the Minister of Agriculture, Food and Rural Initiatives, January
GTZ Deutscher Gesellschaft für Technische Zusammenarbeit (2005a) Data sheets for ecosan projects: 004 ecological housing estate, Lübeck Flintenbreite
GTZ Deutscher Gesellschaft für Technische Zusammenarbeit (2005b) Data sheets for ecosan projects: 008 Gebers collective housing project, Orhem, Sweden
Gustavsson J, Cederberg C, Sonesson U, van Otterdijk R, Meybeck A (2011) Global food losses and food waste: extent, causes and prevention. FAO, Rome
Hermann L (ed) (2011) How energy from livestock manure can reduce eutrophication. Ecoregion perspectives. Fourth Issue: sustainable agriculture in the Baltic Sea region in times of peak phosphorus and global change. CBSS-Baltic21
Hernandez Leal L (2010) Removal of micropollutants from greywater. Ph.D. thesis, Wageningen University, The Netherlands
Hirota R, Kuroda A, Kato J, Ohtake H (2010) Bacterial phosphate metabolism and its application to phosphorus recovery and industrial bioprocesses. J Biosci Bioeng 109(5):423–432
Hoekstra AY, Chapagain AK (2007) Water footprints of nations: water use by people as a function of their consumption pattern. Water Resour Manage 21:35–48
Kalmykova Y, Harder R (2012) Pathways and management of phosphorus in urban areas. J Ind Ecol (forthcoming)
Kalmykova Y, Karlfeldt Fedje K (2012) Phosphorus recovery from municipal solid waste incineration fly ash. Under review
Karlsson P, Aarsrud P, de Blois M (2008) Återvinning av näringsämnen ur svartvatten—utvärdering projekt Skogaberg. Svenskt Vatten Utveckling, Report 2008-10
Kvarnström E, Emilsson K, Richert Stintzing A, Johansson M, Jönsson H, af Petersens E, Schönning C, Christensen J, Hellström D, Qvarnström L, Ridderstolpe P, Drangert J-O (2006) Urine diversion: one step towards sustainable sanitation. Report 2006-1, EcoSanRes
Lamprea K, Ruban V (2008) Micro pollutants in atmospheric deposition, roof runoff and storm water runoff of a suburban Catchment in Nantes, France. In: 11th International conference on urban drainage, Edinburgh, United Kingdom
Lamprecht H, Lang DJ, Binder CR, Scholz RW (2011) Animal bone disposal during the BSE crisis in Switzerland—an example of a "disposal dilemma". Gaia 20(2):112–121
Lienert J, Bürki T, Escher B (2007) Reducing micropollutants with source control: substance flow analysis of 212 pharmaceuticals in feces and urine. Water Science Technology 56(5):87–96
Liu Y, Villalba G, Ayres RU, Schroder H (2008) Global phosphorus flows and environmental impacts from a consumption perspective. J Ind Ecol 12(2):229–247
Lundqvist J, de Fraiture C, Molden D (2008) saving water: from field to fork. Curbing losses and wastage in the food Chain. SIWI Policy Brief. Stockholm International Water Institute, Sweden
MacDonald JM, Ribaudo MO, Livingston MJ, Beckman J, Huang W (2009) Manure use for fertilizer and for energy: report to congress. Economic Research Service, United States Department of Agriculture, June

Malley DF, Ulrich AE, Watts PD (2009) Food and water security in the Lake Winnipeg Basin—transition to the future

Matsubae K, Kajiyama J, Hiraki T, Nagasaka T (2011) Virtual phosphorus ore requirement of Japanese economy. Chemosphere 84:767–772

Meinzinger F (2010) Resource efficiency of urban sanitation systems: a comparative assessment using material and energy flow analysis. Ph.D. thesis, Technischen Universität Hamburg-Harburg, Germany

Mesure S (2008) The £20 billion food mountain: britons throw away half of the food produced each year. The Independent: Sunday, 02 Mar

Monte MC, Fuente E, Blanco A, Negro C (2009) Waste management from pulp and paper production in the European Union. Waste Manage 29:293–308

Morf LS (2012) Phosphor aus Klärschlamm—Strategie des Kanton Zürichs und der Schweiz (Phosphorus from sewage sludge—the strategy of the Canton of Zürich and Switzerland), vol 45. Essner Tagung Wasser- und Abfallwirtschaft, 14–16 März 2012 in Essen (English version)

Ohtake H (2010) Biorecycle of phosphorus resource for sustainable agriculture and industry. OECD workshop on biotechnology for environment in the future: science, technology and policy, 16–17 Sept 2010

Ohtake H (ed) (2011) The coming phosphorus crisis (Rin shigen kokatsu mondai toha nanika). Handai Livre Publishing, Osaka

Olde Venterink H (2011) Legumes have a higher root phosphatase activity than other forbs, particularly under low inorganic P and N supply. Plant Soil 347(1–2):137–146

Organisation for Economic Co-operation and Development (2008) OECD environmental data: compendium 2006–2008. OECD, Paris

Ott C, Rechberger H (2012) The European phosphorus balance. Resour Conserv Recycl 60:159–172

Parfitt J, Barthel M, Macnaughton S (2010) Food waste within food supply chains: quantification and potential for change to 2050. Philos Trans R Soc B Biol Sci 365(1554):3065–3081

Petzet S, Cornel P (2010) Recycling of Phosphorus from Sewage Sludge—Options in Germany, Sonderausgaben Wasser und Abfall 1:34–36

Petzet S, Cornel P (2011) Towards a complete recycling of phosphorus in wastewater treatment—options in Germany. Water Sci Technol 64(1):29–35

Petzet S, Peplinski B, Cornel P (2012) On wet chemical phosphorus recovery from sewage sludge ash by acidic or alkaline leaching and an optimized combination of both. Water Res. doi:10.1016/j.watres.2012.1003.1068

Phosphorus Recycling Promotion Council of Japan (2008) Establishment of the phosphorus recycling promotion council of Japan. Phosphorus Recycling Promotion Council of Japan, Tokyo

Pierzynski GM, McDowell RW, Sims JT (2005) Chemistry, cycling, and potential movement of inorganic P in soils. In: Sims JT, Sharpley AN (eds) Phosphorus: agriculture and the environment. Agronomy Monograph, vol 46, pp 53–86

Remy C, Jekel M (2008) Sustainable wastewater management: life cycle assessment of conventional and source-separating urban sanitation systems. Water Sci Technol 58(8):1555–1562

Richards I, Dawson C (2008) Phosphorus imports, exports, fluxes and sinks in Europe. In: Society PotIF (ed). International Fertiliser Society

Scholz RW (2011) Environmental literacy in science and society: from knowledge to decisions. Cambridge University Press, Cambridge

Schoumans OF, Rulkens WH, Oenema O, Ehlert PAI (2010) Phosphorus recovery from animal manure: technical opportunities and agro-economical perspectives. Alterra report 2158, Alterra, Wageningen, The Netherlands

Selman M, Greenhalgh S, Diaz R, Sugg Z (2008) Eutrophication and hypoxia in coastal areas: a global assessment of the state of knowledge, eutrophication and hypoxia in coastal areas: a global assessment of the state of knowledge. World Resources Institute, Washington

Shakhramanyan N, Schneider UA, McCarl BA, Lang DJ, Schmid E (2012) The impacts of higher mineral phosphorus prices and externality taxation on the use of organic phosphorus sources in US agriculture. Working Paper IETSR-1, Institute of Ethics and Transdisciplinary Sustainability Research, University of Lüneburg, Germany

Shiroyama H, Yarime M, Matsuo M, Schroeder H, Scholz RW, Ulrich AE (2012) Governance for sustainability: knowledge integration and multi-actor dimensions in risk management. Sustain Sci 7(1):45–55

Smit AL, Bindraban PS, Schröder JJ, Conijn JG, van der Meer HG (2009) Phosphorus in agriculture: global resources, trends and developments. Report to the Steering Committee Technology Assessment of the Ministry of Agriculture, Nature and Food Quality, The Netherlands, in collaboration with the Nutrient Flow Task Group (NFTG), supported by Development Policy Review Network (DPRN). Wageningen Plant research International B.V., Wageningen

Someus E (2009) PROTECTOR—Recycling and upgrading of bone meal for environmentally friendly crop protection and nutrition. Final Report

Sustainable P Initiative (2013) Sustainable phosphorus research coordination network. Sustainable P Initiative. http://sustainablep.asu.edu/prcn. Accessed 19 Aug 2013

Sustainable Sanitation Alliance (2010) Urine and faecal wastewater separation at GTZ main office building, Eschborn, Germany. http://www.susana.org/docs_ccbk/susana_download/2-63-en-susana-cs-germany-eschbornhouse-1-2009.pdf

Tanum Kommun (2008) Urine separation. http://www.tanum.se/vanstermenykommun/miljo/toaletterochavlopp/urineseparation.4.8fc7a7104a93e5f2e8000595.html

Trencher G, Yarime M, McCormick KB, Doll CNH, Kraines SB (2013) Beyond the third mission: exploring the emerging university function of co-creation for sustainability. Science and Public Policy. doi:10.1093/scipol/sct044

U.S. Geological Survey (2012) Mineral commodity summaries 2012. U.S. Geological Survey

Ulrich AE, Malley DF, Voora V (2009) Peak phosphorus. Opportunity in the making. International Institute for Sustainable Development, Winnipeg

United States Food and Drug Administration (2008) Substances prohibited from use in animal food or feed. 21 CFR Part 589

Vinnerås B (2001) Faecal separation and urine diversion for nutrient management of household biodegradable waste and wastewater. Licentiate thesis, Swedish University of Agricultural Sciences, Uppsala

Vinneras B, Jonsson H (2002) The performance and potential of faecal separation and urine diversion to recycle plant nutrients in household wastewater. Bioresour Technol 84(3):275–282

Wajima T, Haga M, Kuzawa K, Ishimoto H, Tamada O, Ito K, Nishiyama T, Downs RT, Rakovan JF (2006) Zeolite synthesis from paper sludge ash at low temperature (90 °C) with addition of diatomite. J Hazard Mater 132:244–252

Willer H, Kilcher L (2012) The world of organic agriculture—statistics and emerging trends 2012. Research Institute of Organic Agriculture (FiBL), Frick, and International Federation of Organic Agriculture Movements (IFOAM), Bonn

Winker M (2010) Are pharmaceutical residues in urine a constraint for using urine as a fertilizer? Sustain Sanitation Pract 3:18–24

Winker M, Vinnerås B, Muskolus A, Arnold U, Clemens J (2009) Fertiliser products from new sanitation systems: their potential values and risks. Bioresour Technol 100(18):4090–4096

World Resources Institute (2012) Eutrophication and hypoxia: nutrient pollution in coastal waters. World Resources Institute, Washington. http://www.wri.org/project/eutrophication. Accessed 27 May 2012

Yarime M, Trencher G, Mino T, Scholz RW, Olsson L, Ness B, Frantzeskaki N, Rotmans J (2012) Establishing sustainability science in higher education institutions: towards an integration of academic development, institutionalization, and stakeholder collaborations. Sustain Sci 7(1):101–113

Zalouk S, Barbati S, Sergent M, Ambrosio M (2009) Disposal of animal by-products by wet air oxidation: performance optimization and kinetics. Chemosphere 74:193–199

Chapter 7
Trade and Finance as Cross-Cutting Issues in the Global Phosphate and Fertilizer Market

Olaf Weber, Jacques Delince, Yayun Duan, Luc Maene,
Tim McDaniels, Michael Mew, Uwe Schneidewind,
and Gerald Steiner

Abstract This chapter provides an overview of trade and finance issues in the global phosphate and fertilizer market. First, we analyze global trade dynamics affecting fertilizers and their raw materials. Secondly, we present factors that influence fertilizer prices. Based on these analyses, we infer that prices for raw materials, energy and transport costs, supply and demand, subsidies, trade and finance, the supply chain, regional influences, the food price, and fertilizer

O. Weber · Y. Duan
University of Waterloo, 200 University Avenue West, Waterloo, ON N2L 3G1, Canada
e-mail: oweber@uwaterloo.ca

J. Delince
AGRILIFE Unit (IPTS—JRC, EC), European Commission, Joint Research Centre,
Edificio Expo, c/Inca Garcilaso 3, 41092 Sevilla, Spain
e-mail: Jacques.Delince@ec.europa.eu

L. Maene
Former Director, General-International Fertilizer Industry Association (IFA),
16 bis rue Jeanne d'Arc 78100 Saint-Germain-en-Laye, France
e-mail: maeneluc1@gmail.com

T. McDaniels
The University of British Columbia (UBC Vancouver), Vancouver Campus,
2329 West Mall B.C., Vancouver V6T 1Z4, Canada
e-mail: timmcd@interchange.ubc.ca

M. Mew
CRU International-Consultant, Polignac, Route de Fources, 32100 Condom, France
e-mail: michaelmew@wanadoo.fr

U. Schneidewind
Wuppertal Institute for Climate, Environment and Energy, Döppersberg 19,
Postfach 100480, 42004 Wuppertal, Germany
e-mail: uwe.schneidewind@wupperinst.org

G. Steiner (✉)
Harvard University, Weatherhead Center for International Affairs (WCFIA),
1737 Cambridge Street, 02138 Cambridge, MA, USA
e-mail: gsteiner@wcfia.harvard.edu

substitutes all influence fertilizer prices. Our analyses also show that, since 2007, the volatility of commodities significantly increased and strongly affected fertilizer purchases for crop production. Finally, we propose case studies to analyze challenges and opportunities related to phosphate and fertilizer markets and their sustainability implications.

Keywords Global phosphate trade · Price dynamics of phosphate rock · The 2008 price peak · Food commodity prices

Contents

1	Background..	276
2	Global P and Fertilizer Trade..	277
3	The Price of Phosphate and Fertilizer...	279
4	Influences on Price and Trade...	282
5	Draft Model..	290
6	Work in Global TraPs..	291
	6.1 Knowledge Gaps and Critical Research Questions	291
	6.2 Role, Function, and Kind of Transdisciplinary Case Process.............................	292
	6.3 Suggested Case Studies	292
References..		293

1 Background

This chapter summarizes the influence of trade and financial issues on phosphate and fertilizer prices. The first part of the chapter (Sects. 1–3) gives an overview about the current literature and results in this field. The second part of the chapter (Sects. 4–6) presents a data analysis and conclusions based on the results of this research.

Fertilizer demand has historically been influenced by dynamic and interrelated factors such as population and economic growth, agricultural production, fertilizer prices, and government policies (Food and Agriculture Organization of the United Nations 2008). The global sector changed significantly over the last years. For instance, the United States became a net nitrogen importer being a net exporter for a long time. At present, China is the leading producer of ammonia, phosphate rock (PR), urea, and diammonium phosphate (DAP). The production of potash is highly concentrated in Canada that produced 15,586 kt in 2010 and exported nearly half of it to the United States. The United States is still the global leader in sulfur production. Hence, to date, China plays a major role in the market for fertilizers and their raw materials. Generally, price changes for fertilizers are explained by energy prices, increased crop production to meet biofuels demand, exchange rate movements, food price, and global demand and supply conditions (Oehmke et al.

2008; Ott 2012). However, three developments since 2006 distinguish the current state of agricultural markets from past fluctuations (Food and Agriculture Organization of the United Nations 2008).

1. The hike in global prices in 2007 concerns nearly all major food and feed commodities (Oxfam 2011)
2. Record prices emerged because of numerous influences such as high demand, changes in food price, etc. (Ott 2012)
3. There seem to be stronger linkages between agricultural commodity markets and other markets than in the past (Food and Agriculture Organization of the United Nations 2008).

These phenomena were already manifest in 2006 and strengthened in 2007—a year that was characterized by persistent market uncertainty, record prices and unprecedented volatility in grain markets (Food and Agriculture Organization of the United Nations 2008; Saravia-Matus et al. 2012) as well as by the beginning of a global financial crisis. Though United States nominal prices of nitrogen, phosphate, and potash fertilizers already increased as early as 2002, they increased sharply and reached historic highs in mid-2008 (Huang et al. 2009). The price of DAP increased by five times—from $262 to $1,218/t—from January 2007 to April 2008, but had fallen to $469/t in mid-December (Hargrove 2008). In contrast to DAP, potash was the only fertilizer ingredient whose price was rising in 2008. Standard grade muriate of potash, the most common source of potassium, cost $172/t in January 2007 and $875/t in mid-December 2007 (Hargrove 2008). As a consequence, the USDA Index of Prices Paid by Farmers for Fertilizer, which reflects all types of fertilizer and has a base period of 1990–1992, increased by 216 points between 2006 and 2008 (Kenkel 2010). However, it seems that the significant price increase in 2008 was of a short-term "spike" nature compared to underlying production cost increases.

The magnitude and nature of these changes have led some observers to refer to a paradigm shift in agriculture away from decreasing real food prices over the past 30 years. A consequence could be that increasing food prices will cause an increase in agricultural production and an increase in the price of fertilizers as well (Food and Agriculture Organization of the United Nations 2008; Ott 2012).

Because it is expected that PR and fertilizer prices will remain volatile, this chapter examines the key factors that influence the price of phosphorus and mineralized fertilizer, and it considers possible consequences of potential increase in price volatility.

2 Global P and Fertilizer Trade

This chapter begins with an overview about the global trade of fertilizers and their basic commodities. The overview is based on the Global Fertilizer Trade map

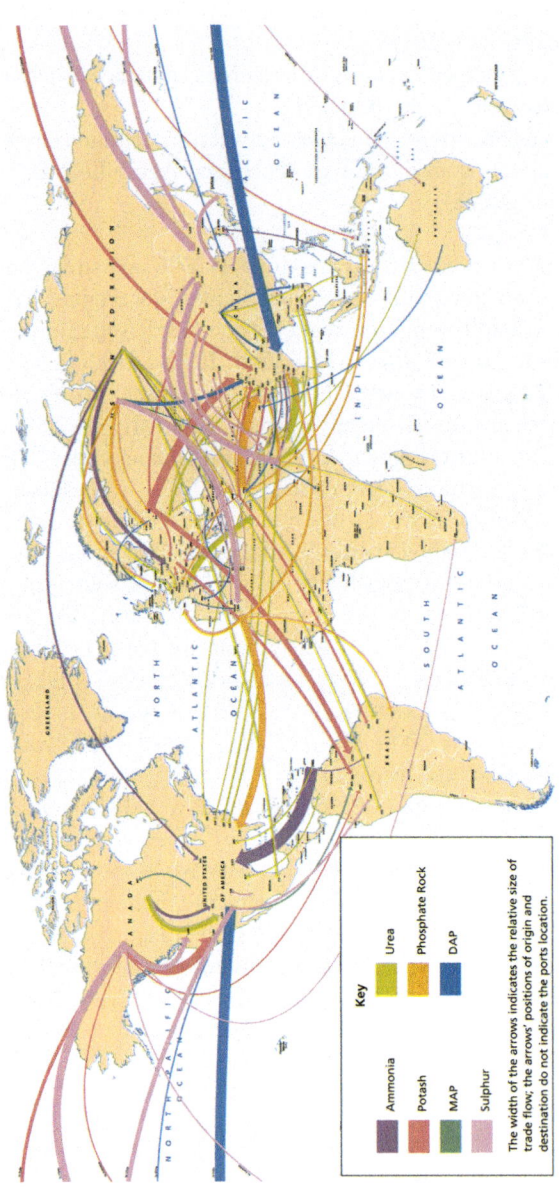

Fig. 1 Global fertilizer trade and production (*Source* ICIS) http://www.icis.com/resources/fertilizers/trade-flow-map-2014/

produced by ICIS in Fig. 1 (see www.icis.com). We will mainly concentrate on PR and DAP. As the top fertilizer-processing country, China contributes about 25 % to the world fertilizer production and consumption; the top three fertilizer consumers (China, United States, and India) accounted for 58 % of the world consumption in 2010 (Liu et al. 2008). African countries with the exception of South Africa do not import much fertilizer.

Phosphate rock. In 2010 according to ICIS, the biggest producers of PR, globally, are China, Morocco, the United States, Russia, Tunisia, and Jordan. Morocco is the largest exporter of PR in the world (Malingreau 2012). In 2010, it supplied 2,575 kt of phosphate rock (PR) to North America (USA and Canada), 2,090 kt PR to West Europe, 920 kt PR to India, 735 kt PR to Mexico, 635 kt PR to East and Central Europe, 555 kt to New Zealand, and 535 kt to Brazil. After China, Morocco was the second biggest producer of PR with 26 Mt PR in 2010.

China is the largest producing country of PR. It produced 69,100 kt in 2010, but only exported 655 kt to South Korea according to Fig. 1. The rest was consumed domestically. With 25,245 kt, the United States is the third biggest producer for PR in 2010, having been the second biggest in 2009. In addition to that it receives PR from Morocco. The second biggest exporter of PR is Jordan. In 2010, it provided 2,935 kt of PR to India and 685 kt to Indonesia. Globally, the biggest six PR producers are responsible for nearly 80 % of the global production.

In continental terms, North Africa is the biggest supplier of PR. However, the continent does only purchase little PR, DAP or other fertilizer products from other countries. Clearly, Sub-Saharan uses less fertilizers than the world average (Malingreau et al. 2012) though African countries belong to the biggest of PR.

India is the biggest importer of PR in the world. Its suppliers are Jordan, Morocco, Egypt, Togo, and Vietnam. The total amount imported was approximately 5,860 kt in 2010.

China is the largest producer of DAP while the United States is the largest exporter before China. The other main producers in 2010 were India, Morocco, Russia, and Tunisia. Global DAP production in 2010 was 31,438 kt according to Fig. 1 and 32,800 kt according to International Fertilizer Industry Association (IFA). The biggest six DAP producers are responsible for nearly 85 % of the global production.

In sum, the primary global producers and exporters of PR and DAP are China, the United States, and Morocco.

3 The Price of Phosphate and Fertilizer

This section discusses the dynamics of phosphate prices and prices of other commodities. All data are taken from publicly available sources. As valid for other markets, mainly supply and demand influences the price of fertilizers. The demand for fertilizers is mainly influenced by food, fiber, and biofuel production. Higher

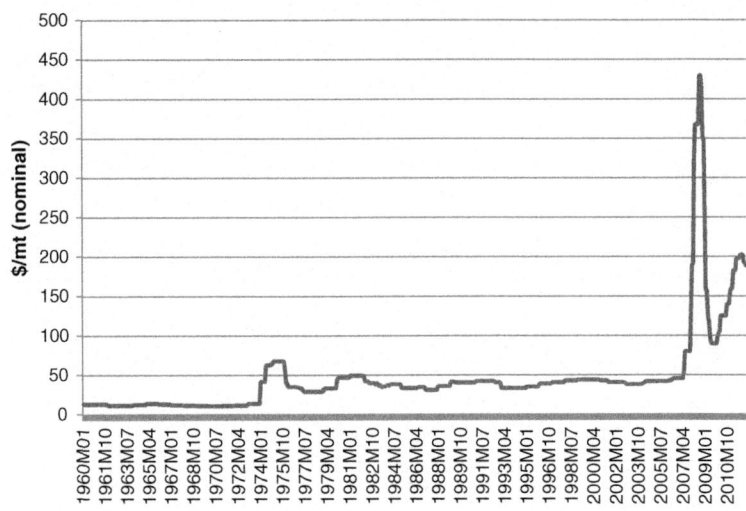

Fig. 2 Phosphate rock prices from January 1960 to April 2012 [*Source* World Bank (www.econ.worldbank.org)]

prices for these products influence the demand for fertilizers positively. On the other hand, supply factors influence the price of fertilizers. These are, among others, the material supply and prices, installed production capacity and its utilization, or trade and distributions costs.

After a long phase of relatively stable prices for PR (USD 30–50), by the end of 2008, the price peaked at USD 430 for RP and over USD 1,100 for P (see Fig. 2). The 2008 peak lasted only for a limited amount of time. Prices dropped sharply in late 2008, recovered slowly in 2009, increased through 2010 and the first 9 months of 2011. Prices fell in late 2011 and early 2012 and have recovered again recently (al Rawashdeh and Maxwell 2011). Simultaneously, world fertilizer prices doubled in 2007 and reached all-time highs in April 2008. Prices began dropping dramatically in October and November 2008 (Hargrove 2008).

Figure 2 clearly shows the price peak in 2007 after a very long period of relatively stable prices. Figure 3 shows the price index for fertilizer during the same time and presents nearly identical results in nominal terms. These results show that the prices for fertilizer and for PR correlate significantly ($r = 0.97$, $r^2 = 0.93$, $p < 0.00001$).

By using the annualized volatility of fertilizer prices based on real price data standardized around the year 2000 value, the data demonstrate that volatility has increased from 2008 on, but not to the 1973 level (Fig. 4).

This data raises two main questions about phosphorous and fertilizers:

1. Which factors influence the price of commodities such as phosphorous and fertilizers as basis for agriculture and thus for the global food production?

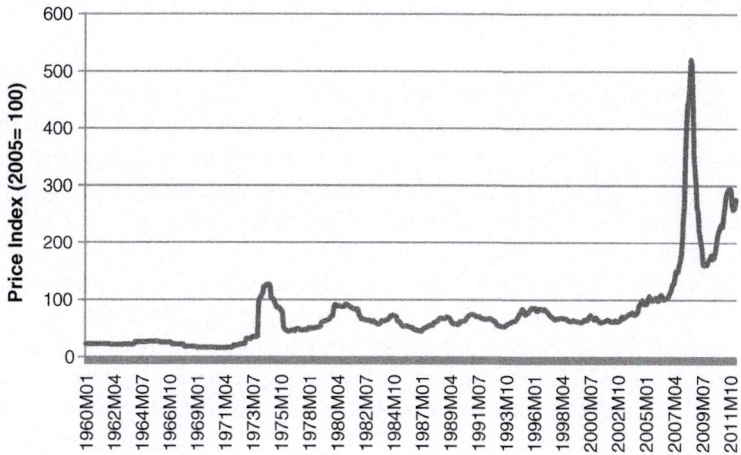

Fig. 3 Fertilizer price index from January 1960 to April 2012 [*Source* World Bank (www.econ.worldbank.org)]

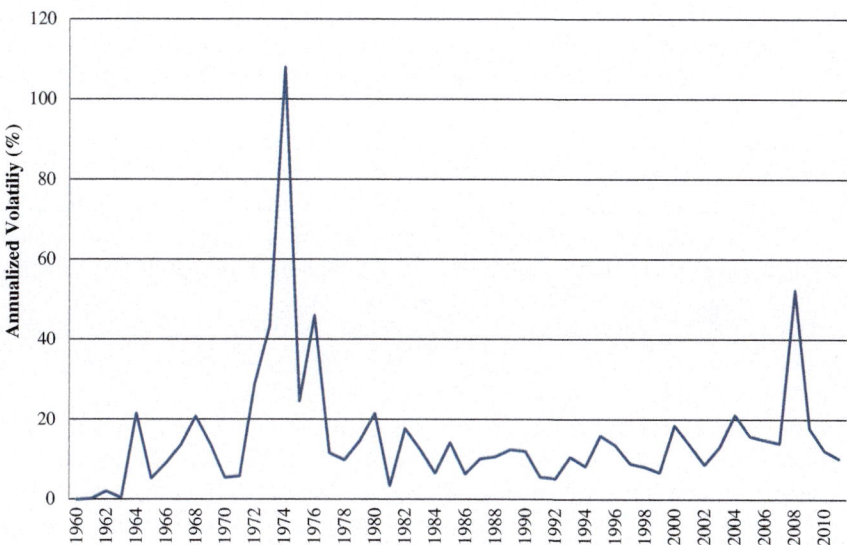

Fig. 4 Annualized volatility of real prices of fertilizer from 1960 to 2011 and standardized around 2000 (*Source* Calculation by European Commission/Agrilife Unit; data World Bank)

2. How does the price of the commodities influence its use and thus its impact on sustainable development?

In the following, we address these two questions by reviewing and summarizing relevant literature and data analyses.

4 Influences on Price and Trade

What influences fertilizer prices? A report from the United States Department of Agriculture mentions higher prices for raw materials, the inability of the US fertilizer industry to quickly adjust to surging demand or sharp declines in international supply, production cutbacks, and decreasing supplies from fertilizer imports are influencing fertilizer prices in the United States (Huang 2009).

The world economy shrank in 2007. However, countries and economies in transition continued their strong economic performance with only a mild reduction in 2007-growth rates, compared to previous years. For example, China and India sustained high though slightly decreasing growth through increasing South–South trade and financial linkages. However, developing countries could not avoid the effects of the global economic slowdown and the volatility of international commodities and the financial market (Food and Agriculture Organization of the United Nations 2008).

Global growth is seen as remaining sufficiently robust to sustain the demand for food in emerging economies thereby strengthening demand for fertilizers (Food and Agriculture Organization of the United Nations 2008; Malingreau et al. 2012).

During 2007 and 2008, a rapid increase in fertilizer prices appeared, followed by lower prices in late 2008 (Huang et al. 2009). Explanation for this price volatility could relate to an imbalance between supply and demand (Hargrove 2008) as well as the economic downturn.

Huan et al. (2009) explain the record fertilizer prices in early 2008 by strong United States and global demands for fertilizers in conjunction with low fertilizer inventories, and the unwillingness of the US fertilizer industry to adjust production levels because of costs. Furthermore, high export tariffs imposed by China caused less urea and DAP availability in the market. Eventually, fertilizer demand reached a level that supply could not match (Hargrove 2008).

The underlying causes of fertilizer price volatility relate to supply and demand factors and the structure of the fertilizer industry. US total fertilizer production capacity declined 42 % between 2000 and 2008 (Kenkel 2010). As fertilizer demand increased in 2007, the US nitrogen inventory fell by 15 % to 0.88 Mt by the end of 2007. The US phosphate inventories fell 27 % to 0.59 Mt in late 2007. Potash inventories in North America (including Canada) dropped by 1 Mt (49 %) to 0.9 Mt at the end of 2007. Domestic and foreign fertilizer producers were not able to quickly adjust their production as inventories dwindled (Huang et al. 2009) because in the short-run production is determined by installed capacity and plants need to operate above 80 % capacity and most run at close to 100 % of design capacity. In the long run, there is a 3–5-year lag from decision to production for new capacity. Additionally, the long supply lines in international fertilizer trade impede quick replenishment of stocks when demand increases unexpectedly. In the US market, the reliance on imports for nitrogen supply has caused periodic shortages since supply exceeded 50 % from imports. Domestic supplies of phosphate fertilizers and imports of potash from Canada are rarely affected by supply line.

Total fertilizer prices continued to increase in early 2008 and were 26 % higher in August than in April. But by late 2008, monthly average prices had fallen, particularly for nitrogen fertilizer (Huang et al. 2009). This decrease in prices was explained by the unwillingness of farmers to pay much higher prices than in early 2007, by the collapse of the global credit market and by the economic slowdown. Thus, demand decreased significantly and caused a significant price decrease (Hargrove 2008).

Global fertilizer demand weakened in response to the record-high fertilizer prices and declining crop prices. According to Huang et al. (2009), the decline in fertilizer prices might be attributed to several factors:

- Weak global fertilizer demand in reaction to the fertilizer price surge and declining crop prices
- A shortened window for US application of fertilizer in fall 2008, caused by wet weather that delayed spring plantings and fall crop harvests,
- An increase in fertilizer supplies from overseas to the United States
- Tighter credit availability, making debt-financed fertilizer purchases more difficult
- Congested distribution supply chains due to farmers postponing purchases.

The increasing demand for agricultural commodities in developing countries put pressure on land resources to create higher yields, which keep the demand of fertilizers high as well (Terazono and Farchy 2012). While demand for basic food crops, high value crops, animal products, and crops capable of being used to produce biofuels is likely to remain strong, it is expected that increased fertilizer consumption required to support higher level of production will be adequately catered for by growing supply worldwide in 2011 and 2012 (Food and Agriculture Organization of the United Nations 2008).

At the global level, it is anticipated that—with new capacity coming on stream—in coming years there will be ample supply of all three major fertilizer nutrients with surpluses of nitrogen, phosphate, and potash. Africa remains a major phosphate exporter and will increase nitrogen exports but will still be a minor player in nitrogen exports compared to the Middle East while importing all of its potash. America will be the net importer of nitrogen, and a primary supplier of potash. Asia is expected to produce a rapidly increasing surplus of nitrogen and continue to import phosphate and potash.

According to Terazono and Farchy (2012), global fertilizer consumption is positively influenced by high agriculture commodity prices. This corresponds to FAO's claim that the high commodity prices experienced over the recent years led to an increased production and correspondingly greater fertilizer consumption, as reflected in high demands and higher fertilizer prices at the beginning of 2007 (Food and Agriculture Organization of the United Nations 2008; Piesse and Thirtle 2009).

Between January 2007 and mid-2008, corn prices increased by 100 %, wheat prices rose by 83 %, and soybean prices were up 112 %. At the same time, growth in worldwide biofuel production diversified the use of grains, sugarcane, soybeans, and rapeseed and contributed to higher prices for biofuel feedstock, particularly corn (Huang et al. 2009). Thus, high agricultural commodity prices encouraged producers to expand total crop acres, adjust the mix of crops planted, and increase fertilizer use to increase yields, all of which led to an increased global fertilizer demand (Huang et al. 2009).

Some groups believe that speculation from the financial sector influenced commodity prices in the period 2007–2008 (Terazono and Farchy 2012). The growing speculative investment in raw materials by financial market players and especially the strong increase in derivatives and over-the-counter trades (Maslakovic 2011) should be taken into account as influencing prices. After the collapse of the mortgage market in the United States, institutional investors and others looked for less riskier places to invest their money. To the extent that food prices have risen, they directed their capital to the futures market pushing the price of grains upwards (Piesse and Thirtle 2009; Oxfam 2011). However, the direction of influence of the financial markets on commodity prices is still unclear (Domanski and Heath 2007) and some even see benefits of the participation of financial market players (Sanders et al. 2008). Because of this influence, commodity markets have become similar to financial markets. They probably will be more liquid but prices will be harder to predict.

As of 2012, some analysts estimate that a significant part of financial investment in the commodity sector is based on financial instruments that are not connected with the real use of the commodities but are driven by the financial industry (Gilbert 2008; Abbott et al. 2011). As institutional investors invest mostly into commodity funds, prices of commodities could be influenced without any connection to supply and demand. However, because many commodity indices and financial products based on these indices do not have fertilizers or phosphorous in their portfolio, the influence of the financial sector on the price of these commodities is probably low. Furthermore, there is no long-term data available to test the influence of financial market speculation on fertilizer prices (Ott 2012).

Even though prices for other commodities also appear to correlate well with each other, this may not be explained by supply and demand. Instead, commodity index-based financial products could be one of the reasons for this correlation. Figure 5 shows the relation between the prices of commodities used for fertilizer, other commodities, and the food price index between 1981 and 2011. All values are standardized and thus directly comparable. Figure 5 shows that the peak of PR is the highest peak among all other depicted commodity indices and commodities, and that towards the end of 2011, its price is increasing more strongly again than that of all other commodities. Furthermore, Fig. 5 suggests that the prices for all analyzed commodities correlate significantly and could be influenced by external factors.

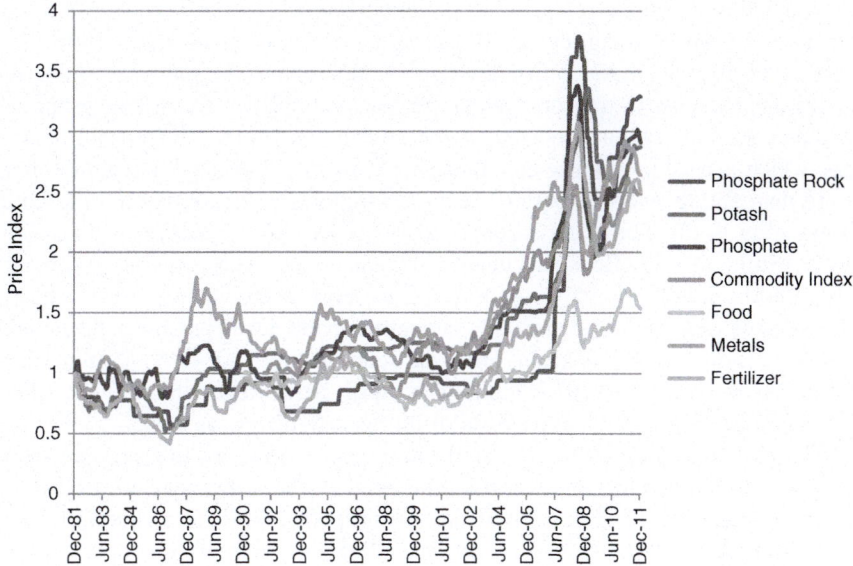

Fig. 5 Prices for commodities used for fertilizer, other commodities, and the food price index

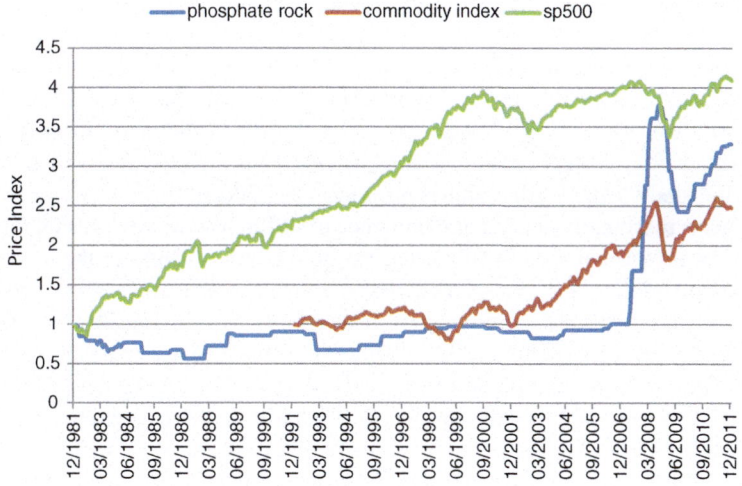

Fig. 6 Development of PR, commodity prices, and the stock market

Furthermore, it is interesting that the strong increase in commodity prices is paralleled by a decrease in the equity index S&P 500 at the beginning of the financial crisis (Fig. 6).

High oil prices contributed to price increases for most agricultural crops by raising input costs on the one hand, and by boosting demand for agricultural crops used as feedstock in the production of alternative energy sources (biofuel) on the other hand (Food and Agriculture Organization of the United Nations 2008).

The oil price increase, which doubled in 2007 and 2008, influences fertilizer prices and the cost of transport related to the food system. Rising energy prices also increased the cost of producing and delivering fertilizers. The price of natural gas, which is used to produce ammonia (the basis for all nitrogen fertilizers), rose more than 550 % over the past 10 years. Natural gas is the second most volatile commodity in the world and accounts for 90 % of nitrogen fertilizer production costs (Smith 2010). Between June 2007 and June 2008, natural gas prices increased by more than 65 %. As a result, the cost to produce nitrogen fertilizer increased as well (Huang et al. 2009). However, recent developments in the natural gas price show a decrease again. From the spike in July 2008 of $172.56/mmbtu, the price decreased to $102.38/mmbtu in April 2012 (World Bank 2012). This should influence the production costs of nitrogen fertilizers as well.

The cost of transporting fertilizers from the US Gulf Coast to crop production regions within the United States is also influenced by energy prices. During the three years ending in January 2008, US rail rates to transport ammonia from the Gulf Coast to the Mid-west increased by 63 %, and an additional 44 % fuel surcharge was added to US rail transport costs in July 2008 because of high fuel prices (Kenkel 2010).

Prices of PR, sulfur, and ammonia—raw input materials used to produce DAP and other fertilizers—increased between January 2007 and early 2008 (Huang et al. 2009). Phosphate prices were influenced by a strong increase in demand and prices for sulfur, vital for producing the popular DAP and other high-analysis phosphate fertilizers. Supply of quality PR also became tight (Hargrove 2008). Moroccan PR contract prices tripled during 2007, international contract prices of sulfur increased by more than 170 %, and Tampa prices of ammonia doubled (Kenkel 2010). However, as Fig. 7 suggests, production costs for phosphorous rock were relatively stable for many years while purchasing prices vary much more.

To more thoroughly analyze the influence of different raw materials on fertilizer prices, we performed a multivariate regression analysis to analyze the association between the price for PR, phosphate, DAP, urea, and potash and fertilizer price. We split the data into two phases, phase 1 from December 1981 to May 2007 and phase 2 from June 2007 to December 2011. While the second phase represents times of turmoil, prices in the first phase were relatively stable. A breakpoint test that was conducted to test whether the two phases were different was significant. Overall (combining phase 1 and 2), the multivariate regression revealed a highly significant association between raw materials and fertilizer price ($p_{\text{phase 1,2}} < 0.00001$). However, we observed differences in these associations by time period (see Table 1).

While in phase 1 all commodities but potash were positively and significantly associated with fertilizer price, in phase 2, the association between DAP and fertilizer price became negative and urea and PR were no longer significantly associated with fertilizer prices. Further, the association between phosphate and fertilizer price appeared much stronger in period 2 ($\beta_{\text{phase1}} = 0.10$, vs. $\beta_{\text{phase2}} = 0.76$). Taken as a whole, these data suggest that, while the relation between the price for basic fertilizer commodities and fertilizer appears less

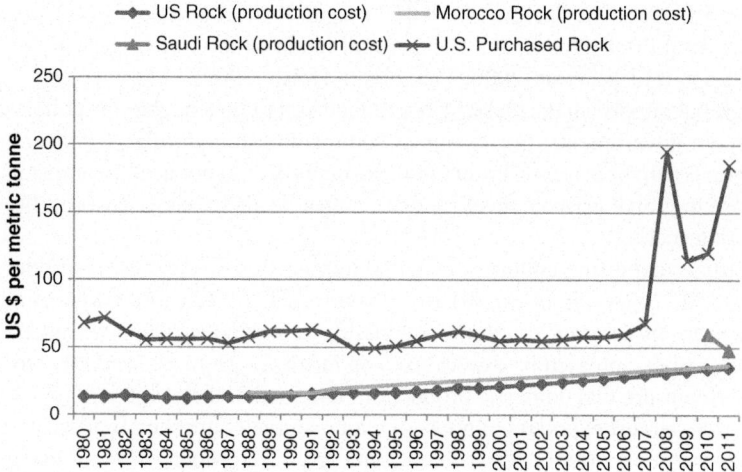

Fig. 7 Production costs of phosphate rock in different world regions and price of purchase rock (*Source* FERTECON Research Center)

Table 1 Regression coefficients and *p*-values from Wald statistic to study association between raw materials and fertilizer prices in two different time periods

Variable	Phase 1 (1981–2007)		Phase 2 (2007–2011)	
	β-estimate	p value	β-estimate	p-value
Phosphate rock	0.264	<0.00001	0.116	0.197
DAP	0.136	<0.00001	−0.320	0.039
Potash	0.028	0.059	0.299	0.001
Phosphate	0.099	<0.00001	0.756	<0.00001
Urea	0.29	<0.00001	0.034	0.609
Constant	0.081	<0.00001	−0.050	0.79

significant in recent years (2007–2011) compared to a time period where prices overall remained relatively stable (1981–2007), particularly the association between PR and fertilizer price has strengthened after the economic downturn in 2007.

Other important factors in determining fertilizer prices are the value of the US dollar and China's imposition of high tariffs on fertilizer exports (Hargrove 2008). World trade has expanded rapidly since 2005, largely driven by trade increases of both oil and non-oil commodities as well as capital goods (Food and Agriculture Organization of the United Nations 2008).

In 2007, imports accounted for 49 % of the nitrogen fertilizer supply in the United States and 85 % of the US potash supply. The value of the US dollar relative to currencies of other major nations supplying fertilizer to US farmers, except Mexico and Trinidad and Tobago, has declined since 2003. For example, relative to the Brazilian national currency, the Real, the US dollar dropped 48 % from January 2003 to January 2007 (Huang et al. 2009). Similarly, in 2008, the Moroccan national currency, the Dirham, also achieved an all-time high compared to the US Dollar. The decline of the dollar against most other currencies has made imports from the United States cheaper and lessened the true impact of the rise in world price (Huang et al. 2009; Piesse and Thirtle 2009).

A global supply chain makes US producers also dependable on policies made by foreign governments. For example, some countries subsidize fertilizers or cap domestic fertilizer prices. Farmers in these countries do not need to adjust usage in response to global price changes, which, in turn, leads to increased volatility in the remaining market.

Fertilizer-exporting countries can also impose tariffs or export restrictions. For example, in 2008, China imposed an export tariff of 185 % on MAP and DAP products to ensure that domestic production remained in China (Kenkel 2010). Overall, China increased its export taxes on fertilizers from 35 % in 2007 to 135 % in 2008 to ensure that domestic production remained in the country. China is one of the world's largest exporters of urea—a major source of nitrogen fertilizer—and the second largest exporter of phosphate. China provided roughly 17 % of the urea and 18 % of the phosphate traded globally in 2007. Despite export restrictions and taxes, China's phosphate fertilizer exports increased by 63 % from 2007 to 2011. In 2011, it supplied 25 % of the phosphate fertilizer traded globally.

Freight rates have become a more important factor in agricultural markets than in the past because of rising transportation costs (Food and Agriculture Organization of the United Nations 2008). Given that in 2007, 58 Mt of fertilizers were shipped to US agricultural producers (Huang et al. 2009), the influence of such cost increases is significant.

Transportation costs account for more than 20 % of the cost of ammonia shipped from Trinidad and Tobago to the US Gulf Coast, and for more than 50 % of the cost of ammonia sourced from Russia to the US Gulf Coast (Kenkel 2010). In addition, the cost of transporting fertilizers from the US Gulf Coast to farmers throughout the Midwest rose dramatically. Specifically, between 2005 and 2008, US rail rates to transport ammonia increased by 63 %, and an additional 44 % fuel surcharge was added to US rail transport costs in July 2008 because of high fuel prices (Huang et al. 2009). Because logistics have a significant influence of fertilizer prices in Africa as well, they could increase the fertilizer price on this continent.

Increased shipping costs influenced the geographical pattern of trade as countries choose to source their import purchases from nearer suppliers to save on transport costs. The impact of transportation costs on fertilizer prices will grow as fertilizers are being produced in fewer localities, which tend to be close to raw materials and ample energy availability (Food and Agriculture Organization of the United Nations 2008). An exception is Saudi Arabia that provides a large low-cost supply source to the world's largest importer, India.

The world population is currently increasing at a rate of 75 million per year creating additional food demand (Kenkel 2010). The strong global growth in average incomes, particularly in developing countries, has increased food and animal feed demand as well (Huang et al. 2009). The continuing population growth contributes to food demand as well. It is anticipated that 50–70 million people will be added annually to the world population until the mid 2030s. Almost all of this increase is expected to take place in developing countries. More food and fiber will be required to feed and clothe these additional people (Food and Agriculture Organization of the United Nations 2008). Over the long run, population and income growth will continue to put upward pressure on demand for fertilizers (Kenkel 2010).

The value of total agricultural output has almost tripled in real terms since 1961. Latin America and South Africa have seen a small increase, while East Asia and the Pacific have more than doubled the agricultural value added per capita over the last four decades (Food and Agriculture Organization of the United Nations 2008). Furthermore, consumers in developing countries not only increased the consumption of staple foods but also diversify diets to include more meats, dairy products, and vegetable oils. This, in turn, amplified rising demand for the feed grains and oilseeds used to produce these foods (Huang et al. 2009).

For example, fruit and vegetable production is increasing rapidly particularly in China where there is a large availability of cheap labor for labor-intensive crops. The area planted to fruit and vegetables in China is currently equivalent to almost 40 % of the area cropped to cereals. With average fertilizer application rates for fruit and vegetables being about double those of cereals, it is estimated that fruits and vegetables are responsible for about half of the increase in fertilizer demand (Food and Agriculture Organization of the United Nations 2008).

Total meat production more than quintupled in developing countries, from 27 to 147 million tons between 1970 and 2005. Although its growth pace has slowed down, global meat demand is expected to increase by more than 50 % by 2030 (Food and Agriculture Organization of the United Nations 2008). Increased meat and aquaculture production will require more feed. Conversion of grain areas to vegetable and fruit production will translate into higher fertilizer demand as average application rates for the latter are about double as those for grain crops (Food and Agriculture Organization of the United Nations 2008). The above trends support continuing and increasing demand for mineral fertilizers to restore and enhance the fertility of the world's agricultural land for higher yields and improved produce quality (Food and Agriculture Organization of the United Nations 2008).

Increasing oil prices, which more than doubled by mid-2008, compared to 2006 led to increasing investment in the production of alternative fuels such as those of plant origin. Governments in the United States, the European Union, Brazil, and other countries have subsidized production of agro-fuels in response to the scarcity of oil and global warming. Biofuels are being promoted as contributing to a wide range of policy objectives, rather than based on market fundamentals such as oil and feedstock prices. The predicted impact of increased biofuel production on world fertilizer demand is expressed in two ways, as percentage age of fertilizer consumed by bioenergy crops and, in absolute terms, as total fertilizer used for feedstock production (Food and Agriculture Organization of the United Nations 2008).

On the one hand, increased demand for fertilizers to produce biofuels in the United States, Brazil, and Europe diminished fertilizer stocks. On the other hand, increased livestock production created increasing demand for grain and thus for fertilizers. Grain reserves became historically low and prices rose sharply (Hargrove 2008).

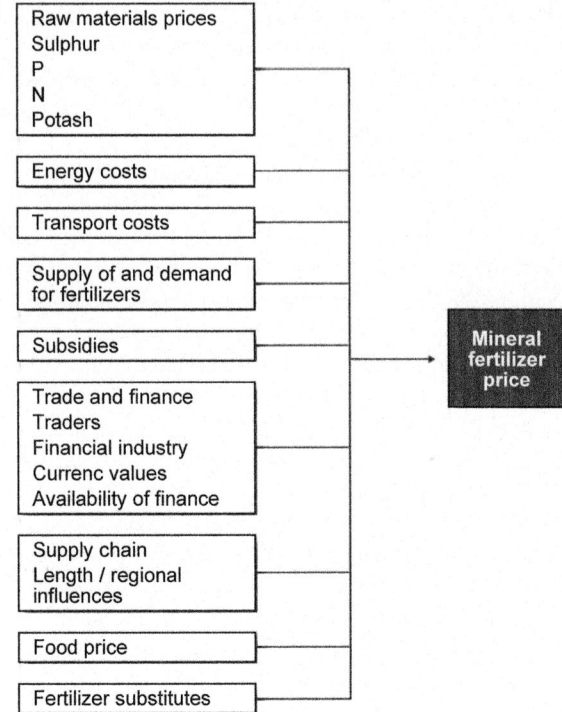

Fig. 8 Factors influencing the price of mineral fertilizers

Green fuel production can lead to food shortages and increasing international food prices, which will encourage farmers to expand planted acreage used for biofuels (Chen et al. 2010). To give just one example, in 2007 in the United States 20 % of the total cereal harvest was used to produce ethanol and it is estimated that this figure will reach 33 % in the next decade. This has increased the derived demand for global fertilizers and increased fertilizer prices (Chen et al. 2010).

5 Draft Model

Based on literature research, data analyses, and workshop discussions, we constructed a relatively simple draft model of factors that potentially influence the price of mineral fertilizers. The model will be refined and further developed based on further analyses and inputs.

A summary of preliminary variables that feed into the model is presented in Fig. 8.

This work will be completed in a series of interviews and meetings, followed by efforts to develop a more refined model.

During the research process, we will also explore how systems-based models can be applied to gain insight into key decisions by relevant groups that can shape the resilience and performance of the worldwide phosphate industry.

6 Work in Global TraPs

Trade and finance have never been subject of transdisciplinary processes related to sustainability before, at least not in the field of commodity trade. Reasons, among others, could be the potential for price-rigging or the competitiveness in the sector. Therefore, any collaboration among different business agents on questions of sustainable transitioning of the phosphorus trade and finance has to be reflected from the perspective of national or international antitrust authorities (Tarullo 2000). Because of this, the basic imperative of the Global TraPs project has to be applied and questions from the precompetitive domain can be dealt with. Further, with a very few exceptions in finance, sustainable entrepreneurship, sustainability in general, or environmental protection is not in the basic frame of the business goals of trading and financial agents (Dean and McMullen 2007). However, as we have shown in this chapter, understanding the phosphorus market and the impact certain regulations, actions, programs, etc. may have, e.g., for getting access for phosphate fertilizer, is important.

Against this background, it is remarkable that one of the big international phosphorus traders and the IFA agreed to participate in the Global TraPs project. In a first round of conversation, three basic questions emerged: (1) How does the Global P market function? (2) What are the features of (un-)sustainable P trade? Are there any? (3) What role may micro-finance systems play?

6.1 Knowledge Gaps and Critical Research Questions

Based on discussions in four Global TraPs workshops as well as the analysis presented in this chapter, the following critical questions will be looked at:

How does the global fertilizer market function? What types of actors are involved? How is the market structured compared to other markets like gas, energy or other markets? How does the financial sector influence the market?

How do different market players influence sustainable phosphate use and what role can they play in fostering sustainable phosphate use? To answer this question, a definition of sustainable phosphate use needs to be developed. Based on this definition, research on current phosphate use and trade and on how to foster sustainable phosphate use should be conducted. Specifically, tools could be developed that support sustainable phosphate use.

Which factors influence the price of commodities used for fertilizer such as PR and the food price? The chain between PR and food has to be analyzed and to be understood. Further, specific analyses could be done on the volatility of prices and interactions with other markets and with different market players. Thus, for instance, case studies that compare regulated and non-regulated fertilizer markets in different regions could be conducted.

How can negative impacts on sustainable phosphate use, resulting from phosphate prices, be avoided? Our results suggest that the volatility of phosphate price negatively impacts sustainable phosphate use. Prices are only partly dependent on supply and demand but are influenced by speculation and general financial market influences. Proposals should be developed to avoid these negative impacts on sustainable phosphate use.

What instruments of micro-finance can be developed to support the sustainable use of phosphate and fertilizers for farmers in developing countries? Existing instruments like microfinance and subsidiaries will be analyzed with regard to their usability to guarantee a sustainable food production.

6.2 Role, Function, and Kind of Transdisciplinary Case Process

As mentioned above, transdisciplinarity meets new frontiers when approaching the field of sustainable trade and finance. There is a multitude of stakeholders from different fields (from international mining companies and traders to local smallholder farmers, traders, and financial institutions) and scales which are involved in the P supply–demand chain. The inclusion of certain stakeholders in the discussion of the research questions and in the case studies depends on the topic/question. Many of these stakeholders have different views and a different understanding of sustainable P use. Some may be interested in increasing prices, some may be interested in decreasing prices. Some concentrate on the positive impact of fertilizer use for the food supply in developing countries, some concentrate on negative environmental impacts of fertilizer overuse. Whereas some P produces or P users may prefer stable prices, others may be interested in volatility.

From a classical economics perspective, markets are efficient without regulations. On the contrary, we have to acknowledge that presumably no industry receives as many regulations (including subsidies) as agronomy. Thus, one part of mutual learning of the transdisciplinary project will be to describe and understand different "models of P trading" from the perspectives of different key actors (traders, producers, users, farmer organizations, etc.). Based on this, criteria of sustainable P trade and finance may be developed. Though, this is not a typical case study as it does not deal with a specific region, process, we list it under the list of case studies.

6.3 Suggested Case Studies

Phosphorus Trading (T&F Case Study 01). P is taken as a case for a commodity. Key experts from the P-trading business (managers and consultants of trading companies, mining companies, etc.), scientists from the field of sustainable resources management, sustainable finance, and business identify key actors, market mechanisms, regulations and constraints of the P market. This case study will be led by key actors from the P trade and finance market and science.

The vehicles of mutual learning are small (non-public) expert workshops with key consultants of key stakeholders which may provide a description of the global

P market and relations/mechanisms that may affect its functioning. On this basis, scenarios of markets may be constructed with the help of Master theses.

Fertilizer Subsidies (T&F Case Study 02): Access to fertilizer by farmers in different markets. In order to analyze sustainable fertilizer use, a multiple-case study (including at least two studies) should be run to compare markets in different regions. Influences like subsidies, the availability of finance, cost structures, and the supply chain could be analyzed. Furthermore, models of cooperation between financing institutions such as microfinance organizations and fertilizer producers and traders could be explored in order to guarantee the fertilizer supply in developing countries for increasing their food production. As the map in the beginning of this chapter demonstrates, getting access to mineral fertilizer by African farmers seems to be difficult. Therefore, an African country should provide one case. On the other hand, India is a big fertilizer consumer and uses subsidies and would be an ideal candidate to be integrated into the case study to conduct a comparative study.

Fertilizer Subsidies (T&F Case Study 03): Phosphate mining in different world regions. This (multiple) case study compares the financial flows and mechanisms in phosphate mining in North Africa and in the United States. It will analyze the influence of financial capital and regulations on phosphate mining, how and why the industries were developed and what impact long-term versus short-term strategies have on mining. The goal of this study is to analyze the sustainability of current business models and to develop models for a sustainable phosphate supply.

Acknowledgments We thank D. Ian Gregory, Melih Keyman, and Roland W. Scholz for important comments on earlier drafts of the paper.

References

Abbott PC, Hurt C et al (2011) What's driving food prices in 2011? Farm Foundation, Oak Brook, IL, p 50
al Rawashdeh R, Maxwell P (2011) The evolution and prospects of the phosphate industry. Mine Econ 24(1):15–27
Chen PY, Chang CL et al (2010) Modeling the volatility in global fertilizer prices. http://hdl.handle.net/1765/20377
Dean TJ, McMullen JS (2007) Toward a theory of sustainable entrepreneurship: reducing environmental degradation through entrepreneurial action. J Bus Ventur 22(1):50–76
Domanski D, Heath A (2007) Financial investors and commodity markets. BIS Q Rev 53–67
Food and Agriculture Organization of the United Nations (FAO) (2008) Current world fertilizer trends and outlook to 2011/12. FAO, Rome
Gilbert CL (2008) How to understand high food prices. Discussion paper, University of Trento, Trento, Italy
Hargrove T (2008) World fertilizer prices drop dramatically after soaring to all-time highs. http://www.eurekalert.org/pub_releases/2008-12/i-wfp121608.php. Cited 2 Feb 2012
Huang W-Y (2009) Factors contributing to the recent increase in U.S. fertilizer prices, 2002–2008. United States Department of Agriculture, Washington
Huang W-Y, McBride W et al (2009) Recent volatility in US fertilizer prices. Amber Waves 7(1):28–31

Kenkel P (2010) Causes of fertilizer price volatility. Oklahoma State University, Oklahoma, p 2

Liu Y, Villalba G et al (2008) Global phosphorus flows and environmental impacts from a consumption perspective. J Ind Ecol 12(2):229–247

Malingreau J-P, Hugh E et al (2012) NPK: Will there be enough plant nutrients to feed a world of 9 billion in 2050? Foresight and horizon scanning series. European Union, Brussels, p 32

Maslakovic M (2011) Commodities trading. TheCityUK, London

Oehmke JF, Sparling B et al (2008) Fertilizer price volatility, risk, and risk management strategies final report. Canadian Fertilizer Institute, George Morris Centre, Guelph, ON, p 19

Ott H (2012) Fertilizer markets and its interplay with commodity and food prices. Joint Research Center Technical Report. European Union, Sevilla

Oxfam (2011) Not a game: speculation vs food security. Oxfam issue briefing. Oxfam International, Oxford

Piesse J, Thirtle C (2009) Three bubbles and a panic: an explanatory review of recent food commodity price events. Food Policy 34(2):119–129

Sanders DR, Irwin SH et al (2008) The adequacy of speculation in agricultural futures markets: too much of a good thing? Marketing and outlook research report, vol 2. Urbana-Champaign, Department of Agricultural and Consumer Economics, University of Illinois at Urbana-Champaign

Saravia-Matus S, Gomez y Paloma S et al (2012) Economics of food security: selected issues. Bio-based Appl Econ 1(1):65–80

Smith J (2010) Fertilizer price volatility likely to remain. Southwest Farm Press

Tarullo DK (2000) Norms and institutions in global competition policy. Am J Int Law 94(3):478–504

Terazono E, Farchy J (2012) Fertilizer sales soar as farm product prices surge. Financial Times, London

World Bank (2012) GEM commodities (Pink Sheet). http://data.worldbank.org/data-catalog/commodity-price-data. Cited Jul 2012

Appendix: Spotlight 10

Phosphorus and Food Security from a Greenpeace and Indian Smallholder Farmer View

Reyes Tirado and Vijoo Krishnan

Phosphorus is an essential element for all living things, needed (in the form of phosphate, PO_4) in cells for construction and renewal of DNA and RNA, of phospholipids, and of energy transduction molecules such as ATP. In vertebrates, PO_4 is a main component of the mineral apatite in bones. Thus, human health depends on an adequate dietary supply of P (Dietary Reference Intake (USA): 580 mg for an adult, 1,055 mg for children and youth). The body tightly regulates P homeostasis, primarily by modulating PO_4 excretion in the kidney, closely in concert with calcium (Ca) due to their joint role in bone formation.

Mining Phosphorus, at What Price?

The only industrial source of phosphorus, apart from recycling organic forms of it, currently comes from mines in a handful of countries. Mined phosphorus is by far the main source of phosphate fertilizer used worldwide. Vulnerability to trade speculation, influence of fossil fuel price volatility and increasing demand in emerging economies: all these factors put pressure on mineral phosphorus prices. In 2008, international PR prices increased by 800 %. International prices went down quickly, but never to the pre-peak values: prices are now in 2012 about four times higher than they were before 2006. This volatility makes phosphate import-dependent countries, and low-income farmers more vulnerable and financially insecure. Affordable and sustainable phosphate access is an imperative to ensure food security of nations and livelihood security of the small and marginal farmers.

R. Tirado
Greenpeace Research Laboratories, University of Exeter, Exeter,
EX4 4RN, 01392 247920, UK
e-mail: reyes.tirado@greenpeace.org

V. Krishnan
All India Kisan Sabha (AIKS), 4 Ashoka Road, New Delhi 110001, India
e-mail: vijookrishnan@gmail.com

Phosphate Rock Environmental Hazards

Some PRs contain low levels of radionuclides, including uranium. When these PRs are processed for eventual conversion to phosphate fertilizers, a majority of the radionuclides and other contaminants are concentrated in the phosphogypsum by-product. The resulting phosphogypsum stockpiles present a serious environmental problem, with potential local hazard for human health and pollution of the groundwater. In addition, some phosphate fertilisers contain small amounts of the heavy metal cadmium. Because cadmium is highly toxic to humans, there are concerns about its accumulation in agriculture soils and transfer through the food chain.

A Very Leaky Phosphorus Flow

On a worldwide scale, we are mining five times the amount of phosphorus that humans are consuming in food. Overall, if we simply relate human intake to the agroinputs, about 90 % of the P entering the system is lost into the environment. For example in agriculture, only between 15 and 30 % of the applied P fertilizer in farmlands is actually taken up by annually harvested crops with the rest remaining in the soil. If soil erosion is an issue, the phosphorus lost ends up in water systems causing widespread pollution in lakes, rivers and coastal areas, algal blooms and dead zones in the oceans (together with nitrogen). Thus, ironically phosphorus represents both a scarce non-renewable resource for living beings and a pollutant for living systems.

Phosphorus Applied to Soils Ends up in Water

As noted above, when phosphorus fertilisers are applied, only a small proportion of it is immediately available to plants, the rest is stored in soils in varying degree of availability. It is common for commercial farmers to apply phosphorus as well as other nutrients in excess so not to limit yield or in the case of phosphorus to try to compensate for a specific soil condition such as cool temperatures. However, this increases the risk of significant phosphorus losses if run-off, leaching or soil erosion events occur with excess P ending up in lakes, rivers and oceans. This results in financial losses for the farmer and environmental damage to the soil and surface waters. Ameliorative measures need to be developed to rectify this anomaly.

Too much of a Good Thing

Excess nutrients in water systems, eutrophication, is a major and common problem worldwide driven mostly by overuse of phosphorus and nitrogen fertilisers. Excess phosphorus and nitrogen are causing widespread damage to the Earth systems, especially water systems.

Global studies of phosphorus imbalances found that phosphorus deficits covered 29 % of the global cropland area, while 71 % of the area had overall phosphorus surpluses (MacDonald et al. 2011). On average, developing countries had phosphorus deficits during the mid twentieth century, but current phosphorus fertiliser use may be contributing to soil phosphorus accumulation in some rapidly developing areas, like China, together with relatively low phosphorus use efficiency. Even the idea of most African soils being phosphorus depleted is contested by new analysis; there are vast areas where phosphorus excesses are more common although inefficiently used for food production (MacDonald et al. 2011).

Solutions for a Broken Phosphorus Cycle

Ensuring phosphorus remains available for food production by future generations and preventing pollution of water systems is possible by working towards restoring the phosphorus cycle. This requires strong actions in two main areas: reducing phosphorus losses, especially from agricultural lands, and increasing phosphorus recovery and reuse to agricultural lands from all sources, including livestock wastes, food waste and human excreta. Closing the broken phosphorus cycle should follow two main drivers:

1. Stop or minimise losses, by increasing efficiency in the use of phosphorus, mostly on arable land and the food chain. Additionally, sustainable phosphorus use will benefit from shifting to plant-rich diets that are more efficient users of phosphorus (and other resources) than meat-rich diets, and from minimizing food waste.
2. Maximise recovery and reuse of phosphorus, mostly of animal and human excreta, and thus minimise the need for mined phosphorus.

Organic fertilisers, when locally available are generally cheap, and make farming more secure and less vulnerable to the problems of external inputs' access and price fluctuation, should be promoted as part of an integrated soil fertility management. Natural nutrient cycling and nitrogen fixation can contribute to soil fertility, and at the same time reduce farmers' expenses on inputs and provide a healthier soil, rich in organic matter, better able to hold water and less prone to erosion. Conscious efforts to improve efficient nutrient use and to encourage judicious use of ecological methods of nutrient fixation must be made.

Evidence shows that farming without synthetic fertilisers can still produce enough food for all. This is especially true if we consider a vision aimed at farming with biodiversity, closing nutrient cycles, recycling nutrients from non-conventional sources (sewage, food waste, etc.) and with more sustainable diets. Many scientists, institutions like FAO, UNEP and farmers associations are documenting remarkable success from ecological farming in achieving high yields and fighting poverty in low-income regions.

Real Holistic Solutions

To bring about real effective change, there is a need for convergence in agriculture, livelihoods, energy and sanitation initiatives, especially in low-income regions where agriculture productivity is low and farmer capital and infrastructure are lacking. We suggest two guiding principles for potential alternatives to future sustainable phosphorus use:

A Holistic Approach to Address Rural Livelihoods and Agriculture Issues

Work in sectorial isolated silos will not produce the much-needed effective changes. A people-centred, multi-institutional and transdisciplinary approach will be required. With regards to phosphorus in rural areas, this might mean integrating energy needs, eco-sanitation and fertilisers for food production in setting goals on initiatives related to phosphorus research.

Research and Funding in Agroecological Systems and Holistic Solutions

For decades, research has been directed to an agriculture model that is intensive in external inputs and aimed at increasing yields in a few staple grains, while often detrimental to the environment (Foley et al. 2011). However, less attention has been placed on research for scaling up low-input local practices and agroecological solutions that improve overall food production, nutrition and livelihoods at the local scale. Many examples exist, what is lacking is the research and development to scale up and adapt these solutions to different local realities (3).

The Global TraPs initiative offers an opportunity to focus research and development on a new holistic model of agriculture centred on people, not agrochemicals or other expensive inputs, which can increase food production where it is most needed, and at the same time help in rural development and protection of the environment (IAASTD 2009; Schutter 2009; Scialabba 2007).

References

Foley JA, Ramankutty N, Brauman KA, Cassidy ES, Gerber JS, Johnston M, Mueller ND, O'Connell C, Ray DK, West PC, Balzer C, Bennett EM, Carpenter SR, Hill J, Monfreda C, Polasky S, Rockstrom J, Sheehan J, Siebert S, Tilman D, Zaks DPM (2011) Solutions for a cultivated planet. Nature 478:337–342

IAASTD 2009. International Assessment of Agricultural Science and Technology for Development. Island Press. www.agassessment.org

MacDonald GK, Bennett EM, Potter PA, Ramankutty N (2011) Agronomic phosphorus imbalances across the world's croplands. Proc Natl Acad Sci 108:3086–3091

Schutter OD (2010) Agroecology and the right to food. UN Special Rapporteur on the right to food. http://www.srfood.org/images/stories/pdf/officialreports/20110308_a-hrc-16-49_agroecology_en.pdf

Scialabba NE-H (2007) Organic agriculture and food security. FAO. ftp.fao.org/paia/organicag/ofs/OFS-2007-5.pdf

Erratum to: Sustainable Phosphorus Management: A Transdisciplinary Challenge

Roland W. Scholz, Amit H. Roy and Deborah T. Hellums

Erratum to: Chapter 1 in: R. W. Scholz et al. (eds.), *Sustainable Phosphorus Management*, DOI 10.1007/978-94-007-7250-2_1

In the artwork of Fig. 7, the terms "Mineral fertilizer + Manure" and "Mineral fertilizer" should be replaced with "Phosphate rock + Manure" and "Phosphate rock", respectively. Hence, Fig. 7 should be displayed as:

The online version of the original chapter can be found under
DOI 10.1007/978-94-007-7250-2_1

R. W. Scholz (✉)
Fraunhofer Project Group Materials Recycling and Resource Strategies IWKS, Brentanostrasse 2, 63755 Alzenau, Germany
e-mail: roland.scholz@isc.fraunhofer.de

ETH Zürich, Natural and Social Science Interface (NSSI), Universitaetsstrasse 22, CHN J74.2, 8092 Zürich, Switzerland

A. H. Roy · D. T. Hellums
International Fertilizer Development Center (IFDC), P.O. Box 2040, Muscle Shoals, AL 35662, USA
e-mail: aroy@ifdc.org

D. T. Hellums
e-mail: dhellums@ifdc.org

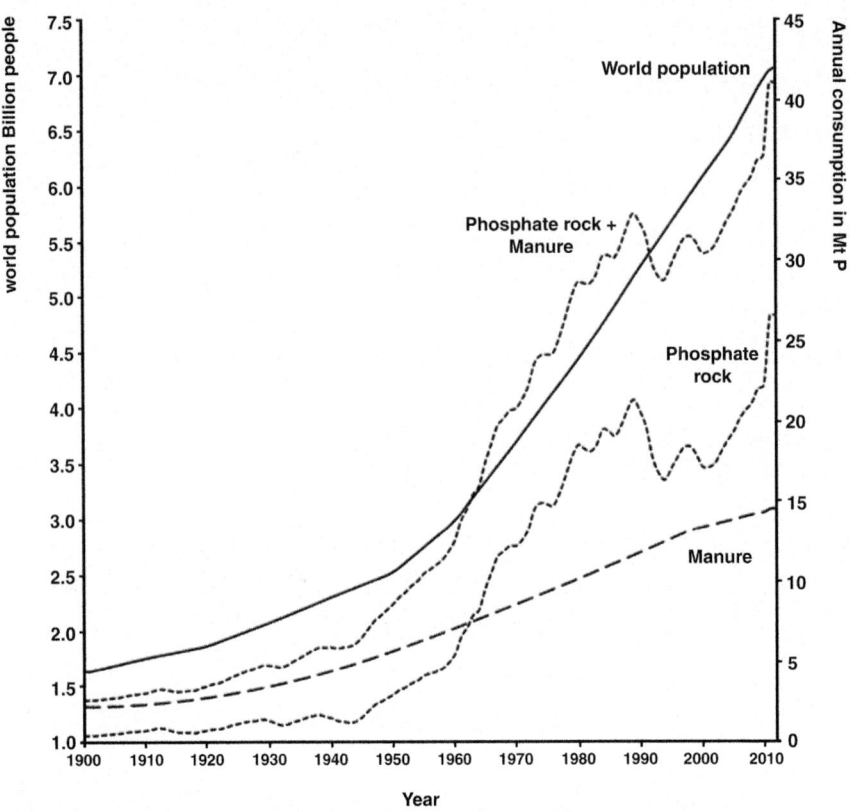

Fig. 7 The evolution of phosphorus fertilizer use, where mineral phosphorus plays an increasingly important key role (phosphorus data—80 to 85 % for fertilizer use—from USGS, presented as moving mean with a 3-year sliding window; rough estimation of manure data based on literature, see Sect. 4.8.2; population data from (USCB 2009); manure data extrapolated from different data on annual manure production, see text)

In the artwork of Fig. 28, the term "Mineral fertilizer" should be replaced with "Phosphate rock". Hence, Fig. 28 should be displayed as:

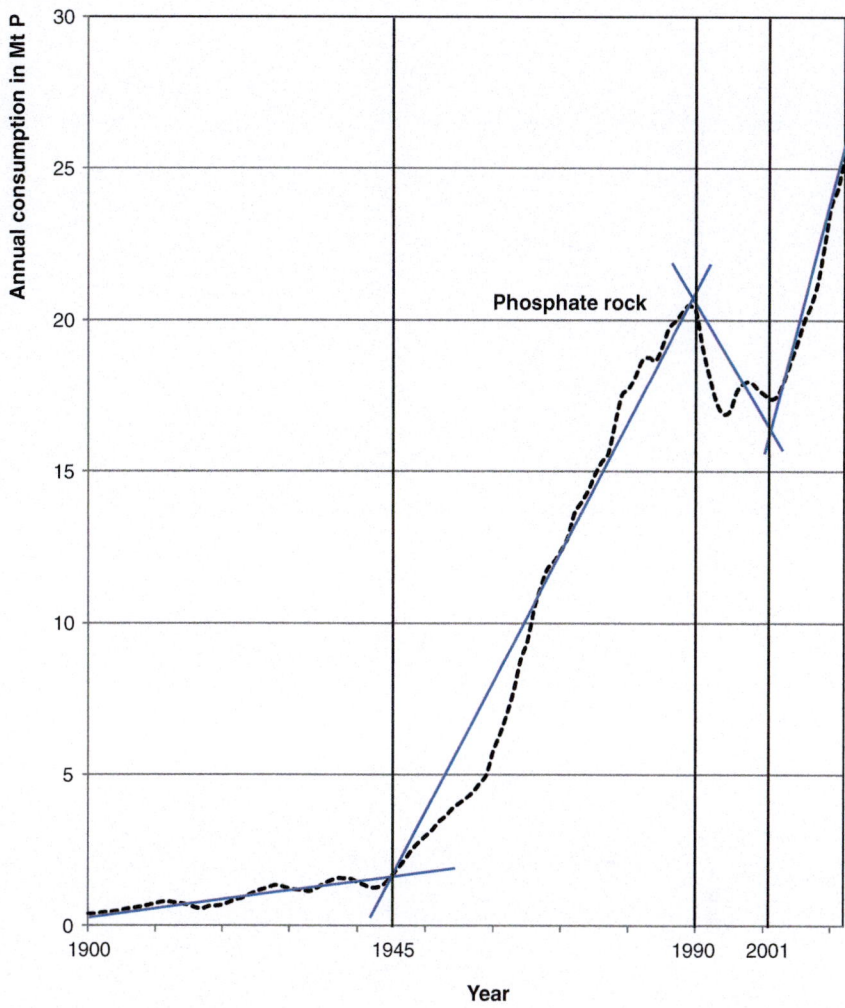

Fig. 28 Different trends of phosphorus use in different periods (*x*-axis: Mt P, data from USGS Mineral Commodity Summaries; the graph is generated by unweighted moving average statistics to smooth annual fluctuations using a five-year time window)

In the published volume, Fig. 28 was mistakenly placed in "Spotlight 2" (chapter appendix) near Fig. 29. It should be moved to page 100 (in chapter text).

The manufacturer's authorised representative in the EU is Springer Nature Customer Service Centre GmbH, Europaplatz 3, 69115 Heidelberg, Germany. If you have any concerns regarding our products, please contact ProductSafety@springernature.com

Printed and bound by CPI Group (UK) Ltd, Croydon, CR0 4YY
25/03/2026
02078169-0005